Guidebook to Real Time Electron Dynamics

This practical book presents an overview of the various approaches developed to understand the dynamics of irradiation of electronic systems in physics and chemistry. It also illustrates typical application examples, namely atoms, molecules, and clusters such as nano objects.

There exist many theories adapted to specific physical situations (both in space and time), but there is not yet a common theory for all possible dynamic scenarios. This book provides a general perspective on the topic, supplying the readers with a guide to navigate the wide spectrum of approaches and select the one that best suits their purpose.

This book provides an overview of available theories to address various problems in the irradiation of finite systems, discussing the possibilities and limitations of the available theories to help readers understand the applicability of a given theory, or set of theories, to address a given physical or chemical situation.

It is an ideal guidebook for graduate students and researchers in physics and chemistry.

Key Features:

- Presents a critical survey of available theoretical tools to help readers choose the appropriate method or approach for any given physical or chemical situation in an irradiation scenario
- Accessible, with an emphasis on avoiding details of formal and technical difficulties
- Provides an extensive guided tour based on typical examples starting from the actual physical or chemical situation down to the tools to be used to describe it

Jorge José Kohanoff is a "Beatriz Galindo" Distinguished Researcher in the Instituto de Fusión Nuclear "Guillermo Velarde" at Universidad Politécnica de Madrid, Spain.

Paul-Gerhard Reinhard is an Emeritus Professor at the Institute for Theoretical Physics at the University of Erlangen-Nuremberg, Germany.

Lorenzo Stella is a Senior Lecturer in the Centre for Light-Matter Interactions (CLMI) at Queen's University Belfast, United Kingdom.

Eric Suraud is a Distinguished Professor at Paul Sabatier University, France. He is also Honorary Professor at Queen's University Belfast, United Kingdom.

Guidebook to Real Time Electron Dynamics

Irradiation Dynamics From Molecules to Nanoclusters

Jorge José Kohanoff, Paul-Gerhard Reinhard,
Lorenzo Stella and Eric Suraud

CRC Press
Taylor & Francis Group
Boca Raton London New York

CRC Press is an imprint of the
Taylor & Francis Group, an **informa** business

Designed cover image: Clarisa Siperman

First edition published 2024
by CRC Press
2385 NW Executive Center Drive, Suite 320, Boca Raton FL 33431

and by CRC Press
4 Park Square, Milton Park, Abingdon, Oxon, OX14 4RN

CRC Press is an imprint of Taylor & Francis Group, LLC

© 2024 Jorge José Kohanoff, Paul-Gerhard Reinhard, Lorenzo Stella and Eric Suraud

---

Library of Congress Cataloging-in-Publication Data

Names: Kohanoff, Jorge José, author. | Reinhard, P.-G. (Paul-Gerhard),
1945- author. | Stella, Lorenzo, author. | Suraud, Eric, author.
Title: Guidebook to real time electron dynamics : irradiation dynamics from
molecules to nanoclusters / Jorge José Kohanoff, Paul-Gerhard Reinhard,
Lorenzo Stella and Eric Suraud.
Description: First edition. | Boca Raton, FL : CRC Press, 2024. | Includes
bibliographical references and index. | Summary: "This practical book
presents an overview of the various approaches developed to understand
the dynamics of electronic systems in physics and chemistry." -- Provided by publisher.
Identifiers: LCCN 2023039216 | ISBN 9780367651268 (hbk) | ISBN
9780367648220 (pbk) | ISBN 9781003127949 (ebk)
Subjects: LCSH: Electrons--Emission. | Irradiation. | Molecules. |
Particles (Nuclear physics)
Classification: LCC QC793.5.E627 K64 2024 | DDC
539.7/2112--dc23/eng/20231117
LC record available at https://lccn.loc.gov/2023039216

ISBN: 978-0-367-65126-8 (hbk)
ISBN: 978-0-367-64822-0 (pbk)
ISBN: 978-1-003-12794-9 (ebk)

DOI: 10.1201/9781003127949

Typeset in CMR10
by KnowledgeWorks Global Ltd.

Publisher's note: This book has been prepared from camera-ready copy provided by the authors.

Dedication

To our families.

Contents

Preface

What do an oncologist working in a cancer therapy center and an engineer working on the electronics for autonomous cars have in common? The answer is that they both try to optimize and control the effects of radiation, or more precisely the effects of radiation damage. The oncologist will try to maximize them in the most localized way possible to avoid destroying healthy tissues around the tumor [Sol17, Obo19]. The oncologist uses dedicated radiation, for example gamma rays emitted by radioactive nuclei in radiotherapy or proton and carbon beams in hadrontherapy. In turn the engineer will try to minimize the effects of radiation damage by shielding electronic devices to avoid their intermittent malfunctioning [AM15]. The engineer does not control the origin of radiation which is to a large extent due to cosmic rays. He has to deal with various particles such as in particular protons, neutrons or muons depending on the altitude and thus on how much cosmic rays have interacted with our atmosphere. Radiation damage thus covers major all-day life societal issues with numerous applications from health to industry [Int].

Oncologists and engineers have a vast experience in how to deal with radiation effects [DRAJ$^+$92, EBP05]. One has now reached a remarkable level in the control of radiation delivery in oncology treatments. And one is also able to robustly protect most sensitive electronic circuits, in spaceships, planes and at sea level. Some issues remain of course and there is still a sizable margin of progress in the optimization of these processes. However, even if a huge experience has been gathered over decades on how to deal with radiation effects, this was attained on a mostly empirical basis, in terms of technical and technological know how. A deeper understanding of irradiation processes, and, above all, of underlying mechanisms, thus remains essentially incomplete and, to a large extent, uncertain. Even if a qualitative analysis is sometimes at hand, there remains a huge gap before reaching fully quantitative access. We thus still lack a complete, quantitative understanding of irradiation phenomena and underlying mechanisms, based on a well-founded theoretical framework.

Obtaining such an understanding is, of course, a formidable task. Irradiation of matter, namely deposition of energy by radiation, such as by light or charged particles, may take place over very short length scales (the atomic scale) and time scales (the electronic time scale). But the manifestations of irradiation occur on human length and time scales. It is thus a process whereby microscopic mechanisms directly impact the macroscopic world. Furthermore, understanding irradiation requires us to describe far off equilibrium processes, involving electron transport and emission, as well as strong coupling to ionic degrees of freedom. Large correlation effects, including electron dissipation leading to thermalization are also expected to play an important role, in relation to the far-off equilibrium nature of the considered situation. It is a highly challenging problem which requires the analysis and understanding, experimentally and, even more so, theoretically, of the interaction of matter (at a fully quantum level down to electrons) with potentially strong electromagnetic fields. The huge accessible phase space and time scales have led to a diversity of tailored approaches which blurs a complete coherent picture.

To the best of our knowledge, and in spite of its timeliness and importance in particular for applications to radiation damage, there exists, to date, no available global theory of irradiation dynamics, or at least a reference theory in specific cases from which to validate approximation schemes. A proper theory of irradiation is today plagued by major bottlenecks concerning many-body correlations, coupling to ionic motion and the insufficiencies of multi-time/multiscale modeling to link microscopic to macroscopic worlds, to name a few.

This book is designed as a guidebook into the realm of irradiation dynamics in finite many-electron systems with a focus on the most demanding part, the description of electron dynamics. As, in many cases, irradiation takes place locally both in time and space, an appropriate description of irradiation in finite systems from atoms to molecules and nanoclusters is a key issue to analyze and understand underlying microscopic mechanisms. It is thus the choice we have made here to restrict our discussions to such finite systems and focus on electronic aspects which are crucial for the whole process. Electrons are light and react immediately to irradiation while preserving numerous quantum mechanical features in spite of the fact that they often drive far-off equilibrium dynamics. The task is thus complex and difficult and confining the topic as we did is compulsory.

The book is organized in 5 chapters. After a long introduction providing numerous examples of irradiation dynamics in Chapter 1 we make a rapid survey of available theories in Chapter 2 and of theoretical tools of access physical observables in Chapter 3. Chapter 4 is devoted to detailed discussions of typical examples of applications of the theories introduced in Chapter 2. Finally, Chapter 5 provides a summary of today's achievements and open challenging questions. In the spirit of a guidebook, we have avoided entering technicalities in Chapter 2 but discussed in detail hotspots of Chapter 4. In the same spirit, we tried, in particular in Chapter 5, to provide the reader with a practical tool to know what can be achieved (and what cannot) with a given theory for a given physical situation.

1 The Physics of Irradiation, from Molecules to Nano-objects

Irradiation processes cover a huge range of beams and targets, from the most elementary particles to the macroscopic world, e.g., laser welding of materials or sunburn of the human skin. In this book, we will focus on irradiation phenomena in atoms, molecules and clusters up to nano-particles with a bias on high energy impact. This is an interesting topic in itself and is also a crucial preparatory step to understand irradiation at macroscopic scales because even there the actual irradiation may occur on a rather limited spatial range. Such small systems are especially demanding because they exhibit the largest quantum and detailed many-body effects, which require the most elaborate methods. Once developed and tested on those finite quantum systems, these methods provide reference theories from which one can deduce approximations for larger systems and/or more energetic situations, where classical approaches become increasingly valid. At some places in this book, we will thus also sketch the connection of the elementary irradiation processes limited to small spots into a larger environment of surrounding matter.

The term *irradiation* covers a broad range of possible projectiles from photons and light-charged particles (electrons, muons) to ions (protons, alpha particles, heavy-ions) and neutrons. All impact the target via the electromagnetic interaction. The case of neutrons deserves a special comment, though. These interact first with nuclei in materials via nuclear processes leading mostly to the production of ions which, in turn, cause an electromagnetic perturbation in the material. We thus confine the discussion to electromagnetic excitations, either by photons, typically from a laser, or by charged particles. The latter deliver usually a short electromagnetic pulse with unspecific frequency content, i.e. covering all frequencies. Lasers provide much more versatile pulses allowing to shape frequency, intensity, and pulse duration, up to dedicated pulse sequences in pump-and-probe setups. Modern developments in atto-physics allow to shape even extremely short pulses with broad frequency content similar to collisions with charged particles.

Another issue is time scales which also range from atto-seconds to macroscopic times. We shall focus here on the "early" stages of the irradiation processes where the impinging field (laser, ion, electron) couples directly to a finite many-electron system (atom, molecule cluster). This means times up to a few hundred of femto-seconds. Thermal relaxation processes and energy transport into surrounding matter are thus not included. They require a

DOI: 10.1201/9781003127949-1

different, more macroscopic, modeling, which we shall only marginally address when considering the coupling of electrons to ions and/or environment.

1.1   AN APPETIZER

Even within the above-described focus, the field of irradiation dynamics in finite quantum systems at the nanometer scale remains huge, covering numerous complementary directions of investigation. Before better going into details and to collect actual material for the forthcoming discussions we present a few examples. This will also help to figure out a "generic scenario" of irradiation exhibiting major effects such as the interplay between time and spatial scales or the importance of correlations between constituents of the system. We shall consider here three examples taken from recent investigations including numerous aspects thereof which have been studied for several years.

1.1.1   PLASMONIC EFFECTS IN CLUSTER IRRADIATION

The first example is an Ag cluster irradiated by a dual laser pulse. A dual laser pulse is a double pulse composed of two separate pulses following each other with a tunable delay time [FMBT+10]. One often calls such a setup a pump and probe experiment to exemplify the fact that the first laser pulse pumps the system into an excited state, while the second laser pulse probes the de-excitation of the system.

In the present case the target system is a large Ag cluster (size around 2000 atoms). Ag clusters are soft metal clusters. When exposed to strong laser fields, clusters lead to the formation of a transient nanoplasma [FMBT+10]. A sizeable fraction of electrons are extracted from their mother atom by the laser. A few of them are immediately ejected from the cluster, but the resulting net charge is sufficient to trap inside the cluster the electrons extracted from the atomic cores later on. These electrons thus remain bound to the cluster as a whole but not to their parent atom anymore. Together with the ionic background, they thus form a nanoplasma. This process is called inner ionization. It is a transient phenomenon as the long-term evolution of the system is usually a complete explosion due to the huge accumulated charge. Indeed, explosion involves ions which set the pace thereof. It can be rather long, depending on ion masses and net charge, but in any case it is much larger than electronic times and the typical timescale associated with the early direct ionization furbishing the early net charge of the system.

One usually separates the inner ionization from the outer ionization labeling electrons which have already left the cluster. The existence of such a transient nanoplasma is also favoring the huge energy absorption observed in irradiated clusters. Indeed, at variance with atoms in the gas phase, the metastable nanoplasma has a lifetime which allows to deposit huge amounts of energy within the cluster before it explodes. The effect can even be enhanced by properly tuning the laser pulse shape. This huge energy deposit is converted into the emission of X-rays and energetic ions and electrons. Properly understanding these mechanisms is highly relevant in applications for pulsed X-ray generation and energetic particle sources and several investigations have thus been led on such systems since the mid 1990s, see e.g., [FMBT⁺10].

The existence of a nanoplasma furthermore points out the importance of plasmon excitation in such systems, namely the collective oscillation of the electron cloud against the ionic background. Inner ionized electrons constituting the nanoplasma behave as a quasi-free electron gas, a bit similar to a metallic sphere. This situation is well described by Mie theory [Mie08] which delivers the eigenfrequency of oscillations of the electron cloud against the ionic background as

$$\omega_{\mathrm{Mie}} = \sqrt{\frac{e\,\varrho_{\mathrm{ion}}}{3\varepsilon_0 m_{\mathrm{e}}}} \tag{1.1}$$

where ϱ_{ion} is the ionic density, e and m_{e} are the electron charge and mass respectively and ε_0 the dielectric constant. The Mie plasmon frequency Eq. (1.1) is the basic ingredient of the fast-developing field of plasmonics (section 4.3). The existence of Mie plasmon frequency is crucial in such a scenario as it introduces a reference frequency, which may interfere with the laser frequency possibly leading to the resonant coupling between the laser and the system.

This irradiation scenario is analyzed in Figure 1.1. Several aspects thereof are considered mixing both experimental results and theoretical interpretation. The theoretical modeling used here relies on a classical Molecular Dynamics (MD) description, see section 2.4.2, treating both electrons and ions as classical particles following Newton equations of motion. Interactions between the particles are simply the Coulomb interaction. This is thus a full many body description. The MD approach accounts for all correlations between particles but only at the classical level. The approximation is justified here because of the large excitation energy deposited in the cluster.

Gross properties of the irradiation scenario are analyzed in the right panel of Figure 1.1 for a given delay time Δ_{delay} of 1 ps (1ps $= 10^{-12}$s) between pump and probe pulses. This panel is itself divided into 3 pieces. The upper right panel displays energy absorption (red curve, right scale) together with the external electric field delivered by the laser (E_{laser}, gray) and the effective field inside the cluster (E_{center}, blue). Note that the latter E_{center} is the field responsible for electron cloud oscillations against the ionic background, namely the source of the plasmonic response Eq. (1.1). The lower right panel distinguishes both inner and outer ionization while cluster size (radius) is shown in the middle panel. The three pieces together provide a clear illustration of the irradiation scenario.

The pump pulse leads to a small initial inner ionization (about 1 electron per atom) and a very small outer ionization, but sufficient to allow a sizable increase of inner ionization during the second half of the pump pulse. Little occurs then at the electron side up to the onset of the probe pulse.

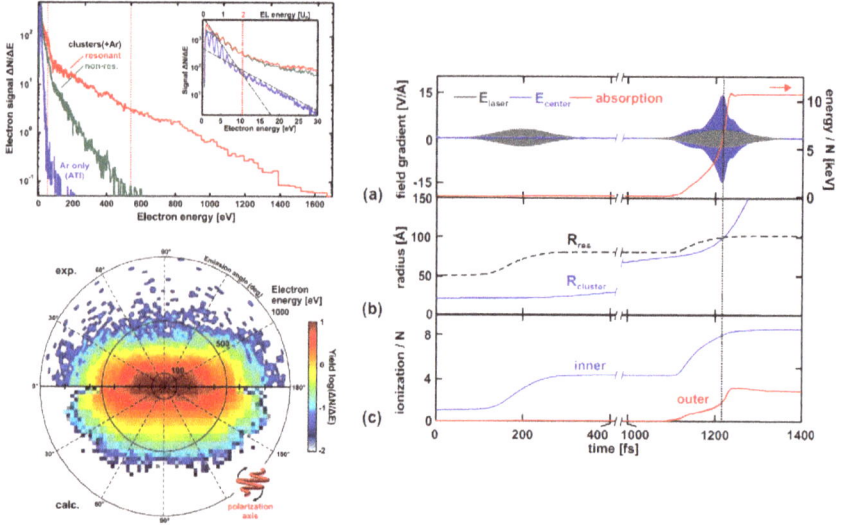

Figure 1.1   Combined figure merging experimental results in Ag clusters (size around 2000) to MD simulations in Ag$_{2500}$ clusters. Upper left panel: experimental Photo Electron Spectrum (PES, section 1.2.3) along the laser polarization axis. Dual laser pulses (pump and probe) of 150 fs duration and intensity of 0.9 10^{14} W. cm^{-2} are used. The case of Ar gas (blue curve) is also plotted in comparison to Ag clusters for two optical delays of 0.4 and 1.2 ps corresponding to non-resonant (green) and resonant (red) cluster excitations. The inset provides a zoom in the low-energy domain. Lower left panel: experimental (upper part) versus theoretical (lower part) Photo Angular Distribution (PAD, section 1.2.3) at optimal delays which correspond to maximum plasmonic couplings. The experimental (1.2 ps) and theoretical (1 ps) values qualitatively match. Right panels: theoretical summary of the irradiation scenario for the optimal delay of 1 ps, as a function of time. (a) Total energy absorption per atom (right axis) and electric field amplitudes (left axis) of the laser E_{laser} and the effective field in the cluster center E_{center}. (b) Cluster radius $R_{cluster}$ calculated from the ionic root-mean-square radius together with the estimated resonant radius R_{res} for the actual inner cluster charge state. (c) Evolution of the inner and the outer ionization. From [PIT$^+$12].

But, because of the net charge, ionic expansion starts and the cluster radius increases steadily ($R_{cluster}$, see panel (b)). In turn, inner ionization and radius expansion change the ionic density ϱ_{ion} and Mie plasmon frequency ω_{Mie} according to Eq. (1.1). This is illustrated in panel (b) by the "resonant radius" R_{res} namely the radius equivalent to Mie plasmon frequency. The delay $\Delta_{\mathrm{delay}} = 1$ ps has precisely been chosen here so that R_{cluster} matches R_{res} during the probe pulse. As a consequence the probe pulse hits the cluster at resonance. This explains the associated huge energy absorption and sudden and large increase of inner and outer ionization as well as cluster expansion. The effect is especially large here precisely because of this resonant effect. Using another delay time which misses the resonant condition would lead to a much softer response.

The above resonant scenario is illustrated in experimental results in the left panels. In the upper left one the experimental signal is the Photo Electron Spectrum (PES, see also section 1.2.3), namely the kinetic energy spectrum of emitted electrons (outer ionization). Three cases are plotted for comparison. The blue curve (Ar only) corresponds to the signal as obtained from a gas of Ar atoms, not clusters. The two other curves correspond to a resonant probe (red curve) and a non-resonant probe (green curve). The red curve corresponds to the resonant scenario schematically described in the right panels of the figure. The figure delivers two key messages. First, the high capability of energy absorption in clusters as compared to atoms is directly visible from the spread of energies of the emitted electrons. Clearly, electrons emitted from clusters then attain much larger energies than when emitted from atoms. Mind that the energy integrated PES provides the total number of emitted electrons, which complies with its enhanced value in the resonant case. Second, the emission spectra are different in the resonant and non-resonant cases. Again, the resonant condition leads to much larger kinetic energies of emitted electrons as compared to the non-resonant case. This confirms the resonant enhancement of emission at the level of electron energies, following plasmonic coupling. Finally, the Photo-electron Angular Distribution (PAD) of the emitted electrons (see also section 1.2.3) is also shown in the left lower panel of Figure 1.1 with a comparison to theoretical prediction. There is a clear angular dependence in emission with a maximum along the laser polarization axis (horizontal axis).

This example is interesting in many respects. The first aspect to be mentioned is of course the scientific interest of such a setup which exhibits the surprising capability of energy storage by clusters, as compared to their constituent atoms. Clusters clearly provide an invaluable laboratory for testing energy deposition in matter. This is crucial for a better understanding of underlying microscopic mechanisms. But it is also worth mentioning here potential applications in terms of X-ray or high energy particle sources, both resources being not so easy to access.

The second interesting aspect is the actual setup used to perform the experiment. The pump and probe scenario illustrates the high versatility of lasers in such situations. The setup indeed allows a remarkable tunability of the dynamical scenario up to the use as manipulation tool. To be noted as well, in this respect, is the somewhat surprising sensitivity of the system to the setup.

The delay between pump and probe suffices to strongly impact the system's response.

The third important point to be mentioned concerns our experimental access to the system's response. In the present case experimental signals are purely electronic. They could have been complemented by ionic signals as done in former studies [KSK+99]. But the focus on electrons is already quite telling as they are key ingredients in the system's response. Their analysis in terms of kinetic energy and angle of emission, namely PES and PAD, provides indirect but instructive signals. We have seen for example how much can be extracted from the PES.

The last two important aspects we would like to point out concern actual physical mechanisms. To a large extent they rely on underlying modeling, allowing to point out the most striking features. The first mechanism to be mentioned is the role of the plasmon, which is a system's specific property. The irradiation scenario thus directly connects to an intrinsic system's characteristics, actually the global behavior of the electron cloud with respect to the ionic background illustrates the key coupling between electrons and ions. In this specific case the coupling is mediated by the plasmon but the ionic expansion stems from electron emission and conversely plasmon response reflects both net charge and ionic expansion.

1.1.2   ELECTRONIC RELAXATION DRIVEN BY ULTRASHORT LASER PULSES

Having identified a few interesting key mechanisms in the example of cluster irradiation let us now consider another example in a different regime. Indeed the Ag clusters were large clusters, typically objects of nanometric size and their dynamics have been explored on rather long time scales in the picosecond range (see section 1.1.1). The emerging aspects are thus rather "global", both spatially and in terms of time scales. This is, by the way, reflected by the strong interplay between electronic response and ionic expansion. Furthermore, the kind of scenarios we have considered correspond to "strong damage" as they lead to a complete destruction of the system. As a consequence of the high excitation energies considered modeling through classical MD is acceptable and qualitatively sound. It is thus now interesting to consider a much different situation in terms of spatial and time extensions and in terms of dynamical regime. To this end, we take an example in the other extreme limit of space and time relying on an UltraFast (UF) setup in a small organic molecule. In this dynamical domain the classical description is no longer acceptable and the irradiation scenario will point out various crucial quantum features.

Photo induced ultrafast dynamics [CPHL+17, YUG+18] has become a major scientific field following laser technology developments [BK00, KI09]. It is now possible to explore sub-fs (1 fs $= 10^{-15}$ s) dynamics down to the attosecond (1 as $= 10^{-18}$ s) hence opening the door to fully time-resolved analysis of irradiation at the electronic level. Ultrafast processes are involved in many mechanisms such as vision, photosynthesis or radiation damage of bio-molecules. Understanding them is essential, for example, to control chemical reactivity but also in astrochemistry or even for solar cells.

Up to now most investigations focused on ionization of atoms and only recently on small molecules. Most theoretical works were accordingly performed in simple systems dominated by single electron effects and/or using simplified modeling, but leaving several questions open and few possibilities of extension to multi-electron systems. Elaborate methods are, in turn, restricted to (very) small systems and more complex systems are still little explored. Forthcoming experiments will address complex molecules (C_{60}, ...) to explore how excitation may induce ultrafast processes (charge dynamics, energy flows...) at the many electron level. First results in moderate size molecules start to be published and we consider, in the following, one such example.

We consider again a pump and probe scenario but this time with much shorter pulses and on a small size organic molecule, Naphthalene (Naph, $C_{10}H_8$). An initial Infra Red (IR) laser pulse of frequency 1.55 eV and duration below 30 fs duration is split into a pump and a probe pulse. The pump pulse is converted by High Harmonic Generation (HHG) into an eXtreme Ultra Violet (XUV) pulse with a set of frequencies between 17 and 35 eV, centered around 26 eV. Mind that the Ionization Potential (IP) of Naph is 8.12 eV, well below the frequency span of the XUV pulse, whose duration is also below 30 fs. The second part of the original IR pulse is kept as is but delayed with respect to the XUV pulse.

The XUV pulse ionizes the Naph molecule and populates excited cationic states with energies near the double ionization threshold at 21.5 eV. This is a spectrally dense region with states involving strong multi-electronic effects. The setup is thus ideal to test many body effects within the system and study hole migration. The probe pulse then provokes a second ionization and the dynamics of the many body quantum cationic states can be analyzed in real time (by varying the IR pulse delay) in terms of Photo Electron Spectra (PES). This finally gives access to the lifetime of the excited states created after the pump pulse. A word of caution is, nevertheless, necessary here. One has to subtract from the measured PES signal single and double ionization components due to the pure XUV (no IR probe pulse). Practically this can be achieved by focusing on the low energy part (below 1eV) of the PES.

Figure 1.2 gathers the major results of this investigation. The upper panel provides a schematic picture of the experimental setup. The pump probe leads to $Naph^+$ and after a time delay Δt the probe pulse leads to $Naph^{++}$. The PES is analyzed here through the associated Velocity Map Imaging (VMI) measurements, see also section 1.2.3, in terms of momenta components p_y along the laser polarization axis and p_x, perpendicular to it.

The lower panels provide an analysis of experimental results. The right panel displays the time resolved PES at energies below 1 eV. This points out the appearance of 3 major structures, respectively around 0.4, 0.64, and 0.88 eV. These can be associated with the ionization of 3 well-identified orbitals of

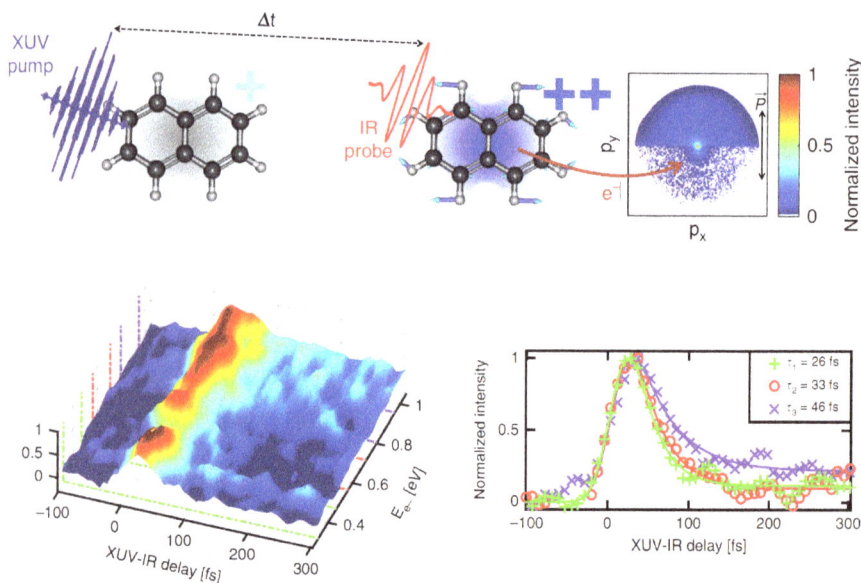

Figure 1.2 Upper panel: schematic picture of the ultrafast relaxation dynamics of a naphthalene cation. A short XUV pump pulse leads to Naph$^+$ while an IR probe pulse with delay time Δt leads to Naph^{++}. The light polarization is set parallel to the detector. The emitted photoelectrons are collected with a Velocity Map Imaging spectrometer (VMI, see section 1.2.3), delivering the electronic momenta as a function of Δt (right upper panel). Left lower panel : resulting PES (kinetic energies of emitted electrons) as a function of delay time Δt. Three structures appear that can be related to the ionization from specific orbitals of the neutral molecule (shown in green, red, and purple, along the vertical axis). Right lower panel: time-dependent electron signal for the three structures identified in the lower left panel. The extracted time constants, obtained from a simple fitting procedure are indicated. From [MDL$^+$19].

the neutral molecule. The analysis of the PES as a function of delay Δt (left panel) allows one to access the time evolution of the electronic signal. After a maximum, the signal slowly decreases and, by fitting the signal, it is possible to extract the lifetimes. These correspond to values around 24±5 fs, 33 ±6 fs, and 46±7 fs, as it can be seen from the lower right panel.

The lower right panel suggests another interesting feature. When simulating in great detail with the help of dedicated quantum modeling, the decay of the structure around 0.4 eV, one observes long living oscillations which could be attributed to coherent vibrational dynamics of a low frequency (ionic) mode around 0.06 eV. This is an interesting aspect. While XUV-induced coherent dynamics is known for low-excited molecular states, it had not been observed before in complex highly excited states of a rather large molecule.

The example described above is interesting in many respects. As a first remark we should again point out the importance of several tools for analyzing electron response. These are, in particular, PES derived from a VMI analysis of electronic emission. Another common aspect of the former example (section 1.1.1) is the use in both cases of pump and probe laser setups. These double laser irradiation allows it to first excite the system and then follow in time its de-excitation. While the case of section 1.1.1 corresponded to the deposit of huge excitation energy leading to a brute force breaking of the system, the present test case explores a lower excitation domain in which detailed quantum features are accessible. Interestingly enough we observe again a coupling between electronic and ionic degrees of freedom. In the plasmonic case ionic expansion was the leading mechanism while in the ultrafast case vibronic coupling was observed at least for one specific excited state. More generally one can also say that these experiments allow to analyze details of excitation and relaxation mechanisms. The analyzed effects are rather sensitive to "details" (resonant coupling in the plasmonic example, actual molecular excited state in the ultrafast case).

Some aspects, though, are new as compared to the former plasmonic case. The major, global one, is what one could call the level of analysis. At variance with the Ag cluster case, the Naphthalene case represents a rather small molecule, studied and analyzed at short spatial and time scales. Clearly one is addressing physics at a much more detailed level than in the plasmonic scenario. For example the PES is measured in a time resolved manner allowing to access rather short time scales in the few fs range. This is of course linked to the fact that the actual excitation is much softer in the Naphthalene case, even if it allows to access rather complex excited states. In turn, this level of detail requires a full quantum treatment which could be overlooked in the more energetic plasmonic case.

Both cases thus show that it is possible to access irradiation dynamics with a certain degree of subtlety. We have not discussed in detail the underlying theoretical descriptions which in both cases were reasonably working. We shall come back to this aspect later on (section 3.1). Both scenarios also focused on electron response in isolated systems (gas phase), even if coupling to ionic degrees of freedom was a key ingredient to understand the response of the system. Both cases, finally, relied on laser manipulation (excitation and analysis). It is thus interesting to address the complementing case of irradiation scenarios with a focus on ionic observables using excitation via a highly charged projectile. This will also be the opportunity to explore the possible role of the environment. An example is presented in the next section.

1.1.3 ION-INDUCED FRAGMENTATION OF URACIL: THE ROLE OF ENVIRONMENT

Understanding irradiation of biological material is a central issue for oncology [Sol17, Obo19]. It is also an important task to protect human beings in space or to understand how life may have appeared on Earth. The interaction of ionizing radiation, in particular via charged particles, causes serious structural and chemical alterations in the cells. This leads to gene mutations and cellular deaths. At the microscopic level, radiation damage mostly stems from single and double DNA (DesoxyriboNucleic Acid) strand breaks induced either by direct ionization or by secondary particles such as low-energy electrons, ions or Free radicals. The overall irradiation scenario is thus complex and proceeds in several steps. It involves both biological molecules and their, mostly water, environment.

Gamma-ray therapy has been for decades a tool of choice in oncology [EBP05]. The onset of hadron therapy using protons and multiply charged ions as beam provides an interesting alternative with a priori better properties in terms of locality of dose deposit as compared to γ-ray therapy. Analyzing the irradiation of biological molecules by charged particles is thus a key issue [DMRH20]. For the sake of realism these molecules need to be considered in their natural environment, namely water.

The present investigation focuses on Uracil $(C_4H_4N_2O_2)$ which is one of the nucleobases in RiboNucleic Acid (RNA) and which plays a fundamental role in coding, decoding, regulation and expression of genes. In order to analyze the effect of environment Uracil is considered both as a single molecule and as an embedded molecule, either in an Uracil cluster or as in an Uracil-water cluster. The projectiles are $^{12}C^{4+}$ ions at a kinetic energy of 36 keV. This is a rather high energy. Ions move at about 15 a_0/fs which means that the time Carbon ion needs to "cross" the molecule is in the sub-fs range. It extends into the fs range for clusters, not more.

The high charge of the projectile strips a few electrons (1 or 2 mostly) from the target system and deposits some energy. Altogether this leads to a fragmentation of the target, which is a long-time process, because it involves target ions. The irradiation process is thus extremely rapid but the measurements focus on the long-term evolution of the system, namely its fragmentation. Fragments are recorded with help of a Time of Flight (ToF) apparatus which records the time at which a fragment reaches an electrode at a certain distance from the reaction chamber. This immediately provides the mass over charge ratio m/z of the emitted fragment, which is a key quantity to analyze results.

A summary of the results obtained in this experiment is shown in Figure 1.3. The lower panel (a) corresponds to the case of pure Uracil, the middle panel (b) to the case of Uracil clusters and the upper panel (c) to the case of Uracil-water clusters. Spectra are split into regions labeled 1-8 (upper axis) which can, roughly speaking, be interpreted as corresponding to the number of "heavy atoms" (C, N, O) in the observed charged fragments. It furthermore provides a simple reading pattern of the spectra. In the case of the pure Uracil molecule the fragmentation spectrum is dominated by intact parent ions $(C_4H_4N_2O_2^+)$, namely singly ionized Uracil molecules $(m/z \sim 112)$. There are of course other fragments present. Among the most important, and well-known

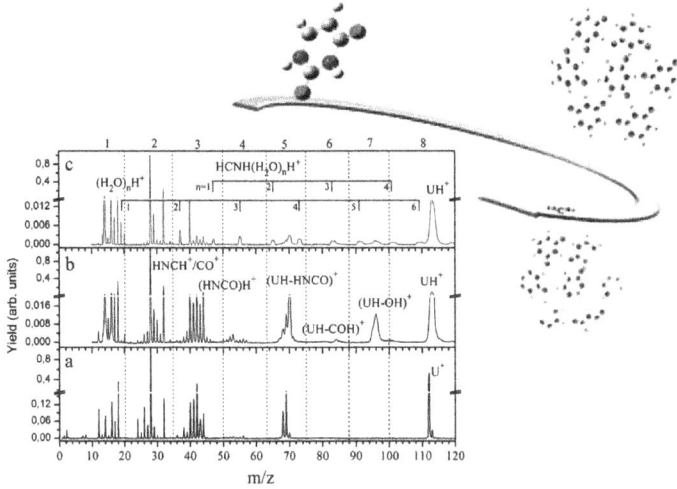

Figure 1.3 Illustration of the role of the environment in the ion-induced fragmentation of uracil. Projectiles are $^{12}C^{4+}$ ions at 36 keV kinetic energy. The figure compares the ion-induced mass spectra obtained for three cases in the charge over mass m/z region up to the monomer: pure uracil molecules, lower panel (a); uracil clusters, middle panel (b); nano-hydrated uracil clusters, upper panel (c). The three spectra have been normalized to 1. The regions labeled 1 to 8 (upper horizontal axis) can be roughly assigned to fragments containing, respectively, 1 to 8 "heavy" atoms, namely C, N, and O. A schematic representation of the three species (uracil molecule, pure uracil cluster, and nano-hydrated uracil cluster) is also given in the upper and left part of the figure. From [MBC+16].

patterns, are the loss of an HNCO group ($m/z \sim 69$) and the appearance of HNCO$^+$ cations ($m/z \sim 43$). Many other patterns can be identified such as low-intensity peaks in the $m/z < 10$ region which can also be attributed to atomic di-cations C^{2+}, N^{2+}, and O^{2+}. But one can also identify molecular fragments such as $C_2H_3N_2^{2+}$ or $C_3H_3NO^{2+}$ and more, in the $20 < m/z < 35$ range. It is interesting to note that these fragmentation patterns are in rather good agreement with quantum structure calculations. The most interesting aspect, though, is to analyze the impact of the environment.

At first glance, all spectra may look rather similar. The presence of the environment, nevertheless, leads to the formation of broader peaks due to the larger kinetic energy of the fragments or successive loss processes within the cluster. But more important is the fact that the relative weights of the various fragmentation channels have been modified. Even more so, some fragmentation channels have disappeared while new ones have appeared. As examples one can look at regions 6 and 7 where clearly new fragmentation peaks pop up in clusters, while they did not show in the pure molecular case. This can be traced back to the formation of an internal bonding network within the cluster, where water molecules also take their share. Globally speaking, though, the cluster environment allows a redistribution of the charge and of the internal energy. It thus acts as a buffer which tends to protect the cluster, to some extent, against fragmentation. For example there is no example of the formation of doubly charged fragments in the case of cluster targets apart from doubly charged atomic fragments in the low m/z range (not shown in Figure 1.3). Small fragments which occur from many-body dissociation channels are also particularly protected by this buffer effect generated by the cluster.

Finally, it should also be noted that, in the case of nano-hydrated clusters, one observes a series of hydrated fragments, which is clearly specific to the presence of a water environment. These fragments are labeled $HCNH(H_2O)_n H^+$ ($n = 1$ to 4) and span over regions 3 to 7 of the spectrum (upper panel). Details of formation mechanisms are not yet fully understood. Altogether the important point nevertheless remains the protective effect of the environment which significantly reduces damage and the yield of low-mass fragments by one to two orders of magnitude.

This last example brings interesting new notions. It should first be noted that it focuses on ionic data, exclusively. This is at variance with former examples which focused on electronic data, even if ionic degree of freedom clearly played a major role. This may be the occasion to discuss a bit the theoretical access. Fragmentation is a complex process taking place for very long time. It deals with long-term evolution of the system and is thus rather indirect, at least what concerns underlying mechanisms. A direct theoretical modeling of the dynamics of such a process is still far away, both formally and practically. One has thus to rely on structural approaches which at least allow to identify energetically favorable fragmentation channels. But this overlooks details of the dynamics. There is thus space for further improvement in the field.

The major interest of this example, though, is to demonstrate the role of the environment. The environment is present in a bunch of physically relevant situations and can often not be overlooked. The notion of an environment itself is interesting. This may not be so obvious in the case of hydrated Uracil clusters as one may be tempted to naturally separate water molecules from Uracil ones and thus potentially define an environment of water around Uracil. But in the case of pure Uracil clusters, we are facing a situation in which irradiation, which is primarily taking place locally, here by electron stripping, has to be understood within the cluster constituents. And there is no difference between the Uracil molecule from which the electron was stripped and the surrounding molecules. This points out the fact that one may have to specify, somewhat arbitrarily, what is the environment in terms of modeling. We shall come back to that question later on, see section 2.5.5.3.

1.2 GENERAL FEATURES

1.2.1 SYSTEMS, TIME AND SPATIAL SCALES

1.2.1.1 Learning from Examples

The three cases discussed in section 1.1 have provided interesting examples of irradiation scenarios in quite different systems. The first case of Ag clusters demonstrated violent explosions driven by a high-charge state. It occurred in a rather large object and could be understood, to a reasonable extent, by means of simple classical modeling. The second example dealt with a much smaller Naphthalene molecule and focused on lower energies, well above the perturbative regime but still in a domain in which quantum features dominate. In both cases we concentrated the analysis on electronic degrees of freedom while also looking at the interplay between electrons and ions. Finally, the last example concerned Uracil as a typical molecule of biological relevance and we concentrated, in that case, on the impact of the environment, in particular the ever-present water. The experimental analysis was performed at a much coarser level focusing on purely ionic data. But even in that case, we could point out subtleties in the actual mechanisms underlying the whole dynamical evolution of the system.

To summarize, the messages delivered by the above examples allow to identify several key features which should necessarily be accounted for an understanding (both at experimental and theoretical levels) of irradiation dynamics. The first important point is the strong interplay between electrons and ions. Putting the system far off equilibrium does not break the constitutive relation between electrons and ions inside a molecule or a cluster. The same holds true when the irradiated system is embedded into an environment. A subtle interplay between all electronic and ionic degrees of freedom remains and, to some extent, drives the ensuing dynamics. This strong coupling between electrons and ions implies to cover both electronic and ionic spatial and temporal scales in their own range. While electrons move at the fs pace, ions may explore dynamics in the hundreds of fs range and even in the ps domain. In space, electrons may remain localized, or explore rather large spatial domains within the cluster. The spatial hierarchy is even enhanced in connection with an environment. This altogether means that we face, with irradiation dynamics, a multi-scale problem which requires to combine several ways of analysis.

Experimentally speaking it means that one should ideally access both ionic and electronic signals, if possible simultaneously. This is not a simple issue and most experiments focus on one or the other aspect, complementing missing information by qualitative reasoning or modeling. The task becomes even more complicated when measurements are indirect.

The situation is not better from the theoretical point of view. Indeed multi-scale modeling is an awfully demanding task. Most theories are restricted to a certain range of validity. A full description of an irradiation process over all ranges of time and space in a molecule, for example, remains at the edge of modeling possibilities, except for very simple situations. In turn, numerous theoretical approaches exist which each deal with a certain spatial and/or temporal range. This is extremely useful to understand key steps of a process and crucial sub-systems as we will discuss in the many examples later. There remains the difficulty to develop validated interfaces between regimes, even more so if coupling to an environment comes into play. Solution strategies are still in the developing stage. We will discuss in that context some presently available hierarchical approaches, see section 2.5.5.3.

1.2.1.2   Systems

In order to clarify the situation let us circumscribe a bit better the typical systems and situations we will address in the following. As already discussed in the preamble to Chapter 1, we focus in this book on the irradiation of finite quantum systems. Even within this limitation, the examples discussed in section 1.1 cover already a wide range of physically relevant situations. This suggests that a proper understanding of elementary irradiation processes, in general, will benefit from the analysis of "small" finite quantum systems. They allow to disentangle the various mechanisms at work. And, as it is possible to continuously pass from small to larger and larger systems, one may hope to extend the understanding in a constructive manner step-wise. Our primary target systems will thus be atoms, molecules and clusters, first as isolated objects in gas phase and later with a few steps toward systems in contact with the environment. The term environment itself has many meanings. It can label embedding the system in or on a different material. It can be the successive layers of water molecules coating a biological molecule. And it can also refer to separating large systems into a strongly active region (hotspot) which requires a high level of modeling and a less active region which allows a lower level of theoretical description.

The chosen target systems also determine the typical theories one may have access to. Very small systems, such as atoms and small molecules, allow highly elaborate theoretical approaches including high levels of quantum details and correlations. Larger systems, on the other hand, require dedicated approximations especially in the account of correlations. This is all the more true in the case of dynamical scenarios which naturally are more demanding and thus impose more severe limitations. We will discuss the available theories in Chapter 2 and the topic of correlations specifically in sections 2.3 and 5.2.1. The situation is even more complex if an environment is involved which, again, requires further dedicated modeling and approximations, see section 2.5.5.3.

1.2.1.3   Scales

Figure 1.4 gathers in a visual and compact way the typical spatial and temporal scales we shall address in this book. Some energy values are also added and several keywords are indicated for completeness. Let us discuss these

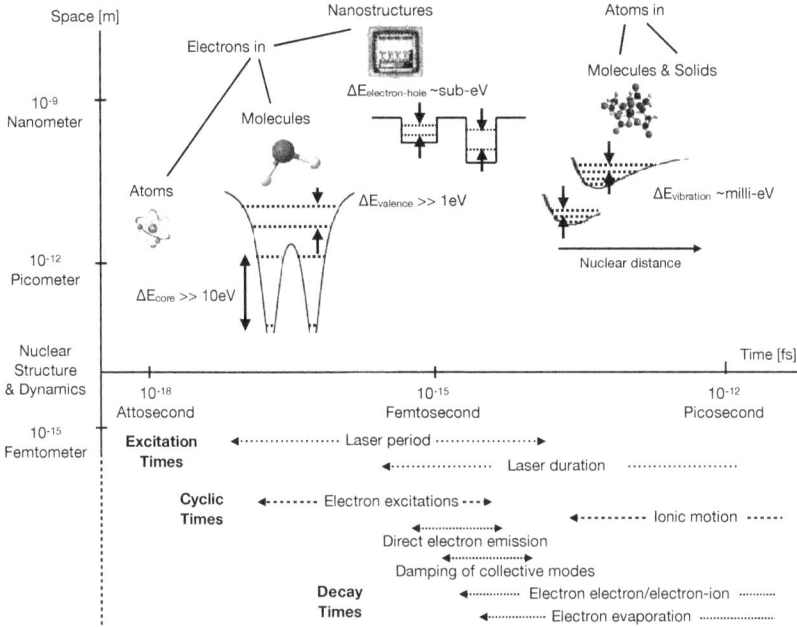

Figure 1.4 Illustration of time and spatial scales in a schematic manner (adapted from [KI09]). The scales are restricted to systems of interest in this book namely from atoms to molecules and nano-objects. Typical energy scales are also indicated for completeness in the various systems of interest. The lower part of the figure gathers time scales relevant in irradiation dynamics. Actual values have been estimated in the case of Sodium to provide quantitative indications. Time characteristics (period, duration) of typical "femtosecond" lasers are indicated in the upper part (excitation times). The other time scales are separated in cyclic ones (cyclic times) and the ones associated to decay times (lower part). Both classes of times include electronic and ionic times.

various scales in more detail. We start with the typical target systems which are schematically represented in the upper part of the figure and located in a time-space map. The relevant spatial scales spread below nanometers (10^{-9} m), which cover nanostructures and clusters, down to molecules and atoms. Atomic radii are of order a few Bohrs (1 $a_0 = 0.529 \, 10^{-10}$ m). Nuclear radii are of the order a few 10^{-15} m but even if nuclear processes may be involved in irradiation processes we shall not address them here, as outlined in the preamble to chapter 1. The typical system's sizes we shall consider here thus range between 10^{-10} m and some 10^{-9} m at most. Exceptions might be considered on the upper limit when dealing with an environment, whose description will usually be considered in a schematic way thus somewhat generous with spatial details.

The time scales associated with our target systems are indicated along the abscissa axis of Figure 1.4. Let us first focus on the picture delivered by the upper part of the figure, which links time and spatial scales in a schematic way. The typical time scales range from the attosecond (1 as $= 10^{-18}$ s) to the picosecond (1 ps $= 10^{-12}$ s). This covers electronic motion within atoms, molecules and nanostructures. The characteristic time of electronic motion is in the femtosecond (1 fs $= 10^{-15}$ s) range. It can be resolved experimentally down to a fraction thereof, thanks in particular to recent advances in laser technology, which allow to attain trains of attosecond laser pulses. The electronic motion may last on rather long time scales, in particular in relation to collective motion, as attained with plasmons in clusters (section 1.1.1) and reach hundreds of fs up to picoseconds.

This enters the domain of ionic motion. Typical ionic time scales increase with the square root of ionic mass but also depend on the nature of bonding. They can go down to some tens of fs for Hydrogen motion inside a molecule or a cluster and easily reach picoseconds for heavy atoms. The relaxation of ionic motion can require even longer time spans.

Finally, the upper part of Figure 1.4 gives a few typical energy spacings for various situations. Inner atomic transitions may reach large values up to keV. They are associated with the attosecond time range. Typical electronic processes in the valence shell of molecules and clusters lie in the eV range and below, especially for large nanostructures. The motion of ions in a molecule is related to meV transitions.

More details on time scales are given in the lower part of Figure 1.4. Because electrons play a central role in all our considerations, we choose the femtosecond (fs) as the standard time unit, and the eV as the standard energy unit. The lower part of Figure 1.4 is shared in three layers, each corresponding to a certain type of time scale. The upper layer corresponds to excitation mechanisms. We indicate the range of typical laser periods as well as pulse durations (see also section 1.2.3). Usual laser pulses have periods in the fs range, corresponding to IR photons. By harmonic generation (see the example in section 1.1.2) one can for example reach photon energies of several tens of eV which correspond to sub fs periods. The laser pulse duration, in turn, is quite flexible, typically in the supra fs range. Typical "fs" pulses have a duration of a few tens of fs. This can be extended to several hundreds of fs and delay times Δt in pump and probe scenarios span typically hundreds of fs. Irradiation by swift charged particles (see the example in section 1.1.3) is usually very fast. It is difficult to quantify it precisely, as the Coulomb interaction is infinite range so the action of the charged projectile is theoretically endless. Practically, the actual excitation mechanism of course depends on the projectile velocity but the range of interest typically spans between fs and some tens of fs.

The second layer of the lower part of Figure 1.4 corresponds to what one can call cyclic times, to disentangle them from decay times, characteristic of relaxation processes shown in the lower layer. Single electron excitations oscillate rather quickly, typically below 10 fs. In metallic systems, the excitation spectrum is dominated by the plasmon frequency Eq. (1.1) whose period typically lies in the fs range. This is a collective motion whose time scale interferes directly with characteristic single electron motion. Ionic motion explores the 10 - 100 fs range, depending on ionic mass. Time scales can be rather short for very high excitation energies leading to rapid Coulomb explosion or more gentle when ionic motion is driven by the electron cloud, especially in the domain of vibrations.

The bottom layer of Figure 1.4 finally examines a few relaxation times. A dominant mechanism is energy loss by direct electron emission, as observed in our previous examples (section 1.1). Collective excitations such as plasmons, also have a typical decay rate related to Landau damping from coupling of plasmons with energetically close single-particle excitations. Its characteristic times lie in the 10 fs range. We finally indicate three other crucial mechanisms. The first one is electron-ion coupling the major effects of which span a huge range of times from fs to ps. The second one is the electron-electron relaxation processes associated with dissipation and thermalization of the excited electron cloud. This is a central issue in the relaxation of far-off equilibrium systems and requires elaborate approaches to describe properly the necessary dynamical correlations. This relaxation is difficult to quantify experimentally and even though it was most probably present in our above examples, we did not show any experimental evidence thereof. Typical decay times associated with such electronic processes range from a few fs to tens of fs depending on the actual electron energies. Once thermalized, the electron cloud may finally relax via statistical emission governed by the electron temperature. This electron evaporation decay time span values up to ps.

1.2.2   A GENERIC IRRADIATION SCENARIO

From the examples discussed in section 1.1, one can identify a generic scenario characterizing irradiation dynamics. As we learned in section 1.2.1.3 that spatial and temporal scales are intrinsically connected, we shall describe this generic scenario as a function of time, with implicit links to spatial aspects. A basic scenario can be formulated as follows.

Irradiation delivers a (possibly strong) electromagnetic perturbation. Almost immediately respond the electrons which rapidly reach excited states. Part of the deposited excitation energy is quickly released via direct electron emission as seen in all our examples. But the system usually remains highly excited, either because it has not yet fully released the deposited excitation energy or because it continues to be fed with energy, for example by a long lasting laser pulse or both. After that, electrons will start to relax by various mechanisms. The purely electronic mechanisms, e.g., electronic dissipation,

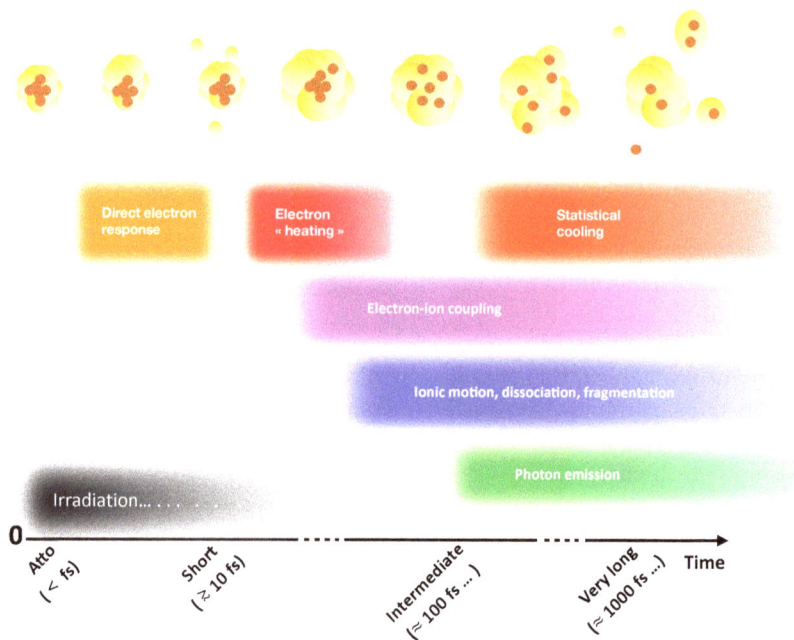

Figure 1.5 Schematic summary of the various dynamical regimes/mechanisms following an irradiation. The lower axis provides a typical time scenario with explicit scales. In the course of time following an irradiation (which may extend over several fs) various phenomena take place like electron heating (short to intermediate times) up to electron cooling via statistical emission (very long times). In parallel electron-ion coupling (intermediate to long and very long times) takes place covering both vibrational (electron-phonon coupling) and explosive scenarios, depending on the actual excitation energy.

pop up first. Coupling to ionic degrees of freedom is the next mechanism to distribute available energy. Mind that ions do, of course, also couple directly to the electromagnetic perturbation. But because of their much larger mass, they react at a much slower pace than electrons and with much less energy intake. They thus come into play much later. Still later follows the reaction of the environment if there is any.

Figure 1.5 provides a schematic summary of this scenario including some actual times. The upper part of the figure delivers a cartoon-like visualization of the excitation and response of an irradiated species. Electrons are represented by a transparent cloud around darker spots associated with ions. The various mechanisms are successively illustrated, e.g., direct ionization, coupling to ionic motion and final explosion. Other less violent scenarios are conceivable, of course. The present "extreme" case has been chosen to cover as many phenomena as possible. The lower arrow contains actual times from attosecond to picosecond, which covers the most relevant scales. As already indicated these time scales should be associated with spatial scales from a few atomic units to possibly infinite expansion. Finally, the middle part of the figure schematically provides major physical mechanisms at play during irradiation. These mechanisms are included in various frames to indicate the typical time scales they are associated with. Note that these frames are blurred on purpose to indicate that the associated time range is more qualitative than quantitative even if some clear differences do exist for example between direct electron emission following irradiation and statistical electron emission during late time cooling. But for example, coupling between electrons and ions may manifest itself at different times because of different types of binding and of different ionic masses. For example, ionic motion takes place at the 10 fs range in a small organic molecule like N_2 while it will pop up in the several hundreds of fs in simple metal clusters such as Na clusters and even later on in Cs clusters. Actual values of time scales are thus to be taken here with a grain of salt.

1.2.3 TYPICAL OBSERVABLES

Now that we have identified a typical irradiation scenario it is time to discuss how it can be addressed practically, namely what are the typical relevant observables which can be measured and computed. The answer depends on the actual response of the system's relevant constituents which gives directions to go for measurements. We have seen that electrons react first because of their small mass while the longer-term response of the system involves both electrons (Figures (1.1) and (1.2)) and ions (Figure (1.3)), and sometimes photons (Eq. (1.1)). As a consequence, relevant signals focus on electrons, ions and photons. The different nature of these measured quantities renders time scales different and so provides various levels of information.

The simplest measurements ignore this aspect and provide only time-integrated signals giving a picture of the response usually well after it is over. Details of specific time scales (electronic versus ionic ones for example) are blurred. However, with the increasing control of temporal details, e.g., by pump and probe scenarios, on can record time resolved signals (see for example Figure 1.2). This adds new, invaluable information. We shall thus consider, in the forthcoming discussion, both time integrated and time resolved signals.

Figure 1.6 sketches presently accessible key observables. The upper panel of the figure recalls the most usual irradiation scenarios, namely with photons (lasers), ions and electrons. Lasers are certainly one of the key tools, because of their availability and versatility. They span a wide range of frequencies and intensities allowing us to access a world of different mechanisms. The time profile of the signals provides another important ingredient. Pump and probe approaches, in particular, are extremely useful for time-resolved analysis both at ionic pace (section 1.1.1) and now even at electronic one (section 1.1.2)).

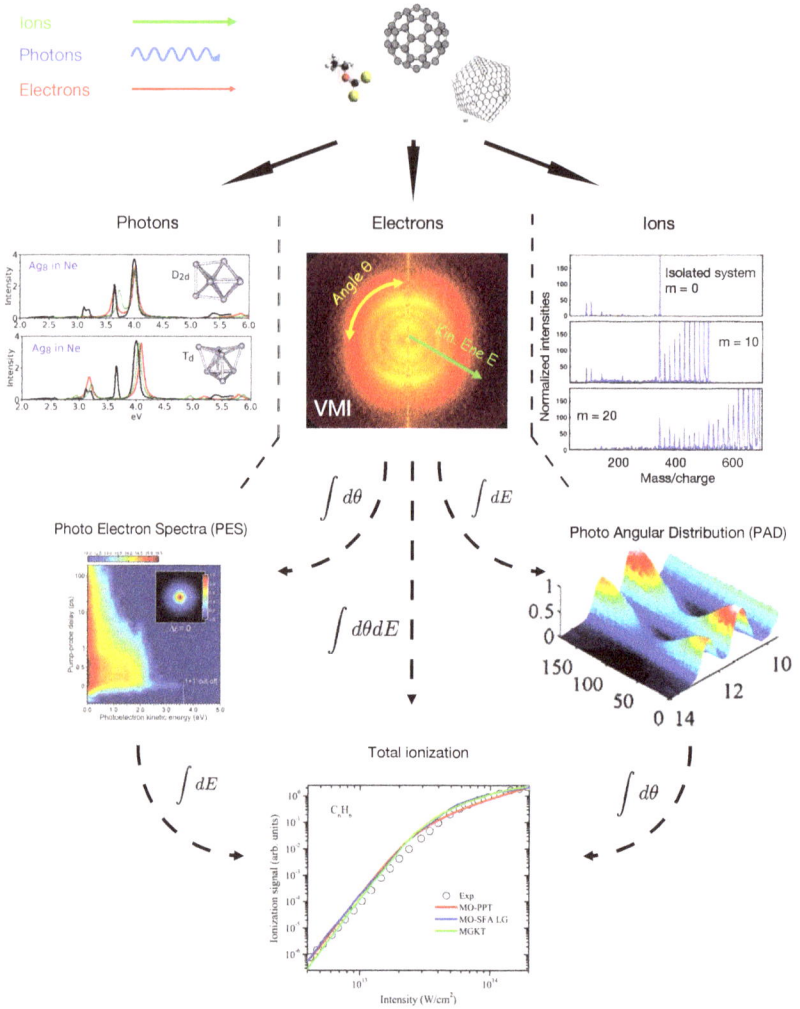

Figure 1.6 Major observables from emission of photons, electrons or ions in the analysis of irradiation. The upper part shows projectiles (photons, ions or electrons) and targets (molecules and clusters). For details see text. The sources for the inserts are: photo-absorption spectrum from small Ag clusters embedded into a Ne matrix in left panel in second row from [SR18]; mass spectrum from fragmentation of a water coated AMP molecule in right panel of second row from [LNH$^+$06]; VMI of emitted electrons from C_{60} in middle panel in second row from [KJJ$^+$10]; time resolved PES from acetylacetone in left panel in third row from [KCS$^+$20]; PAD from C_{60} in right panel in third row from [WGB$^+$15]; total ionization from benzene in lowest panel from [ZLJ$^+$16].

Ion beams require dedicated facilities and are thus more limited in their range of applicability. At variance with lasers, they deliver an electromagnetic perturbation without specific frequency. But ion beams are important in applications both in the medical and industrial domains. The same holds true for electrons which are especially important for understanding irradiation in the biological context, in particular in connection with Dissociative Electron Attachment (section 5.2.3.3). There are other possible beams conceivable such as for example positrons [BBR17]. For the time being they remain marginal in the field and we shall thus not discuss them specifically here.

As said above, the system's response to irradiation can be analyzed in terms of emitted photons, electrons and ions. Let us first consider photons and ions. The photon case is illustrated in the left panel of the second row of Figure 1.6 in terms of the photo-absorption cross section (often quoted optical response) of small Ag clusters (Ag_8) embedded into a Neon matrix. Both theoretical (red curves) and experimental (black curves) cross sections are displayed. In the case of such a small cluster, shape effects are dominant and explain most of the spectral fragmentation of the cross section into several major peaks. The competition between the two close isomers in the gas phase (green curves) is also imprinted in this fragmentation. Last, not least, it is interesting to see the impact of the embedding matrix (red versus green curves) which, in spite of the inert nature of Ne, leads to a shift of dominant peaks.

The ionic case is illustrated in the right panel of the second row in Figure 1.6 on the example of the mass spectra from the fragmentation of the singly charged anion adenosine 5-monophosphate (AMP^-, $C_{10}H_{14}N_5O_7P^-$) coated by a certain number of water molecules. Excitation of the system is achieved here by collisions with neutral Na or Ne atoms. The interest of such experiments is to understand the possible protecting role of water in view of applications in a biological context as AMP is a basic component of the RNA. It is clear that the fragmentation pattern is heavily affected by the number of coating water molecules (see also section 1.1.3).

Finally, electronic observables are documented in the central panel of the second row showing a velocity-map image (VMI) from electron emission. It is the most comprehensive electron observable and contains complete energy and angle-resolved information 3.4.4.5. From that one can deduce by integration reduced, but simpler observables as indicated by the three arrows flowing out of the VMI panel. Let us start with considering the most global value namely total ionization (see also section 3.4.4.3). It is illustrated in the central lower panel where the total ionization of benzene (C_6H_6) is plotted as a function of laser intensity. A comparison is done with several dedicated models and the lasers used for irradiation are considered as being in the strong field domain, namely a regime where the selective impact of frequency is progressively replaced by the dominance of field strength (or intensity, see also section 3.3.2). One also observes a clear scaling of ionization with a power of the number of photons needed to emit one electron, as is well-known in situations where no resonant effect is present [KGD$^+$22], for a more detailed discussion see section 4.1.2.1.

It should be noted that the total ionization is however not necessarily a true electronic observable. It depends on how it is measured, as it can basically be attained in two different ways. One can measure kinematic properties (energy, direction) of emitted electrons and sum them up, or measure the charge states of the emitted ions. When interested only in the total ionization yield the second method, which is in fact an ionic quantity, is obviously simpler to obtain. When interested in more details one needs to recur to a direct measurement of electronic properties.

The highest level of detail is provided by VMI [BPHH96] which delivers an energy and angle-resolved analysis of electron emission. This basically provides a double differential cross section $d^2\sigma/d\theta dE$ where E is the kinetic energy of the emitted electron and θ its angle with respect for example to the laser polarization. The central second upper panel provides an example of such a VMI analysis of laser irradiation of C_{60}. Note that in this case it is indeed the energies of the electrons which have been recorded but one obtains a similar pattern using momenta. The VMI is extremely rich. One can spot in this example both variations of kinetic energy (radial) and of angle with respect to laser polarization (vertical). In turn, it is often useful to consider integrated versions of VMI, which thus deliver Photo Electron Spectra (PES, see also section 3.4.4.5) $d\sigma/dE$ by polar integration (integration of VMI over θ) or Photo-electron Angular Distributions (PAD, see also section 3.4.4.4) $d\sigma/d\theta$ by energy integration (integration of VMI over E).

Both PES and PAD are illustrated in the left and right lower panels and linked to VMI by arrows reflecting angular or energy integration. Mind nevertheless that both PES and PAD can be accessed without reference to a VMI, namely by simply recording either only angles of emission (PAD) or energy of emitted electrons (PES). PES and PAD also deliver extremely rich signals. In the PAD example here, on the right side of the figure, again obtained in C_{60}, an energy content has been preserved which allows to attribute the PAD to a given energy level of C_{60}. One clearly sees in that example variations in the PAD associated with electrons emitted from different levels (here Highest Occupied Molecular Orbital (HOMO) and HOMO-1). The PES example shown on the left side of the figure has been obtained in acetylacetone ($C_5H_8O_2$) using a VUV laser. This is a pump and probe setup which allows to follow in time (vertical coordinate) the evolution of the PES. It is thus a Time-Resolved PES (TRPES) where one clearly sees the evolution of the energetic characteristics of the emitted electrons as a function of delay.

A few complementary words are in place here. First, it should be noted that, most of the time, only one type of observable is recorded in a given experiment. This is clearly due to practical hindrances in experimental setups and in expertise. This clearly induces strong limitations in our experimental access to a given irradiation scenario. This limitation is somewhat less severe from the theoretical side although there are other strong limitations. For example, ionic signals following fragmentation are theoretically an extremely complex quantity to access. They involve very long ionic times hardly accessible to real-time simulations. Further, and more fundamentally, usual simulations do not allow proper treatment of fragmentation dynamics because of the over-whelming amount of fragmentation channels. We shall come back to that later on in section 5.2.3.2.

Another aspect concerns the necessarily limited amount of the observables considered here. Other quantities and/or experimental setups are conceivable while we restrict ourselves here to the most prominent ones. Even so, it is interesting to note that many quantities may have a wider range of applications than expected. For examples PES and PAD are usually associated with laser irradiation but it is obviously possible to record kinetic energies or angular distributions of electrons emitted from a system irradiated by a bypassing swift ion [KGD+22] even if the acronym PES is then misleading.

1.3 SOME CHALLENGES

The above discussions have shown the richness and complexity of the field of irradiation dynamics. It involves a multitude of spatial and temporal scales as well as a variety of signals from photons, electrons or ions. Even richer scenarios unfold if multiple laser pulses are considered in a pump and prove setup. Although a solid basis of understanding has been achieved over the years, there remains a large amount of challenging tasks for theory increasing with the steady improvement of experimental techniques. In the following, we point out a few generic aspects which, we think, could constitute a canvas for more specific approaches. We shall consider a few examples thereof below in sections 1.3.1 and 1.3.2. But before stepping into details, let us briefly explain in more general terms what we have in mind here.

From the experimental point of view a major issue is to have access to a maximum of complementing information. Usual experiments focus on one observable from electrons of ions. The data thus obtained are then by nature incomplete and one is bound to "fill the holes" by complementary assumptions or to deduce underlying mechanisms in an indirect manner. Clearly, even if information is incomplete it is already often quite telling. Still, major progress would consist in correlating observables more systematically, for example, ionic *and* electronic or photonic ones. We shall illustrate this point in specific examples below (section 1.3.1).

It is interesting to make an analogy here with heavy-ion collision experiments in nuclear physics. In this field, one studies collisions between atomic nuclei to learn about nuclear properties. Dynamical effects may be very important and can strongly blur the underlying characteristics of the colliding nuclei [DST00]. Signals are provided by emitted particles which may be neutrons, protons, small nuclei or gamma rays. Measuring so diverse particles simultaneously requires a dedicated combination of different detectors. This is expensive and thus most of the experiments in the field focused on measurements of a selected species like neutrons or massive fragments. It was only in later experiments that correlated detections were considered more systematically. This allows to map the full kinematic of the process. Even more so, the recent generations of detectors are able to record collision products on an event-by-event basis, namely for each individual heavy-ion collision. This is a great achievement for a thorough understanding of the underlying dynamical mechanisms. This stage of a complete measurement has not yet been reached in irradiation experiments on atoms and molecules. The nuclear example shows that any step toward more complete measurements is highly desirable.

The situation at the theory level is different but has also its challenges. One task is that theory might complement missing experimental information. For example, a theory which properly includes both electrons *and* ions could describe ionic dynamics from an experimental knowledge of merely electronic signals, thus "filling the holes" of an incomplete measurement. But that would work only if theories were sufficiently reliable which is not yet always the case, as we shall see.

Still, major progress in that direction has been made over the years and we have now at hand theories which can properly cover some aspects of an irradiation scenario and/or establish relevant connections between electronic, ionic or photonic signals. But, as on the experimental side, there is still room for more complete theories, particularly in the realm of a detailed real-time description of the energetic dynamical processes. We shall discuss a few prominent aspects in section 1.3.2 below. Aspects reaching further into the future will be taken up in section 5.2.

1.3.1 FROM INDIRECT OBSERVABLES TO COMPLETE EXPERIMENTS

As a first example of "indirect" measurement let us consider Transient Absorption Spectroscopy (TAS). TAS was introduced in the early 1950s and aims at measuring the photo-generated absorption energies stored in excited states and associated lifetimes of molecules, materials, and devices. By construction, it is thus based on a photonic signal but provides information on electrons. With the advent of femtosecond pump and probe approaches it has thus become possible to follow, with TAS, electronic response down to some hundreds of fs.

An example is given in Figure 1.7 in which Gold nano-particles have been irradiated by an IR laser of duration of about 50 fs. This pump pulse is complemented by a white light probe pulse with various delays between 100 fs and 10 ps. The interesting quantity is the difference between the ground state photo-absorption signal and the one in the excited state. This is the quantity called ΔA plotted in Figure 1.7. The photo-absorption signal from the ground state has a wide (about 100 nm width) single peak structure plasmon around 520 nm, which reflects the spherical nature of the nano-particles

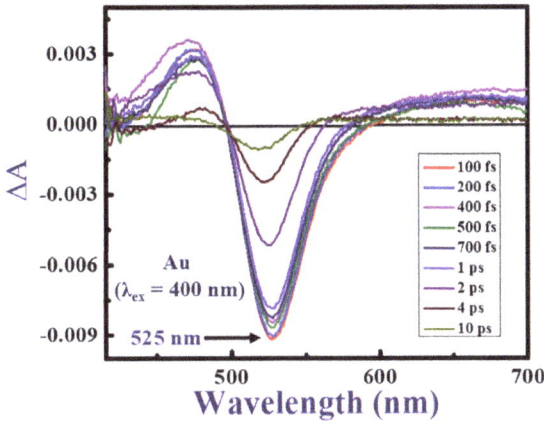

Figure 1.7 Transient Absorption Spectra (TAS, ΔA) of Gold nano-particles in distilled water pumped at 400 nm wavelength at different probe delay times (inset panel with colors). Mind that ΔA represents the variation of the absorption spectrum, whence positive and negative contributions depending on the wavelength. The absorption spectrum of the Gold nano-particles (no pump probe) displays a wide peak in the UVVisible domain, around 520 nm wavelength. From [KCB$^+$19].

used in this work. The ΔA signal is strongly negative around the plasmon peak which could indicate depletion of plasmon electrons. Two positive contributions are observed, one just below 500 nm and a flat one above 600 nm, which could be attributed to absorption through the thermally excited non-equilibrium electron distribution near the Fermi level. With increasing delay between pump and probe both positive and negative contributions in ΔA tend to decrease in absolute value, up to almost disappear for a delay of 10 ps. These measurements allow to extract decay times in the ps range for the electronic relaxation following the pump pulse. The example is interesting for our purpose as it clearly illustrates the indirect nature of the measurements which deal with photons but give indications on electron dynamics.

Figure 1.8 provides a complementing picture of indirect analysis now focusing on ions to access electronic properties. The test case is an I_2 whose dissociation dynamics after charging is followed in time through time-resolved measurement of the kinetic energies of the ions. Ion fragments (I^{p+}) are accelerated and detected in coincidence, which allows to reconstruct their kinetic energies. This is again a pump and probe scheme, but now in the XUV spectral range, as attained by free electron lasers. The laser frequency is 87 eV and the pulse duration is about 60 fs . The intensity of about 10^{14} W . cm $^{-2}$ lies in the multi-photon regime. The first photon produces I_2^{2+} and I_2^{3+} but a second can lead to I_2^{4+} and I_2^{5+}, when two consecutive intense XUV pulses (pump and probe pulses) are considered.

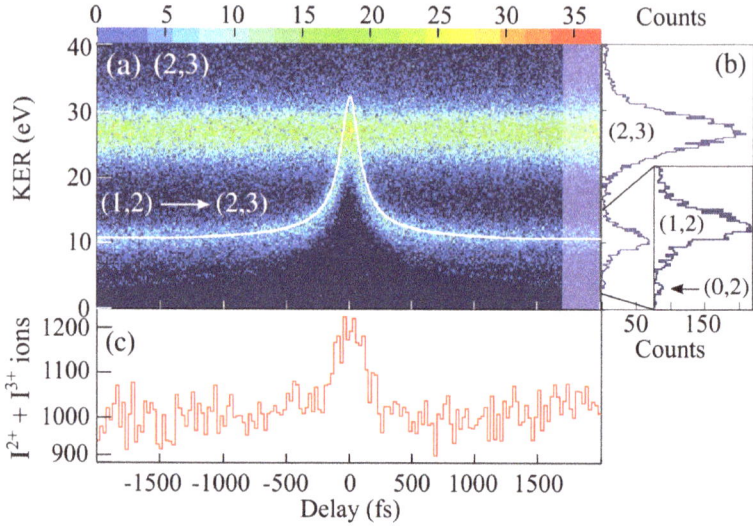

Figure 1.8 Upper left panel (a): Kinetic Energy Release (KER) for all coincident (I^{2+}, I^{3+}) ion pairs as a function of the time delay between pump and probe. The color scale is indicated on top of the panel. The result of a classical simulation via the pathway $I^+ + I^{2+} \rightarrow I^{2+} + I^{3+}$ is superimposed in white. Right panel (b) : Projection of KER's at large delays (white shaded area in panel (a)) onto the KER axis. The inset shows a magnified view of the low energy region. The charge states associated to the observed asymptotic KER's are indicated in brackets (e.g., $(I^{2+}, I^{3+}) = (2,3)$). Lower left panel (c): Projection of all KER's onto the probe delay axis. From [SSK$^+$14].

Figure 1.8 shows the delay-dependent Kinetic Energy Release (KER) for the fragmentation channel $I_2^{5+} \rightarrow I^{2+} + I^{3+}$ as a function of delay time Δt between pump and pulse, and thus as a function of the distance between the two fragments undergoing separation. Note that the delay may take negative values. The picture exhibits two different structures. The one independent of Δt around 27 eV reflects two-photon absorption. The one at lower energy (around 11 eV) characterizes a two step process. For small delays, it overlaps the constant signal as it corresponds to situations in which the two pulses do overlap in time so that the situation is equivalent to one single pulse. The larger delays (positive or negative) do correspond to two well-time-separated pulses and correspond to a two-step excitation process in which iodine attains charge 3+ by the pump and finally 5+ by the probe. This corresponds to a fragmentation scenario

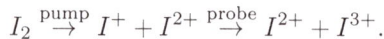

$$I_2 \overset{\text{pump}}{\rightarrow} I^+ + I^{2+} \overset{\text{probe}}{\rightarrow} I^{2+} + I^{3+}.$$

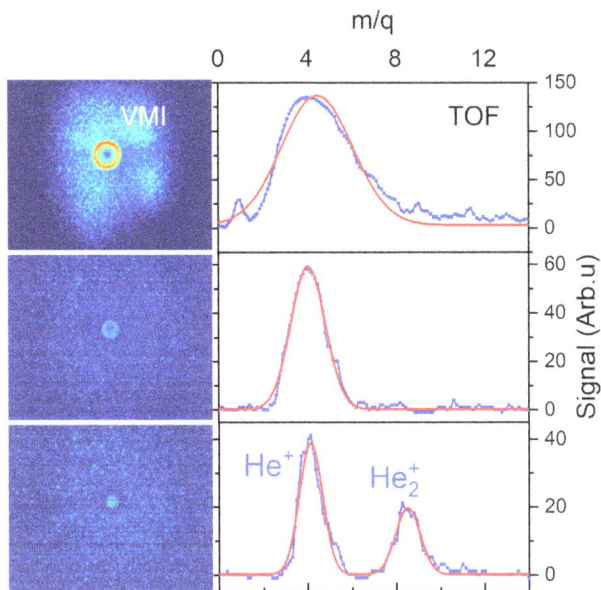

Figure 1.9 Joint electron-ion analysis of the response of Ca-doped He droplets irradiated by a Mid Infra Red laser pulse ($\hbar\omega_{\text{las}} \sim 0.39$ eV, $I = 4.\ 10^{14}$ W.cm^{-2}, duration 50 fs). Both electronic (Velocity Map Imaging, VMI, left panels) and ionic (time-of-flight mass over charge m/q spectra, right panels) have been measured simultaneously. The figure displays clear correlations between ionic and electronic signals. Adapted from [MSR$^+$21].

Further refinements (not shown here) can be envisioned, in particular, the analysis of the competing $I_2^{5+} \rightarrow I^+ + I^{4+}$. In all cases, the analysis can be further supported by classical estimates of the KER. The interesting point for our purpose is that the KER measurement, on ionic fragments, allows to follow in time the charge rearrangement, an electronic process, as a function of time or equivalently as a function of the distance between the two fragments.

An alternative strategy consists of correlating complementary measurements. An example is given in Figure 1.9 in which irradiation of doped Helium clusters is jointly analyzed via electrons and ions. The physical context is the one of a nanoplasma generated by intense femtosecond laser pulses, as considered in section 1.1.1. Nanoplasmas are particularly interesting because they can efficiently convert their particularly strong (resonant) absorption of laser light into fast electrons, highly charged ions, and energetic radiation. This makes them potential candidates for compact light sources or particle accelerators. In the example displayed in Figure 1.9 the target clusters are He nanodroplets doped by Ca atoms which lowers the laser intensity needed to generate a nanoplasma. The laser is in the mid-IR range with a pulse duration of order 50 fs and an intensity in the 10^{14} W.cm^{-2} range. The electronic signal is recorded with a VMI and the ionic one with a standard Time of Flight spectrometer (ToF). The figure thus provides a correlated (electrons + ions) picture of the response of the irradiated system.

One indeed sees a clear correlation between the VMI and the ionic ToF signals. In the upper left panel, for example, one observes a strong low-energy electron component in the VMI (intense central spot) correlated to a wide He$^+$ ionic signal (upper right panel). On the contrary, in the lower left panel, the low-energy electron component in the VMI has been strongly suppressed while the ionic ToF displays two distinct (much narrower) He$^+$ and He$_2^+$ peaks. Low intensity VMI signals thus correlate with low ion kinetic energies and incomplete fragmentation or ion dimerization occurring in a partly ionized nanodroplet The third example (middle panels) corresponds to an intermediate stage between these two extreme cases. Such correlations clearly bring important information to better understand response mechanisms. They are presently not that frequent and should thus be more systematically done, when possible. This is, in our opinion, a major challenge for most experimental investigations where incomplete knowledge may become a source of sterile discussions.

1.3.2 THEORETICAL CHALLENGES

As for experiments, theories are challenged as well. A real-time theoretical description of irradiation dynamics far off equilibrium remains a demanding task, both for formal development and practical implementations. We discuss in this section a few major aspects for theoretical approaches in the field.

1.3.2.1 Specific Difficulties of Far off Equilibrium Dynamics

1.3.2.1.1 *Complexity*

Far-off equilibrium dynamical situations are by nature extremely complex. They cover a huge span of time and spatial scales. Moreover, the most relevant degrees of freedom can change in the course of time. For example, electronic quantum effects dominate initially, often being well localized. Ionic motion and environment become dominant on longer times and on a larger spatial scale. Both aspects, namely detailed quantum description of electrons with high temporal resolution and access to full ionic motion over long times are practically hard to fulfill simultaneously. This means that complementing theories may have to be developed to cover both ranges of scales. The challenge is then to establish proper connections between the two regimes (see section 1.3.2.4).

Another challenge is that far-off-equilibrium dynamics are highly sensitive to many-body correlations (section 2.3). The potentially large energy deposit delivered by the excitation mechanism is only partially released by ionization. The other part of excitation energy initially stored in coherent electronic motion dissipates and thermalizes into intrinsic energy. This can only be achieved by dynamical correlations, going beyond usual mean-field approaches.

1.3.2.1.2   Phase Space

A demanding problem appearing in far-off-equilibrium dynamics is what one could call phase space, i.e. the number of open reaction channels. It grows dramatically with excitation energy. The work-horse for real-time simulations of those processes is the mean-field description TDDFT (TDLDA respectively, section 2.2.3.4) where the mean field averages over all reaction channels thus blurring any detailed information. This can be overcome to some extent by ensemble models with stochastic propagation, see section 2.3.8. Such models in a quantum context are still in a developing stage. And they explore by construction only the dominant channels. Rare reaction channels are even more demanding. They require some a priori expectation of the process and dedicated tailoring of the reaction path. An example is Dissociative Electron Attachment which will addressed in section 5.2.3.3.

1.3.2.1.3   Benchmarking

Another major difficulty, characteristic of far-off-equilibrium situations, is the lack of sufficiently realistic analytical models which could serve as theoretical benchmarks. Along the same line of reasoning, there are usually very few limit cases, if any, against which a theory can be tested. One often considers overly simplified setups such as, for example, the model of one active electron in a 1D molecule. It is better than nothing. But it limits by construction the capability of testing a theory designed for more complex (realistic) cases. Comparison to experiments is, of course, an essential issue and ultimate test, but that should not hide the importance of probing the intrinsic coherence of a given theory. A comparison of a theory to a reference theory, fully under control, is an important validation step, which is unfortunately often missing in scenarios of energetic dynamics.

1.3.2.2   Links to Experiments

Comparison to experiments is of course a key issue. The latter should nevertheless be taken with a grain of salt. Indeed, a direct simulation of experiments is usually complicated for several reasons. Let us consider them successively.

Experimental data often carry uncertainties (e.g., detector acceptance) and are usually post processed. This should be taken into account when comparing theoretical results to data. Ideally, the theoretical observables should be subdued to the same post-processing as experimental ones, before comparison. For example, access to Photo Electron Spectra (PES), see section 1.2.3, can be experimentally limited because of experimental detection thresholds or spectral sensitivity of experimental devices. This should be accounted for, when comparing theory with data, but is rarely done so far (for a nuclear physics example of such a procedure see, e.g., [RRMN21]).

Uncertainties not only come from detectors but also from projectile and target preparation. For example, in a laser irradiation of molecules in the gas phase one does not know the orientation of a given molecule concerning laser polarization which means that signals from various molecules with various orientations are often piled up in the measured observables. This can be cured by applying the same orientation averaging to the theoretical results, although this is a considerable extra computational expense (section 3.4.4.4). Besides orientation, the temperature of the target system is an issue. For example, clusters are often produced in a condensation process involving a finite formation temperature. The latter may considerably alter the shape of the system as it matches the typical ionic vibrations energy range. This in turn may affect numerous quantities such as the optical response in plasmonics (section 2.4.3) or energy or angular resolved electron photo-emission cross sections (section 1.2.3). Finally, the laser itself can be a source of uncertainties due to an incomplete knowledge of the laser pulse. The experimental pulse profile is, in general, known only up to some extent, while the theoretical pulse is modeled by a simple, well-behaved analytical function. Moreover, the laser intensity varies over the pulse diameter which can impact non-linear excitation processes.

1.3.2.3   Computational Aspects

1.3.2.3.1   Some code requirements

Computational aspects are also crucial, especially when developing realistic models which by construction are usually very complex. Practically speaking reliability, flexibility, and robustness are the important criteria, both technically and computationally. Approximations, as unavoidable as they are, limit flexibility and reliability. Avoiding approximations enhances the complexity which, in turn, endangers robustness. One has to find a proper compromise and such compromise can change with the intended applications.

1.3.2.3.2   Difficulties with open source codes

Another issue is the reproducibility, generally speaking, "validity", of computed quantities. Any code behind a model of such a complex scenario as irradiation contains compromises and hidden parameters. This imposes each code limitations in its applicability, beyond the limitations due to the physical description and associated equations. One should keep in mind these intrinsic computational limitations of any code to allow a proper comparison to experiments. The free availability of codes may be dangerous in this respect as users, who have not been directly involved in the code development, may easily overlook such hidden parameters. A compromise is to be found between code access and good practice thereof. The free availability of codes is certainly an important issue, especially in terms of manpower for comparisons with experiments. But it should only be done with caution and awareness.

1.3.2.3.3   Code to code benchmarking

Along this line of reasoning concerning code capabilities and limitations comes as next issue code to-code benchmarking. The equations that model a given problem are usually highly complex. Each code developed for solving the set of equations uses its own algorithms. A key issue is thus to ensure that two different codes deliver compatible outputs in a given physical situation.

Take as a simple example a dimer molecule using a (time-dependent) mean-field approach to describe the ground state and subsequent dynamics. Let us focus first on the ground state of the molecule as an initial condition of the irradiation scenario. From the experimental side, one knows bond length, vibrational frequency, dissociation energy, Ionization Potential (IP) and maybe a few more single-particle energies. All these properties have an impact on dynamics. A real-time TDLDA calculation (section 2.2.3.4) may miss the IP because the self-interaction correction (section 2.2.3.5) was not included, even if dissociation energy and bond length may be correct. The vibrational frequency is a bit more touchy. From one code to the next, depending on the actual functional and pseudo-potential used, one will obtain different values of it, in various relations to experiments.

Now comes the key question of the optimal choice. What is the "best" compromise to be made between the varying quality of the outputs? Is it better to favor structural quantities like bond length, vibrational frequency or ionization potential? In a dynamic context, the choice is disputable. One needs a good IP to properly model direct ionization following the early stages of irradiation. But coupling to ionic motion in the longer term is also a key issue. There is thus a compromise to find and the developers of each code have made their own choices. The question is then to what extent such choices impact dynamical observables. Benchmarking code to code is thus crucial to be able to validate numerical modeling and provide what one could call "numerical" error bars. This should be done ahead of a direct comparison with experiments which by construction will incorporate such hidden effects. Unfortunately, only a few such systematic investigations do exist [OPP+20].

1.3.2.4   Toward Multi-Scale Approaches

As discussed above, each theory has a limited range of applicability due to the approximations made, tailored with some particular scenario in mind. For example, a detailed quantum description of electrons is advisable for the early phases of an irradiation process to account for the immediate coupling of electrons to the electromagnetic perturbations. Macroscopic or classical treatments of electrons at that stage is dubious. On the other hand, the classical

Molecular Dynamics (MD) with appropriate force fields (section 2.5.2) can be very useful to track the long-term evolution of the system, once charge migration has settled down. They can then allow us to follow the dynamics for very long times, hardly accessible to fully quantum mechanical descriptions.

A typical compromise could thus be to develop an expensive quantum approach to the electron response over short times and then, in the longer term, switch to a simpler phenomenological treatment of electrons. Ultimately, once ionization has settled down MD may usually be a good option for very long-term evolution. This is easy to envision, formally, but involved, in practice. Much diligence is required to develop an appropriate interface from quantum regime to classical MD. There exists already some experience for the reduction of a quantum state to an ensemble of classical particles. True problems appear if some fragments or sub-systems remain at or come back to low excitation in the quantum regime. Less clear so far is how to recover a quantum state from a classical phase space distribution. Only a few such interfaces have already been investigated and remain at the prototypical level [NSH+19]. Challenging and excitation tasks thus remain for future development of multi-scale approaches.

2 Theoretical Tools

Chapter 1 has provided a survey of the major phenomena taking place in the dynamics of irradiation of molecules and nano-objects. We have identified several key aspects whose theoretical analysis deserves elaborate tools. An irradiation involves all constituents of the irradiated systems, namely electrons, ions and possibly an extra environment. We thus have first to introduce relevant approaches to describe each of them in a way adapted to the physical context. While structure aspects are rather well under control, the intrinsic dynamical nature of irradiation requires approaches explicitly accounting for the dynamics of all constituents of the system.

The focus of the present chapter is thus on dynamical methods, adapted to scenarios possibly far away from equilibrium, which, to a large extent, eliminates methods based on linear response. Among the various constituents of the considered systems, the most involved part concerns electrons which can, but for exceptional cases, only be dealt with at a quantum mechanical level. To a large extent, ionic degrees of freedom and electronic environment in general, can be treated in somewhat simpler terms. We are thus hitting here the core of the topic of this book, namely the real-time description of, possibly far off equilibrium, electron dynamics. In the spirit of a guidebook, we shall nevertheless not enter technical details and thus refer the reader to proper citations for details. In turn, we shall try to deliver a critical view of the presented methods, with their strengths and fields of applications as well as their limitations.

The most general electron-ion wavefunction describing a many electrons/atoms system reads, in coordinate-space representation, $\Psi(\mathbf{r}, \mathbf{R}, t)$, where $\mathbf{r}$ labels the vector of electronic coordinates, $\mathbf{R}$ the vector of ionic coordinates and t is time. In most general terms $\Psi(\mathbf{r}, \mathbf{R}, t)$ is the solution of the many-body Schrödinger equation:

$$i\hbar\partial_t \Psi(\mathbf{r}, \mathbf{R}, t) = \hat{H}_{\text{tot}} \Psi(\mathbf{r}, \mathbf{R}, t) , \tag{2.1a}$$

$$\hat{H}_{\text{tot}} = \hat{H}_{\text{ions}} + \hat{H}_{\text{el}} + \hat{H}_{\text{coupl}} + \hat{U}_{\text{ext}} , \tag{2.1b}$$

$$\hat{H}_{\text{ions}} = -\sum_{I}^{N_{\text{ion}}} \frac{\hbar^2}{2M_I} \nabla^2 \mathbf{R}_I + \frac{1}{2}\sum_{I \neq J} \frac{Z_I Z_J e^2}{|\mathbf{R}_I - \mathbf{R}_J|} , \tag{2.1c}$$

$$\hat{H}_{\text{el}} = \sum_{i}^{N_e} -\frac{\hbar^2}{2m_e} \nabla^2 \mathbf{r}_i + \frac{1}{2}\sum_{i \neq j} \frac{e^2}{|\mathbf{r}_i - \mathbf{r}_j|} , \tag{2.1d}$$

$$\hat{H}_{\text{coupl}} = -\sum_{I,i} \frac{Z_I e^2}{|\mathbf{R}_I - \mathbf{r}_i|} , \tag{2.1e}$$

$$\text{and} \quad E_{\text{tot}} = \langle \Psi | \hat{H}_{\text{tot}} | \Psi \rangle , \tag{2.1f}$$

DOI: 10.1201/9781003127949-2

where m_e, M_I, Z_I are respectively electron, ion masses, and ion charges and N_e is the number of electrons and N_{ion} the number of ions. The potential $\hat{U}_{ext}(\mathbf{r},\mathbf{R},t)$ is the field delivered by the external perturbation, usually of an electromagnetic nature, such as the one generated by a laser or a bypassing projectile. It may also include constant fields such as the ones created by an external static environment.

Equation (2.1) represents the most general description based on a microscopic many-body Hamiltonian. This is the starting point for all ab initio theories with Hartree-Fock (section 2.2.2) at the lowest level and more advanced approaches as sketched in section 2.3. An alternative, important line of development is Density Functional Theory (DFT) which is explained in section 2.1.5 and used for the majority of applications later on. There, the starting point is the total energy where the electronic part is formulated in terms of the local electron density. There exists no equivalent Hamiltonian. Every step has to be expressed by a variation of the total energy.

As the most involved part are the electrons we will deal with that in several sections. Section 2.1 presents the different theoretical tools to deal with the quantum mechanical many-electron problem. Section 2.2 addresses the various brands of mean-field approximations where electrons are treated as independent particles moving in a common mean-field. These mean-field methods are presently the most efficient, thus widely used, approaches. Section 2.3 presents actual attempts to go beyond the mean-field level, while section 2.4 is devoted to semi-classical methods, relevant for high energies and/or large numbers of particles. Finally, section 2.5 discusses the description of ions and the coupling of electrons to them as well as to an environment in general.

2.1 GENERAL DESCRIPTION OF MANY ELECTRONS SYSTEMS

Sections 2.1, 2.2, 2.3 and 2.4 concern the treatment of the electronic problem. This means that ions, similar to $\hat{U}_{ext}$, appears as an "external" agent acting on the electrons. For the sake of simplicity, we shall use the same notation $|\Psi\rangle$ for the electronic wavefunction as the state characterizing the whole system, now keeping in mind that ionic degrees of freedom are considered as parameters. They will come back into play in section 2.5 in which the coupling with electrons and the dynamics of ions themselves will be directly addressed.

2.1.1 REPRESENTATIONS AND THE MANY-BODY HAMILTONIAN

The starting point of a quantum mechanical description is the state vector $|\Psi\rangle$ in Hilbert space together with operators acting on it, often representing observables. The practical handling is done with respect to a certain set of basis states which allows to represent the wavefunction as an array of complex numbers. This can be, e.g., configuration space representation $\Psi_n(t) = \langle n|\Psi(t)\rangle$ where $|n\rangle$ is some convenient set of basis functions, often oscillator functions or eigenstates of the momentum operator together with spin components. The most common is the coordinate space representation

$$|\Psi\rangle \longrightarrow \Psi(\mathbf{x}_1...\mathbf{x}_N) = \langle \mathbf{x}_1...\mathbf{x}_N|\Psi\rangle \qquad (2.2a)$$

where

$$\mathbf{x} = (\mathbf{r}, \sigma) \qquad (2.2b)$$

is a composite vector of space coordinate $\mathbf{r}$ and spin component σ and the $|\mathbf{x}_1...\mathbf{x}_N\rangle$ are the eigenfunctions of the position and spin operator for the N particles.

Computations in coordinate-space representation will often require "integration" over $\mathbf{x}$. That means in practice

$$\int d\mathbf{x} \equiv \sum_\sigma \int d^3 r , \qquad (2.2c)$$

i.e., integration over the three space components and summation over the spin components.

A typical electronic many-body Hamiltonian in coordinate-space representation is

$$\hat{H} = \sum_{i=1}^{N_e} \left(-\frac{\hbar^2}{2m_e}\nabla_i^2 + \hat{U}_{\text{back}} \right) + \frac{1}{2}\sum_{i\neq j} V(|\mathbf{r}_i - \mathbf{r}_j|) \qquad (2.3a)$$

$$\hat{U}_{\text{back}} = \sum_{i=1}^{N_e}\sum_{I=1}^{N_{\text{ion}}} \hat{U}_{\text{PsP}}(\mathbf{r}_i, \mathbf{R}_I) + \sum_{i=1}^{N} U_{\text{ext}}(\mathbf{r}_i) \qquad (2.3b)$$

where m_e is electron mass, N_{ion} the total number of background ions, ∇_i^2 is the Laplacian operator acting on space coordinate $\mathbf{r}_i$, and $V(|\mathbf{r}_i - \mathbf{r}_j|)$ the Coulomb interaction $V(r) = e^2/r$ between particle i and j. The $\hat{U}_{\text{back}}$ is the one-body operator for the electrons from the ionic background and possible external excitation fields.

The term "ions" means here the nuclei together with some closed shells of core electrons which are energetically well separated from the valence electrons. The interaction between nuclei and electrons is purely Coulombic (Eq. (2.1e)). The intricate interaction between core electrons and valence electrons can be reduced to comparably simple pseudo-potentials (PsP). These are in general non-local operators, but they can be simplified to purely local PsP in some favorable cases (metal ions). A detailed description of PsP goes beyond the scope of this book. We refer to [BHS82, Sza85, GTH96] for examples and details.

Metallic systems often allow a further simplification of the ionic background to the jellium model where the ionic density is replaced by a smooth jellium density. It is advisable to use a jellium with a soft surface zone [MRM94]. The corresponding background potential is the Coulomb potential of the jellium density. That reads for near-spherical systems

$$
\hat{U}_{\text{back}} = \sum_{i=1}^{N_e} U_{\text{jel}}(\mathbf{r}_i) + \sum_{i=1}^{N_e} U_{\text{ext}}(\mathbf{r}_i) \tag{2.4}
$$

with
$$
\frac{\nabla^2}{4\pi} U_{\text{jel}}(\mathbf{r}) = \rho_{\text{jel}}(\mathbf{r}) = \frac{3e^2}{4\pi r_s^3} \left[1 + \exp\left(\frac{|\mathbf{r}| - R_{\text{jel}}}{\sigma_{\text{jel}}} \right) \right]^{-1}, \tag{2.5}
$$

where σ_{jel} characterizes the surface width of the jellium density and the jellium radius R_{jel} is adjusted by the normalization to the total number of ions $\int d\mathbf{r}\, \rho_{\text{jel}} = N_{\text{ion}}$. The central density is determined by the bulk density $\rho_0 = 3/(4\pi r_s^3)$ of the given material of Wigner Seitz radius r_s. The model can easily be extended to account for deformed systems, see e.g., [RS03].

2.1.2 WAVEFUNCTIONS

A quantum mechanically pure electron state can be described by a wavefunction which is, in general, a time-dependent state vector $|\Psi(t)\rangle$ in Hilbert space. Its time evolution is governed by the time-dependent Schrödinger equation

$$
i\hbar \partial_t |\Psi(t)\rangle = \hat{H} |\Psi(t)\rangle \tag{2.6a}
$$

where $\hat{H}$ is the Hamiltonian of the system (an example was given above), which is a Hermitian operator. The time-dependent Schrödinger equation poses an initial-value problem. It determines $|\Psi(t)\rangle$ for given initial $|\Psi(t_0)\rangle$. A compact formulation of the solution can be written in terms of the time-evolution operator $\hat{U}$ as

$$
|\Psi(t)\rangle = \hat{U}(t, t_0) |\Psi(t_0)\rangle \tag{2.6b}
$$

where $\hat{U}$ is solution of

$$
i\hbar \partial_t \hat{U}(t, t_0) - \hat{H} \hat{U}(t, t_0) , \quad \hat{U}(t_0, t_0) = \hat{1} \tag{2.6c}
$$

with $\hat{1}$ being the unity operator. Because $\hat{H}$ is Hermitean, $\hat{U}$ is a unitary operator

$$
\hat{U}(t, t') = \hat{U}^\dagger(t', t) . \tag{2.6d}
$$

Stationary states are determined by the stationary Schrödinger equation

$$
E|\Psi\rangle = \hat{H}|\Psi\rangle \tag{2.6e}
$$

where E is the eigen-energy associated with the state $|\Psi\rangle$. We will concentrate in the following on the time-dependent case as we are mostly interested in theories for electronic dynamics.

In view of the upcoming approximations, it is important to note that the time-dependent Schrödinger equation can be derived from the variational principle of least action

$$0 = \delta_{\langle\Psi|} \left\{ \int_0^t dt' \langle\Psi(t')|i\hbar\partial_{t'} - \hat{H}|\Psi(t')\rangle \right\} . \tag{2.7a}$$

In similar fashion, the stationary Schrödinger equation can be derived from variationally minimizing the total energy leading to the equation

$$0 = \delta_{\langle\Psi|}\langle\Psi|\hat{H} - E|\Psi\rangle \tag{2.7b}$$

where E serves here as the Lagrangean parameter for the normalization condition $\langle\Psi|\Psi\rangle = 1$. Variational principles are the most simple and safe way to formulate approximations at the level of the energy expectation value $\langle\Psi(t)|\hat{H}|\Psi(t)\rangle$ and to derive from that the equations of motion variationally.

Mixed states as they appear, e.g., through thermalization can be described in the wavefunction picture as an ensemble of quantum states

$$\mathcal{E} = \left\{ |\Psi^{(\nu)}(t)\rangle, W^{(\nu)} \right\} \tag{2.8}$$

with a probability $W^{(\nu)}$ associated with the ensemble member Ψ^ν. Such ensembles can be useful in practice, see e.g., the STDLDA approach in section 2.3.8, but are too bulky for formal derivations. This is why one often steps forward to more general objects such as density matrices or Green's functions which will be introduced below.

As outlined in section 2.1.1, the practical handling is done in a specific c-number representation, most commonly the coordinate-space representation for the states $|\Psi(t)\rangle \rightarrow \Psi(\mathbf{x}_1...\mathbf{x}_N, t)$ and similarly for the operators.

The Hilbert space covers the quantum states of a many-body system. The union of Hilbert spaces to all N_e delivers the Fock space as the most general space of many-body systems. The Fock space allows to formulate many-body physics in very general terms where Fermion creation and annihilation operators $\hat{a}_i^\dagger$ and $\hat{a}_i$ are the versatile key objects of representation, see e.g., section 2.2.4.1.

2.1.3   DENSITY MATRICES

The many-body density matrix is an operator in many-body Hilbert space which represents the quantum state of a system. For pure states, i.e. a state which can be represented by one wavefunction, its abstract Hilbert space representation is constructed as

$$\hat{D}(t) = |\Psi(t)\rangle\langle\Psi(t)| \tag{2.9a}$$

and its coordinate-space representation reads accordingly

$$\hat{D}(\mathbf{r}_1...\mathbf{r}_N, \mathbf{r}_1'...\mathbf{r}_N'; t) = \Psi(\mathbf{r}_1...\mathbf{r}_N, t)\Psi^*(\mathbf{r}_1'...\mathbf{r}_N', t) , \tag{2.9b}$$

The density-matrix picture looks at first glance more involved than the wave-functions. However, it is often more elegant for formal derivations of approx-imations as, e.g., in the BogoliubovBornGreenKirkwoodYvon (BBGKY) hi-erarchy [Bal75] and it allows to formulate a semi-classical limit which estab-lishes an immediate link to classical phase-space distributions, see section 2.4. The most obvious advantage is that it allows for a compact representation of mixed states. The density matrix representing an ensemble of wavefunctions Eq. (2.8), reads

$$\hat{D}(t) = \sum_{\nu} |\Psi^{(\nu)}(t)\rangle W^{(\nu)}(t) \langle \Psi^{(\nu)}(t)| \tag{2.10}$$

and analogously in the coordinate-space representation. Changes in the oc-cupation probabilities $W^{(\nu)}(t)$ are related to incoherent processes, often de-scribed by a master equation. Note that the mixed state Eq. (2.10) has to be distinguished from the coherent superposition Eq. (2.51) in MCTDHF with occupation amplitudes c_m which still represents a pure quantum state.

Purely coherent propagation is determined by the generalization of the Schrödinger equation to

$$i\hbar \frac{\partial \hat{D}(t)}{\partial t} = \left[\hat{H}, \hat{D}(t)\right] \tag{2.11}$$

which defines, again, an initial value problem propagation $\hat{D}(t)$ for a given initial $\hat{D}(t{=}0)$. Alternatively, one can express the solution compactly in terms of the time-evolution operator from Eq. (2.6c) as

$$\hat{D}(t) = \hat{U}(t, t_0) \hat{D}(t_0) \hat{U}(t_0, t) \tag{2.12}$$

which, again, expresses the initial-value problem in one compact equation.

The Equation of Motion (EoM) for Eq. (2.11) resembles the Heisenberg EoM for observables. We emphasize that it is different. An observable $\hat{A}(t)$ in the Heisenberg picture evolves in the reverse direction as

$$i\hbar \partial_t \hat{A}(t) = -\left[\hat{H}, \hat{A}(t)\right] \tag{2.13}$$

signified by the minus sign in front of the commutator. One cannot have both. One has to make a decision. In the Schrödinger picture we propagate $\hat{D}(t)$ with Eq. (2.11) and compute the expectation value with a time-independent observable as $\overline{A} = \text{tr}\{\hat{D}(t)\hat{A}\}$ while in the Heisenberg picture, we propagate $\hat{A}(t)$ and keep the state $\hat{D}$ as given initially thus evaluating $\overline{A} = \text{tr}\{\hat{D}\hat{A}(t)\}$.

2.1.4 GREEN'S FUNCTIONS

The Green's function G provides a further alternative to formulate the many-body problem. It expresses the temporal propagation of the wavefunction field.

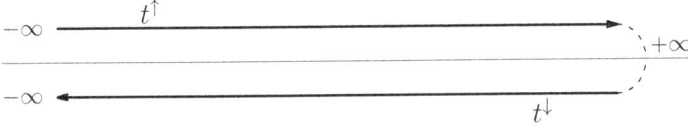

Figure 2.1 Sketch of the Keldysh path along the time axis (faint horizontal) line.

In coordinate space, it reads for example

$$\Psi(\mathbf{r}_1...\mathbf{r}_{N_e}, t) = \int d^3 r_1'...d^3 r_{N_e}' \, G(\mathbf{r}_1...\mathbf{r}_N, t; \mathbf{r}_1'...\mathbf{r}_{N_e}', t_0) \Psi(\mathbf{r}_1'...\mathbf{r}_{N_e}', t_0) \,.$$
(2.14a)

At first glance, it has a similarity with the time-evolution operator. However, the Green's function is usually defined in a restricted time domain. The most widely used is the retarded Green's function describing forward propagation in the domain $t > t_0$. It is defined by the differential equation

$$\left(i\hbar\partial_t - \hat{H} \right) G^r(\mathbf{r}_1...\mathbf{r}_{N_e}, t; \mathbf{r}_1'...\mathbf{r}_{N_e}', t_0) = \delta(t - t_0) \prod_{i=1}^{N_e} \delta^3(\mathbf{r}_i - \mathbf{r}_i') \qquad (2.14b)$$

for $t > t_0$ and $G(\mathbf{r}_1...\mathbf{r}_N, t; \mathbf{r}_1'...\mathbf{r}_N', t_0) = 0$ for $t < t_0$. One often considers also the advanced Green's function G^a which is defined in the complementing domain $t < t_0$ and other combinations, for details see, e.g., [FW71].

The Green's function as defined in Eq. (2.14a) deals naturally with propagating pure quantum states. There is a generalization to mixed states. This is achieved by extending the natural time interval $t = -\infty \longrightarrow t = +\infty$ to the Keldysh path

$$\tau = (t^\uparrow, t^\downarrow) \,, \quad t^\uparrow \in (-\infty \longrightarrow +\infty) \,, \quad t^\downarrow \in (+\infty \longrightarrow -\infty) \,. \qquad (2.15)$$

which is composed out of a forward running branch $t^\uparrow$ complemented by a backward running branch $t^\downarrow$, for an illustration see Figure 2.1. This Keldysh path provides a formally involved, but very powerful, description of non-equilibrium processes. For a detailed presentation see [SvL13].

There is a connection between Green's functions and density matrices, namely the latter are recovered in the limit of $t \to t_0$ as

$$\lim_{t \searrow t_0} G(\mathbf{r}_1...\mathbf{r}_N, t; \mathbf{r}_1'...\mathbf{r}_N', t_0) = D(\mathbf{r}_1...\mathbf{r}_N, \mathbf{r}_1'...\mathbf{r}_N'; t_0) \qquad (2.16)$$

where $\searrow$ means the limit from the domain $t > t_0$.

2.1.5   DENSITY FUNCTIONAL THEORY (DFT)

Density Functional Theory (DFT) is a further formulation of the many-electron problem [Koh99]. It aims at a variational formulation of a many-electron system where the total energy is expressed exclusively in terms of the local electron density $\varrho\,(\mathbf{r})$ which is the diagonal part of the one-body density matrix $\varrho\,(\mathbf{r}) = \hat{\rho}(\mathbf{r}, \mathbf{r})$, see section 2.1.3 and Eq. (2.18b). The first approach in that direction started long ago with the Thomas-Fermi (TF) model of atoms [Fer27, Tho27], where the kinetic energy is modeled as a function of local density yielding together with the direct part of the Coulomb interaction the energy in terms of density alone. The atomic Thomas Fermi model provides reasonable average trends of total energies, which served as a strong motivation for DFT. The next step in that direction was the Slater approximation which expresses the energy of the Coulomb exchange term through the local density [Sla51]. Both the TF model and Slater approximation were intended for immediate practicability, an aspect which we work out in section 2.2. DFT in general is often considered still as an exact many-body theory. The first formal step toward a thorough DFT formulation is provided by the Hohenberg-Kohn theorem [HK64] which states that the total electronic energy E of the ground state can be written as a universal functional of the local density alone, i.e. $E = E[\varrho\,(\mathbf{r})]$. This theorem guarantees the existence of such a functional in principle, and for the ground state, but does not specify it in practice. In fact, the exact density functional must be highly non-analytical to incorporate quantum shell structure and many-body effects.

But already the simpler part of quantum shell structure is lost in simple forms of DFT as TF theory. The problem was overcome in the Kohn-Sham (KS) formulation of DFT [KS65] with the idea to use still the exact kinetic energy of non-interacting particles $E_{\text{kin}} = \sum_i \int d^3r\, \varphi_i^*(\mathbf{r}) \left(-\frac{\hbar^2 \nabla^2}{2m}\right) \varphi_i(\mathbf{r})$ while expressing all interaction energy as functional of local density, i.e. $E^{(\text{KS})} = E_{\text{kin}} + E_{\text{int}}[\varrho\,(\mathbf{r})]$, for details see section 2.2.3.1 and Eqs. (2.26). This energy still depends on the detailed s.p. wavefunctions φ_i via E_{kin} and thus recovers the full quantum structure of the system. As in the Hohenberg-Kohn theorem, the KS formulation guarantees the existence of an "exact" exchange-correlation functional but does not provide any constructive way to reach it. The KS picture, nevertheless, proved to be a very efficient starting point for DFT development and has become over the years the dominating tool in DFT. In the following, we shall thus most of the time use DFT within the KS formulation. A clarification is, nevertheless, in place here: s.p. wavefunctions in the KS picture are merely theoretical tools to compute ground state energy and density. A physical interpretation of them has to be done with caution. We will come back to that in section 2.2.3.5.

DFT can be extended to a Time-Dependent DFT (TDDFT) by virtue of the Runge-Gross theorem [RG84, DG90]. The "exact" functional then becomes conceptually even more involved as it depends on the time-dependent local density $\varrho\,(\mathbf{r}, t)$ and so may include also memory effects. Again, most practical approaches to TDDFT are based on the KS formulation of the theory dealing still with kinetic energy expressed in terms of s.p. wavefunctions. There are several acronyms used for Time-Dependent Density Functional Theory. The generic one is TDDFT which embraces real time treatments as well as linear response computations in frequency space. To emphasize situations where the actual solution is performed as explicit time propagation, some authors use the name real-time TDDFT (rt-TDDFT). For more details on TDDFT and examples see, e.g, the reviews [RS03, MUN06, MMN+12, Ull12]. Note, furthermore, that the acronym TDDFT is often also used for approximate treatments as, e.g., the local-density approximation, see section 2.2.3.

TDDFT relies on a dynamical treatment of the energy-density functional. It can be further extended to include local current distribution in the functional. This leads to Time Dependent Current Density Functional Theory (TDCDFT) with key dynamical variables now being density $\varrho\,(\mathbf{r})$ and current $\mathbf{j}(\mathbf{r})$ [Ull12]. Again TDCDFT is most often used in a Kohn Sham picture. The variational principle then adds a vector potential to the KS Hamiltonian which renders TDCDFT an even more involved theory. The gain is that some dynamical effects can be better included. Unfortunately, practical applications are rare and mostly focus on linear response dynamics, not suited to address far-off-equilibrium situations [DLR+18]. Thus we will not consider applications to TDCDFT in this book and refer the reader to the book [Ull12] for details on TDCDFT.

Both DFT and TDDFT have become over the years major tools of investigations of both structural and dynamical properties of a bunch of physical systems ranging from atoms to solid state. Detailed reviews on DFT can be found in [DG90] for stationary DFT and [MG04] for Time-Dependent DFT (TDDFT). There remains the question of which functional to use (section 2.2.3.2). DFT and TDDFT (with some conditions) theorems as such state the existence of a universal, exact, functional but do not provide an explicit construction principle for it. In practical applications, especially in the time domain, the majority of successful functionals are built on the basis of the simplest approximation called the Local Density Approximation (LDA), which turns out to deliver a robust and reliable mean-field approach. This is the level of (TD)DFT used in most application examples later on in this book. We shall thus discuss LDA in more detail in section 2.2.3.

2.2 MEAN-FIELD DYNAMICS

The term "mean-field model" designates a summary notion for everything which can be formulated in terms of single-particle (s.p.) properties, particularly s.p. wavefunctions. It embraces empirical models based on one-body potentials, Hartree-theory, Hartree-Fock and, in particular, DFT at the level of the Local Density Approximation (LDA) and generalizations thereof. We will address in this section those mean-field models which will be used later on in applications.

Mean-field models are distinguished by two basic features:

1. The system is described in terms of an ortho-normal set of single-particle (s.p.) wavefunctions $\{\varphi_i, i = 1...\Omega\}$ where the number of relevant states is in general $\Omega \geq N_\mathrm{e}$. At zero temperature, we have $\Omega = N_\mathrm{e}$. Finite temperature and/or mixed states require $\Omega > N_\mathrm{e}$.

2. The s.p. wavefunctions are determined by a one-body Hamiltonian $\hat{h}$, coined the mean-field Hamiltonian and, in the case of Density Functional Theory (DFT) also Kohn-Sham Hamiltonian (section 2.2.3). The $\hat{h}$ constitutes the common mean field in which all particles move independently.

In this basic setting, there exist many different approximations which we will outline briefly in this section. We start with the different formal descriptions of an independent-particle state specifying the general objects from section 2.1, namely wavefunction, density matrix, and Green's function, which hold for all mean-field approximations. In subsequent sections, we present the different models for the mean-field Hamiltonian.

2.2.1 DESCRIPTIONS OF AN INDEPENDENT-PARTICLE STATE

Probably the best known way to formulate an independent-particle state is the wavefunction description in terms of a Slater determinant of s.p. wavefunctions φ_i

$$\Psi(\mathbf{x}_1...\mathbf{x}_{N_e}, t) \rightarrow \Phi(\mathbf{x}_1...\mathbf{x}_{N_e}, t) = \mathcal{A}\left\{\prod_{i=1}^{N_e} \varphi_i(\mathbf{x}_i, t)\right\} \qquad (2.17a)$$

where $\mathcal{A}$ is the anti-symmetrization operation. It reads for any function of $f(\mathbf{x}_1...\mathbf{x}_{N_e})$ of N_e-coordinates

$$\mathcal{A} = \frac{1}{\sqrt{N_e!}} \sum_{\mathcal{P}} \text{Sg}(\mathcal{P}) f(\mathbf{x}_{p_1}...\mathbf{x}_{p_N}) \qquad (2.17b)$$

where $\mathcal{P}$ is a permutation $1...N_e \longrightarrow p_1...p_N$ and $\text{Sg}(\mathcal{P})$ its sign (number of commutations to realize $\mathcal{P}$). The pre-factor $(N!)^{-1/2}$ serves to maintain normalization of the total wavefunction (provided the set $\{\varphi_i\}$ was ortho-normalized).

The density matrix Eq. (2.9) for an independent-particle state in coordinate-space representation then becomes

$$\begin{aligned}
D(\mathbf{x}_1...\mathbf{x}_{N_e}, \mathbf{x}'_1...\mathbf{x}'_{N_e}; t) &= \Phi(\mathbf{x}_1...\mathbf{x}_{N_e}, t)\Phi^*(\mathbf{x}'_1...\mathbf{x}'_{N_e}, t) \\
&= \mathcal{A}_\mathbf{x}\mathcal{A}_{\mathbf{x}'}\left\{\rho(\mathbf{x}_1, \mathbf{x}'_1, t)...\rho(\mathbf{x}_{N_e}, \mathbf{x}'_{N_e}, t)\right\} \\
&= \sqrt{N_e!}\mathcal{A}_\mathbf{x}\left\{\rho(\mathbf{x}_1, \mathbf{x}'_1, t)...\rho(\mathbf{x}_{N_e}, \mathbf{x}'_{N_e}, t)\right\} \quad (2.18a)
\end{aligned}$$

where the index $\mathbf{x}$ on $\mathcal{A}$ indicates that this anti-symmetrization acts only among the $\mathbf{x}_i$ coordinates and leaves the $\mathbf{x}'_i$ untouched. The step from the second to the third line exploits the fact that the first anti-symmetrization renders the density matrix already fully anti-symmetric such that the second one may be omitted. The full density matrix D of independent particles is then an anti-symmetrized product of the one-body density matrices

$$\rho(\mathbf{x}, \mathbf{x}', t) = N \int d\mathbf{x}_2...d\mathbf{x}_{N_e} D(\mathbf{x}_1...\mathbf{x}_{N_e}, \mathbf{x}'_1...\mathbf{x}'_{N_e}; t)$$

$$= \sum_{i=1}^{N_e} \varphi_i(\mathbf{x}, t)\varphi_i^*(\mathbf{x}', t). \tag{2.18b}$$

The one-body density matrix is particularly useful. It carries all the information on the independent-particle state in one compact object. For example, it allows to compute the expectation value of any one-body observable $\hat{A}$ as

$$\overline{A}(t) = \int d\mathbf{x}d\mathbf{x}' \, A(\mathbf{x}', \mathbf{x})\rho(\mathbf{x}, \mathbf{x}', t) \equiv \mathrm{tr}\{\hat{A}\hat{\rho}\} \tag{2.19}$$

where $A(\mathbf{x}', \mathbf{x})$ is the coordinate-space representation of the operator $\hat{A}$, and $\hat{\rho}$ is the abstract operator formulation of the one-body density matrix in Hilbert space. The representation of the one-body density matrix in the second line of Eq. (2.18b) is equivalent to the Slater state and it holds for mean-field states at temperature zero.

The advantage of the density-matrix formalism is that it is not confined to temperature zero. It also covers the broader range of mixed states. The most general one-body density matrix reads $\rho(\mathbf{x}, \mathbf{x}', t) = \sum_{i,j=1}^{\Omega} \varphi_i \mathbf{x}, t)\rho_{ij}\varphi_j^*(\mathbf{x}', t)$ with a Hermitian matrix ρ_{ij} and, of course, with $\Omega > N_e$ more states than electrons. This matrix can always be diagonalized which yields the natural orbitals φ_i and the diagonal representation

$$\rho(\mathbf{x}, \mathbf{x}', t) = \sum_i^N \varphi_i(\mathbf{x}, t)n_i\varphi_i^*(\mathbf{x}', t) \tag{2.20}$$

where the diagonal elements $0 \leq n_i \leq 1$ constitute the occupation numbers, the probability with which state φ_i is occupied. Pure states which are also represented by Slater determinants are distinguished by occupation numbers $n_i \in \{0, 1\}$. A limiting case is the thermal equilibrium states where n_i reaches a Fermi distribution.

The Green's function for an independent-particle state factorizes similarly as the density matrix to

$$G(\mathbf{x}_1...\mathbf{x}_{N_e}, t; \mathbf{x}'_1...\mathbf{x}'_{N_e}; t') = (N_e!)^{1/2}\mathcal{A}_\mathbf{x} \{g(\mathbf{x}_1, t; \mathbf{x}'_1, t')...g(\mathbf{x}_{N_e}, t, \mathbf{x}'_{N_e}, t')\} \tag{2.21}$$

where the one-body Green's function reads for the retarded case of a system at temperature zero (pure state)

$$g(\mathbf{x}, t; \mathbf{x}', t') = \sum_{i=1}^{N_e} \varphi_i(\mathbf{x}, t)\theta(t - t')\varphi_i^*(\mathbf{x}', t'). \tag{2.22}$$

The factorization Eq. (2.21) applies also to mixed states, but the one-body Green's function acquires a more involved form on the Keldysh path Eq. (2.15).

2.2.2 HARTREE-FOCK

The prototype self-consistent mean-field theory is the Hartree-Fock (HF) approach. It starts with the ansatz of independent particles for the state of the system whose various formulations are found in section 2.2.1 and derives the mean-field equations from variation of the energy, see Eqs. (2.7), with respect to the s.p. wavefunctions φ_i^*.

The first step is to evaluate an expression for the total energy for an independent-particle state. We give here the result for the generic Hamiltonian Eq. (2.3a) in case of a pure state (temperature zero)

$$
\begin{aligned}
E &= \langle \Phi | \hat{H} | \Phi \rangle \\
&= \sum_i \int d\mathbf{x}\, \varphi_i^+(\mathbf{x}) \left(\frac{-\hbar^2 \nabla^2}{2m} + \hat{U}_{\text{back}} \right) \varphi_i(\mathbf{x}) \\
&+ \frac{1}{2} \int d\mathbf{x}\, d\mathbf{x}'\, V(|\mathbf{r} - \mathbf{r}'|) \left[\rho(\mathbf{x}, \mathbf{x})\rho(\mathbf{x}', \mathbf{x}') - \rho(\mathbf{x}, \mathbf{x}')\rho(\mathbf{x}', \mathbf{x}) \right].
\end{aligned} \quad (2.23)
$$

For the variation, we recall the form Eq. (2.18b) of the one-body density matrix. The variation with respect to $\varphi_i^*(\mathbf{x})$ then delivers the Hartree-Fock Hamiltonian from $\delta E / \delta_{\varphi_i^*(\mathbf{x})} = \hat{h}_{\text{HF}} \varphi_i(\mathbf{x})$ reading in detail

$$
\hat{h}_{\text{HF}} = -\frac{\hbar^2 \nabla^2}{2m} + \hat{U}_{\text{back}} + U_{\text{C}}(\mathbf{r}) + \hat{U}_{\text{x}}, \quad (2.24a)
$$

$$
U_{\text{C}}(\mathbf{r}) = \int d^3 r'\, V(|\mathbf{r} - \mathbf{r}'|)\, \varrho(\mathbf{r}'), \quad \varrho(\mathbf{r}') = \rho(\mathbf{r}', \mathbf{r}'), \quad (2.24b)
$$

$$
\left(\hat{U}_{\text{x}} \varphi_i \right)(\mathbf{x}) = \int d^3 r'\, V(|\mathbf{r} - \mathbf{r}'|)\, \varrho(\mathbf{r}, \mathbf{r}') \varphi_i(\mathbf{r}'). \quad (2.24c)
$$

Note that the exchange term $\hat{U}_{\text{x}}$ is an integral operator with the consequence that the HF Hamiltonian is necessarily a non-local operator which renders the handling of the HF equations rather cumbersome. The time-dependent and stationary HF equations can be written after a proper unitary transformation in the space of occupied states compactly as

$$
\hat{h}_{\text{HF}} \varphi_i = i\hbar \partial_t \varphi_i, \quad (2.25a)
$$

$$
\hat{h}_{\text{HF}} \varphi_i = \varepsilon_i \varphi_i, \quad (2.25b)
$$

with $i = 1...N$, for a detailed derivation see e.g., [MRS10]. It is instructive to write the Time-Dependent HF (TDHF) equation also in terms of density matrices. As the one-body density matrix determines the full state, it suffices to specify its equation which reads (in compact operator notation)

$$
i\hbar \partial_t \hat{\rho}(t) = \left[\hat{h}_{\text{HF}}, \hat{\rho}(t) \right]. \quad (2.25c)
$$

The HF approximation relies on the given microscopic Hamiltonian. It thus belongs to the family of ab-initio theories and represents the lowest-order approach in that family. It provides at once an agreeable description of small atoms and molecules. There are, however, a few limitations:

1. The exchange potential $\hat{U}_x$ is very expensive to compute and the expense increases quadratically with the electron number N_e.
2. The HF approach does not include long-range correlations from collective plasmon modes. This generates a remarkable screening of the Coulomb interaction which becomes increasingly important in large systems.
3. The description of unoccupied s.p. states remains somewhat ambiguous in the HF theory.

All that has led to hefty searches for improvements, either on the side of more efficient approaches or toward more complete models, ideally in both directions. This will be taken up in the following sections.

2.2.3 PRACTITIONERS DENSITY FUNCTIONAL THEORY (DFT)

The HF approach outlined in the previous section renders large many-body systems manageable by virtue of the independent-particle description, see section 2.2.1. However, the explicit computation of the exchange terms can grow very expensive. Moreover, the HF approach misses correlations which are inevitable in Coulomb systems, the more the larger the system. Both ingredients, Coulomb exchange and correlations can be incorporated into an effective mean-field theory while maintaining the independent-particle picture and reducing the effective mean-field to a merely local potential. That can be achieved by the Kohn-Sham (KS) formulation of DFT, see section 2.1.5. However, practical applications must use DFT in connection with manageable approximations of the energy-density functional. These will be outlined in this section. To keep the formalities simple, we ignore the spin in the following.

2.2.3.1 The Local Density Approximation (LDA)

We will now go for approximate energy-density functionals. A key to the robustness of DFT remains the strictly variational formulation. The starting point is the total electronic energy

$$E_{\mathrm{DFT,el}}[\varrho\,(\mathbf{r}),\mathbf{R}] \equiv E = E_{\mathrm{kin}} + E_{\mathrm{C}} + E_{\mathrm{xc}} + E_{\mathrm{back}} \qquad (2.26)$$

where E_{kin} and E_{C} are kinetic and direct Coulomb energy, respectively, E_{xc} is the exchange-correlation energy, and E_{back} stands for the energy from the ionic background potential and possibly external excitation fields (Eq. (2.3b)). The energy $E_{\mathrm{DFT,el}}$ is a functional of the local one-body density and a function of the ionic configuration $\mathbf{R}$ via the background energy E_{back}.

Practical functionals are usually developed on the grounds of the Local-Density Approximation (LDA), which, to quote W. Kohn himself [Koh99], can be considered as the "mother" of almost all DFT methods. The idea is simple. The Coulomb energy is already a local function and needs no further reduction. The a-priori non-local terms in the total electronic energy are E_{kin} and E_{xc}. These are first computed for the interacting, homogeneous electron gas which yields the energy per particle as a function of the density ϱ, e.g, for the xc-energy

$$\frac{E}{N_e} = \epsilon_{\text{xc}}(\varrho). \tag{2.27a}$$

Now one assumes that the local density $\varrho(\mathbf{r})$ in a finite electron system varies only weakly with position $\mathbf{r}$ which allows to construct its energy piece-wise as (again for the example of xc-energy)

$$E_{\text{xc}} = \int d^3r \, \varrho(\mathbf{r}) \, \epsilon_{\text{xc}}(\varrho(\mathbf{r})). \tag{2.27b}$$

The LDA procedure produces smooth effective energies which can be conveniently parametrized and used in practice. The earliest examples are the Thomas-Fermi functional [Tho27, Fer28]

$$E_{\text{kin}}^{(\text{TF})} = \beta_{\text{kin}} \int d^3r \, \varrho^{5/3}(\mathbf{r}), \; \beta_{\text{kin}} = \frac{3}{5}(3\pi)^{2/3}\frac{\hbar^2}{2m}, \tag{2.28}$$

and the Slater approximation to Coulomb exchange [Sla51]

$$E_{\text{x}}^{(\text{LDA})} = \beta_{\text{x}} \int d^3r \, \varrho^{4/3}(\mathbf{r}), \; \beta_{\text{x}} = \frac{3e^2(3\pi^2)^{1/3}}{4\pi}. \tag{2.29}$$

The Thomas-Fermi theory of the atom (section 2.1.5) uses the above simple expressions for kinetic and exchange energy and is able to describe correctly the average trends of atomic binding in terms of energy and radii [LS77, PY89]. However, it totally misses the quantum fluctuations of binding energy and ionization potential which are crucial agents of chemistry.

2.2.3.2   Functionals Beyond LDA: The Jacob's Ladder, Hybrids, and DFT+U

The LDA proved a good approximation, especially in the early times of DFT when it was the only possible and feasible approach. Not surprisingly, it works extremely well for the jellium model for metals (section 2.1.1), in which the electronic density is uniform. It also works very well for simple metals like Na and Al, and quite well for noble metals like Ag and Au. Interestingly, it does work shockingly well for semiconductors and insulators like Si and diamond, and for semimetals like graphite, even if the densities are not uniform. But the local approximation of the density together with the fulfillment of important sum rules regarding the exchange and correlation holes, lead to the cancellation of errors between exchange and correlation contributions, and make the LDA a robust and unexpectedly accurate approximation. Importantly, there are no *flavors* of the LDA. There is only one version, and a few different parametrizations.

Still, there are some aspects that the LDA does not cover sufficiently well. One of them is the inhomogeneity in the density, which becomes important especially when the system exhibits regions with large density variation or very low densities. One emblematic case is that of hydrogen bonds, in which covalent and polarization effects coexist. Such is the case of water, which is a crucial element in chemistry and biology. Corrections for inhomogeneity have been attached through an expansion of the energy density in gradients of the density and the kinetic energy, giving rise to a so-called Jacob's ladder, of increasingly accurate approximations.

The first one is the Generalized Gradient Approximation (GGA), in which the energy density in Eq. (2.27b) is multiplied by an enhancement factor that depends on the density and its spatial gradient, $F_{xc}(\varrho, \nabla \varrho)$. A caveat with the GGA is that it is not unique. There are many different flavors, some of them designed to enforce known exact conditions about the exchange-correlation hole and density scaling, and others fitting parameters to optimize the performance of certain properties of a data set of molecules or crystals. The GGA, with the two favored flavors, one of each kind above, PBE [PBE96] and BLYP [Bec88, LYP88], have been very successful in attracting the chemistry community to DFT.

The next rung in Jacob's ladder includes not only the gradient of the density but also its Laplacian, or the kinetic energy density. The most widely used meta-GGA are PKZB [PKZB99], TPSS [TPSS03], and the recently proposed SCAN [SRP15]. The latter, in particular, improves significantly over lower-rung functionals on geometries and energies of diversely-bonded materials, including covalent, ionic, metallic, van der Waals, and hydrogen bonds, so it is useful also for soft matter.

Another problem encountered by the LDA is the incomplete cancellation of the Coulomb self-interaction by the exchange term, contrary to what happens in the Hartree-Fock approximation where the exchange contribution includes a term that cancels exactly the self-interaction. This limitation of the LDA is shared by the semi-local GGA and meta-GGA approximations (see also section 2.2.3.5).

The observation that the non-local exact exchange contains the ingredients necessary to cancel the self-interaction, and that this latter leads to severe shortcomings like the incorrect long-range exponential decay of the exchange potential (it should be power-law), suggests that a combination of DFT and Hartree-Fock (DFT-HF), could help with problems derived from the self-interaction error, e.g., the ionization limit, the underestimation of HOMO-LUMO energy or bandgaps, the unstable character of most negative ions, or the incorrect behavior of the image potential at metallic surfaces. This has led to the so-called hybrid functionals, which combine semi-local (DFT) and non-local (HF) exchange in a proportion that depends on the DFT exchange approximation. The idea is to tailor the proper asymptotic behavior explicitly and to work out a smooth transition from the DFT in the interior to the differently derived asymptotic potential. For the correlation contribution, a semi-local DFT approximation is retained. As with GGAs, some functionals have been parameterized to reproduce properties in a data set of molecular systems, and others by imposing known limits and scaling properties of the functional. The most popular hybrid functional amongst chemists is B3LYP [Bec93, LYP88], while physicists tend to use PBE0 [AB99]. These are the hybrid extensions of the BLYP and PBE GGA functionals, respectively. A hybrid extension of the SCAN meta-GGA functional is SCAN0 [HC16].

It has also been proposed to combine the semi-local DFT correlation with PT2, a second-order perturbative dynamical correlation part similar to MP2, but using Kohn-Sham orbitals instead of HF ones. These functionals are called *double hybrids*. A further development in hybrid functionals is the range sepa-ration, by which the long-range of the exchange part is treated either exactly as in HF, or by introducing a screening parameter optimized to reproduce certain properties. Short-range and long-range parts combine DFT and HF exchange in different proportions. These are called range-separated hybrid or double hybrid functionals. A reliable extension to large amplitude dynamics has not yet been developed.

A strategy based on the Hubbard model [HF63] has been also extensively explored. In its simplest (one-band) version, the Hubbard model Hamiltonian reads

$$H = -t \sum_{\langle i,j \rangle, \sigma} c_{i\sigma}^{\dagger} c_{j\sigma} + U \sum_{i} N_i^{\uparrow} N_i^{\downarrow}, \qquad (2.30)$$

where t is the so-called hopping parameter, which sets the hybridization be-tween two nearest-neighbor sites, indicated as $\langle i,j \rangle$ in the summation. The operator $c_{i\sigma}^{\dagger}$ creates an electron of spin σ at site i, while operator $c_{j\sigma}$ anni-hilates an electron of spin σ at site j. The parameter U sets the strength of the electron-electron on-site repulsion, while N_i^{σ} is the number of electrons of spin σ at site i. A large value of the Hubbard's U can turn a band metal into a correlated (or Mott) insulator [Kul15].

The DFT+U schemes include a term analogue to the second term in Eq. (2.30) into the DFT energy,

$$E^{DFT+U}[\varrho] = E^{DFT}[\varrho] + \sum_{i} U^{(\mathbf{n})} N_i^{\mathbf{n},\uparrow} N_i^{\mathbf{n},\downarrow} - E^{DC}, \qquad (2.31)$$

along with a term, E^{DC}, which corrects for double-counting [HFdGC14, Kul15]. Note that, in general, the Hubbard's U can depend on the quantum numbers of the atomic orbital to which it is applied. For instance, one can set a different value U (it can also be zero) to the s, p, d, etc. orbitals. In Eq. (2.31) the general dependency on the atomic quantum numbers is indicated by the multi-index $\mathbf{n}$. Those values of U can be either set to reproduce some exper-imental feature, or based on linear response theory [HFdGC14, Kul15]. Re-cently, a DFT+U has been reformulated as a pseudo-hybrid approach so that the values of the Hubbard's U parameters are obtained fully self-consistently [ACBN15].

The DFT+U approach, and in particular LDA+U, has been applied both for solids and molecules to correct the electronic density over-localization of the LDA and other local or semi-local functionals [Kul15]. DFT+U calculations scale more conveniently than hybrid functional calculations, especially in the case of codes based on plane waves. Limitation of the DFT+U schemes comes from the choice of atomic orbitals used to compute the second term in Eq. (2.30). In practice, the results of an LDA+U can depend on the choice of the localized orbitals unless strict requirements are met [HFdGC14].

2.2.3.3 Kohn-Sham Equations for Static LDA

The idea of KS (section 2.1.5) is to use still the exact kinetic energy of a non-interacting system (of single electron wavefunctions φ_i) of the same density Eq. (2.27b) and express only the xc-energy as a (local) density functional. The starting point is then the electronic energy in the form $E = E_{\rm kin} + E_{\rm C} + E_{\rm xc}^{\rm (LDA)} + E_{\rm back}$, where

$$E_{\rm kin} = \sum_i \int d^3r\, \varphi_i^*(\mathbf{r}) \left(-\frac{\hbar^2\nabla^2}{2m}\right)\varphi_i(\mathbf{r}), \tag{2.32a}$$

$$E_{\rm C} = \frac{1}{2}\int d^3r\, d^3r'\, \varrho(\mathbf{r})V_C(|\mathbf{r}-\mathbf{r}'|)\varrho(\mathbf{r}'), \tag{2.32b}$$

and $E_{\rm back}$ is the electron-ion coupling energy from which the coupling potential $\hat{U}_{\rm back}$ in Eq. (2.23) is derived. The energy E still depends on the detailed s.p. wavefunctions φ_i via $E_{\rm kin}$ and thus recovers the full quantum structure of the system. Again, there exists, in principle, a "perfect" xc-functional, but practical xc-functionals are based on LDA for which a great variety is available.

Early constructions focused on incorporating long-range correlations [GL76]. The widely used functional of [PW92] aims at including all correlations being developed from elaborate many-body calculations for the interacting electron gas. Further improvements are achieved by accounting also for the gradient of the local density in GGA [PBE96]. For simplicity, we stay in this presentation with the simple, purely density-dependent functionals.

The ground state is found, similarly to the HF-method, by variation of the energy with respect to the s.p. wavefunctions. This yields the much celebrated Kohn-Sham (KS) equations

$$\hat{h}_{\rm KS}\varphi_i = \varepsilon_i\varphi_i, \tag{2.33a}$$

$$\hat{h}_{\rm KS} = -\frac{\hbar^2\nabla^2}{2m} + U_{\rm KS}, \tag{2.33b}$$

$$U_{\rm KS} = U_C(\mathbf{r}) + U_{\rm xc}(\mathbf{r}) + U_{\rm back}(\mathbf{r}), \tag{2.33c}$$

$$U_{\rm xc}(\mathbf{r}) = \frac{\delta E_{\rm xc}[\varrho]}{\delta\varrho(\mathbf{r})} \tag{2.33d}$$

where the xc-potential is given by functional variation with respect to the local density, U_{back} is the potential from the ionic background, the external fields and the Coulomb potential U_C is given in Eq. (2.24b). The KS equations have much in common with the HF equations (2.25b), except that the involved exchange term is simplified as a local potential in LDA and that in a similarly simple fashion correlations are effectively accounted for.

2.2.3.4 Time Dependent LDA (TDLDA)

Practical realizations of Time Dependent DFT (TDDFT) mostly concern approaches using LDA for the xc functional, in particular in the far-off equilibrium regime. Time-Dependent LDA (TDLDA) thus provides the basic tool for the description of irradiation dynamics. Still, even if most applications use the density functionals in its simplest LDA form, in which case the term TDLDA is appropriate, the usage is often not too precise and TDLDA may be used in cases more general than LDA, such as for example GGA (section 2.2.3.2). Moreover, the name TDLDA covers a lot of different realizations. For example, linearized TDLDA (Sec. 2.2.4.1) is mostly solved in frequency space. To emphasize a solution in real time, one often uses the notion of real-time TDLDA (rt-TDLDA), similar as the term rt-TDDFT brought up already in section 2.1.5.

In TDLDA one simply inserts the instantaneous local density $\varrho(\mathbf{r}, t)$ into the already known LDA functional. This time-local approximation dismisses all memory effects and renders the KS potential a functional of the instantaneous local density. The approximation is often called Adiabatic LDA (ALDA). The correspondingly Time-Dependent KS (TDKS) equations are, again, derived by variation with respect to the s.p. wavefunctions and read

$$\hat{h}_{\text{KS}}\varphi_i = i\hbar\partial_t\varphi_i \tag{2.34}$$

with the same KS Hamiltonian $\hat{h}_{\text{KS}}$ as in the stationary KS equations (2.33). It is only that all potentials now become time dependent, the self-consistent one through the time-dependent density and the external potential $U_{\text{ext}}(\mathbf{r}, t)$ through the time-dependent excitations as, e.g., from a laser pulse as discussed in section 3.2.1.

The time-dependent KS equation poses an initial-value problem. For a given initial set of s.p. states $\{\varphi_i(\mathbf{r}, t=0)\}$, it allows to determine the further time evolution of them and all related observables, for practical aspects see section 3.5. The excitation is brought into the system either by starting from a state different from the ground state or through time-dependent external fields.

2.2.3.5 The Self-Interaction Problem and Corrections to It

DFT at the level of LDA manages to express the action of Coulomb exchange and correlations on the mean field in terms of a local potential $U_{xc}(\mathbf{r}) \equiv U_{xc}(\varrho(\mathbf{r}))$. This potential at a given point $\mathbf{r}$ depends on the local density $\varrho(\mathbf{r})$ exactly at this point and nothing else. Let us call it a local function. That is more confined than what DFT in general asks for, where U_{xc} is a functional of $\varrho\,(\mathbf{r})$ as a whole. This very feature of dealing with a zero-range functional is causing a basic inconsistency in the LDA: the problem of Self-Interaction (SI).

2.2.3.5.1   Status of the Problem

Let us consider typical KS equations without correlation functional but with Coulomb exchange in the Slater approximation Eq. (2.29). To illustrate the SI, we resolve explicitly the sum over s.p. states in the density entering the KS potential as $\varrho\,(\mathbf{r}) = \sum_j |\varphi_j(\mathbf{r})|^2$ which brings the KS equation (2.33) into the form

$$\varepsilon_i\varphi_i = \left(-\frac{\hbar^2\nabla^2}{2m} + U_{\text{back}}(\mathbf{r}) + \sum_j \int d^3r'\frac{e^2}{|\mathbf{r}-\mathbf{r'}|}|\varphi_j(\mathbf{r'})|^2 - \frac{4\beta_x}{3}\Big(\sum_j |\varphi_j(\mathbf{r})|^2\Big)^{1/3}\right)\varphi_i\,.$$

Now consider the term $j = i$. It does not vanish with the consequence that particle i interacts with itself.

Take a mere one-particle system as a limiting case which is interaction free (except, of course, for the external potential). In that system, we have $j = i = 1$ and it becomes obvious that the interaction term in LDA remains active. The one particle interacts with itself. The same SI also persists for larger systems and is considered increasingly irrelevant with increasing particle number N_e because it amounts to a $1/N_e$ effect. However, the painful aspect is that it has long-range consequences. The one particle i if it goes far away from the rest of the system should see the Coulomb field of the N_e-1 remaining particles. But in LDA, the asymptotic Coulomb field becomes e^2N_e/r instead of $e^2(N_e-1)/r$ as it should be.

As a counter check, we consider the exact exchange term from Hartree-Fock. The interaction term in the HF equation for the Coulomb problem becomes

$$U_{\text{HF}}\varphi_i(\mathbf{r}) = \sum_j \int d^3r'\frac{e^2}{|\mathbf{r}-\mathbf{r'}|}|\varphi_j(\mathbf{r'})|^2\varphi_i(\mathbf{r}) - \sum_j \int d^3r'\frac{e^2}{|\mathbf{r}-\mathbf{r'}|}\varphi_j^\dagger(\mathbf{r'})\varphi_j(\mathbf{r})\varphi_i(\mathbf{r'})$$

$$= \sum_{j\neq i} \int d^3r'\frac{e^2}{|\mathbf{r}-\mathbf{r'}|}|\varphi_j(\mathbf{r'})|^2\varphi_i(\mathbf{r}) - \sum_{j\neq i} \int d^3r'\frac{e^2}{|\mathbf{r}-\mathbf{r'}|}\varphi_j^\dagger(\mathbf{r'})\varphi_j(\mathbf{r})\varphi_i(\mathbf{r'})$$

where we see that the $\sum_j$ is identical to the $\sum_{j\neq i}$ because the exchange manages to cancel exactly the SI in the direct Coulomb term. This important feature has been lost in the LDA, and to some extent in more elaborate functionals (section 2.2.3.2).

2.2.3.5.2   Self Interaction Correction (SIC)

The SI error is known since the earliest applications of DFT. Correspondingly, there is a long history of attempts to overcome the SI error. In this section, we

review briefly some pragmatic methods for SI correction (SIC). Thereby, as in
the DFT section (2.2.3), we ignore spin dependence to keep the presentation
simple.

The conceptually simplest SIC approach is just to subtract the offending
terms thus modifying the LDA functional to [PZ81]

$$E_{\text{Cxc}}^{(\text{LDA})} \longrightarrow E_{\text{Cxc}}^{(\text{SIC})} = E_{\text{Cxc}}^{(\text{LDA})}[\varrho] - \sum_i E_{\text{Cxc}}^{(\text{LDA})}[\varrho_i], \quad \varrho_i = |\varphi_i|^2, \qquad (2.35)$$

where the index Cxc stands for direct Coulomb together with exchange-
correlation terms. Variation with respect to φ_i^* yields the modified KS equations for the stationary case as

$$\varepsilon_i \varphi_i = \hat{h}_{\text{KS},i} \varphi_i \qquad (2.36a)$$

$$\hat{h}_{\text{KS},i} = -\frac{\hbar^2 \nabla^2}{2m} + U_{\text{Cxc},i}^{(\text{SIC})}(\mathbf{r}) + U_{\text{back}}(\mathbf{r}) \qquad (2.36b)$$

$$U_{\text{Cxc},i}^{(\text{SIC})} = U_{\text{Cxc}}^{(\text{LDA})}[\varrho] - U_{\text{Cxc}}^{(\text{LDA})}[\varrho_i] \qquad (2.36c)$$

and similarly

$$i\hbar\partial_t \varphi_i = \hat{h}_{\text{KS},i} \varphi_i \qquad (2.36d)$$

for the dynamic case. The subtracted SIC term $U_{\text{Cxc}}^{(\text{LDA})}[\varrho_i]$ is built in the same
way as the KS potential in Eq. (2.33), however using only the s.p. density
$\varrho_i(\mathbf{r}) = |\varphi(\mathbf{r})|^2$ of state i. This looks simple, at first glance. But it has a price:
the new KS potential and with it the KS Hamiltonian depend on the state i
on which they act. This renders $\hat{h}_{\text{KS},i}$ non-Hermitian which, in turn, requires
additional measures to keep the set of self-consistent s.p. states ortho-normal
in the static solution as well as in time evolution.

There are several strategies to cope with that, see e.g., [GU07]. A particularly robust technique for time evolution deals with the simultaneous handling of two connected sets of s.p. states, for a detailed description we refer
to [MDRS08a, MDRS08b]. For the present purposes, it suffices to conclude
that a full SIC can safely be handled in stationary as well as dynamical situations, although it is considerably more expensive than mere LDA because the
Coulomb field implied in the corrective potential $U_{\text{Cxc}}^{(\text{LDA})}[\varrho_i]$ has to be evaluated for each state i separately. That means we need N_e Coulomb evaluations
rather than two (for spin up and spin down). Still, this is much less than the
$N_e(N_e-1)$ evaluations required for the exact exchange term.

2.2.3.5.3 Simplified SIC

There is an extremely simple and robust approximation to full SIC, coined
Average Density SIC (ADSIC), which was used intuitively correctly already
long before the formal background of DFT was developed [FA34]. The idea is
that the local s.p. wavefunctions cover all about the same spatial region such

that one can assume $\varrho_i(\mathbf{r}) \approx \varrho(\mathbf{r})/N_e$. This leads to a SIC functional in the simple form

$$E_{Cxc}^{(LDA)} \longrightarrow E_{Cxc}^{(ADSIC)} = E_{Cxc}^{(LDA)}[\varrho] - N_e E_{Cxc}^{(LDA)}[\varrho/N_e]. \qquad (2.37)$$

This turns back to be a functional of local density alone and the whole machinery of DFT at the level of LDA remains applicable with only a small modification of the KS mean field

$$\begin{aligned}
U_{Cxc}^{(ADSIC)} &= U_{Cxc}^{(LDA)}[\varrho] - U_{Cxc}^{(LDA)}[\varrho/N_e] \\
&= \frac{N_e - 1}{N_e} U_C + U_{xc}^{(LDA)}[\varrho] - U_{xc}^{(LDA)}[\varrho/N_e]. \qquad (2.38)
\end{aligned}$$

The major advantage is that the ADSIC mean-field Hamiltonian is immediately Hermitean which guarantees automatically ortho-normal ground state wavefunctions and unitary time evolution. Another advantage is that it does not require the many Coulomb evaluations of full SIC. All this makes ADSIC a very tempting tool for robust explorations. At first glance, ADSIC looks rather rough. From the assumption $\varrho_i \rightarrow \varrho/N_e$ it is obvious that ADSIC is tailored to metallic systems. But practical experience shows that it performs surprisingly well for many covalent molecules as long as the electron cloud remains singly connected [KDRS13, RS21]. ADSIC is bound to fail in cases of fragmentation or atomic/molecular scattering.

There is a more detailed way to combine the state-dependent KS-SIC potentials into one common (i.e. state independent) effective SIC potential. That is the SIC-Slater approach also invented long before the formal foundation of DFT [SH53]. It is defined as a weighted average

$$U_{Cxc}^{(Slater)} = \sum_i \frac{\varrho_i(\mathbf{r})}{\varrho(\mathbf{r})} U_{Cxc,i}^{(SIC)} \qquad (2.39)$$

which, again, provides a state-independent, thus Hermitean, KS Hamiltonian. There is a drawback too: the approximation is defined at the level of the KS potential. There is no energy functional to it, thus leaving the safe grounds for variational formulation. This may be acceptable for ground-state calculations, but raises a lot of trouble in dynamical applications. For example, energy conservation is not a trivial issue and the zero-force theorem (the requirement that the electron cloud must not leave behind a finite force on its own center of mass) is violated [MKvLR07]. In dynamics, SIC-Slater may be just acceptable for short times and very weak excitations but, in general, it is not a recommended approach for truly dynamical simulations.

There exists an improved version of SIC-Slater proposed by Krieger, Li and Iafrate (KLI) [KLI92, CKLI96] which modifies the SIC-Slater KS potential to

$$U_{Cxc}^{(KLI)}(\mathbf{r}) = \sum_i \frac{\varrho_i(\mathbf{r})}{\varrho(\mathbf{r})} \left\{ U_{Cxc,i}^{(SIC)} - \int d^3r' \; \varrho_i(\mathbf{r}') \left[U_{Cxc}^{(KLI)}(\mathbf{r}') - U_{Cxc,i}^{(SIC)}(\mathbf{r}') \right] \right\}$$

$$(2.40)$$

which then is determined by solving a self-consistent integral equation. The extension was developed and advocated for stationary calculations where it often improves the results. For what dynamics is concerned, SIC-KLI shares the properties of SIC-Slater. It delivers a unitary propagation of the s.p. wave-functions. But energy conservation and zero-force theorem remains an issue of concern [MKvLR07].

Both intermediate approaches, SIC-Slater and SIC-KLI, look here as being motivated ad hoc whereby ADSIC has some intuitive appeal. But both can be derived from formal considerations within another variational approach, the Optimized Effective Potential (OEP) method which will be addressed in section 2.2.3.5.4.

2.2.3.5.4   Combining SIC with Local KS Potentials

A different road to surmount the SI problem is followed by the Optimized Effective Potential (OEP) method. The method was introduced in a more general context long ago [SH53] in order to find an optimal mean field by simplifying for example Hartree-Fock equations at a time at which computational resources were limited. We outline it here in the context of DFT-SIC to which it also perfectly applies. DFT concentrates on finding a reliable functional of the local density alone. The OEP approach starts from the observation that we are at the end happy with having the KS equations employing an effective local one-body potential $U_{KS}(\mathbf{r})$ and tries to optimize that directly by variation of the given energy with respect to a local potential $U_{OEP}(\mathbf{r})$. Given an energy that can be now any functional of the set of occupied s.p. wavefunctions $E = E[\{\varphi_i, i = 1...N_e\}]$, the optimal potential is to fulfill

$$\frac{\delta E[\{\varphi_i\}]}{\delta U_{xc}^{OEP}(\mathbf{r})} = 0 \tag{2.41a}$$

where the s.p. states are solution of the KS equations with the mean-field Hamiltonian

$$\hat{h}_{KS} = -\frac{\hbar^2 \nabla^2}{2m} + U_C(\mathbf{r}) + U_{xc}^{(OEP)}(\mathbf{r}) + U_{back}(\mathbf{r}). \tag{2.41b}$$

It is obvious that the solution of these coupled equations will become rather involved. We refer the reader to the following detailed and comprehensive paper [KK08]. An interesting feature is that SIC-KLI and SIC-Slater approaches (see section 2.2.3.5) can be derived from OEP when starting from identifying the initial energy expression with the SIC energy (2.2.3.5) and dropping a few supposedly less important terms in the OEP potential, see e.g., [MDRS11]. It is the strength of the OEP method that it can deal with a great variety of initial energy functionals $E = E[\{\varphi_i, i = 1...N\}]$. It can be the SIC functional, exact exchange, or hybrid functionals combining exact exchange with approximate correlation functionals. That is all extremely useful for static calculations. However, there are still open questions in long-time dynamical

applications: violation of energy conservation, violation of the zero-force theorem, and sooner or later upcoming instabilities [KMK08].

2.2.4 SIMPLIFIED VERSIONS OF TDLDA AND TDHF

2.2.4.1 Linearized TDLDA (TDHF)

Fully fledged TDHF and TDLDA, also coined real-time TDLDA (rt-TDLDA),
are well capable to deal with dynamical scenarios in the regime of high excitations up to the range where dynamical correlations (see section 2.3) come
into play. They also work, of course, for weak excitations that induce only
small-amplitude oscillations about the stable ground state configuration. The
dynamics here remain in what is called the linear regime and this allows one
to linearize the time-dependent mean-field equations. The emerging formalism
is widely known in many areas of physics as Random-Phase Approximation
(RPA) [Row70, FW71]. The linearization sheds light on the structure of excitations and, at the practical side, often leads to much faster computations,
particularly in situations with symmetry constraints as, e.g., spherical symmetry in the case of atoms or lattice symmetry for solids. It is thus worth to
address it briefly.

Linearization starts from the Thouless theorem [Tho60, FW71] which states
that two Slater determinants can always be connected by the exponential of a
one-particle one-hole ($1ph$) operator. This means applied to the present task
that the full time-dependent mean-field state $\Phi(t)$ can then be related to the
stationary ground state Φ_0 as

$$\Phi(t) = \exp\left(\mathrm{i}\hat{G}(t)\right)\Phi_0 \qquad (2.42a)$$

where $\hat{G}$ is a single particle operator

$$\hat{G} = \sum_{ph}\left(G_{ph}\hat{a}_p^\dagger\hat{a}_h + G_{ph}^*\hat{a}_h^\dagger\hat{a}_p\right). \qquad (2.42b)$$

The $\hat{a}^\dagger$ and $\hat{a}$ are Fermion operators [FW71]. The "hole" states h are states
occupied in Φ_0 and the particle states p are the initially unoccupied ones.
The combination $\hat{a}_p^\dagger\hat{a}_h$ describes a $1ph$ excitation which represents the basic
excitation about a Slater state. Note that we chose here $\hat{G}$ to be Hermitean
which guarantees all-time ortho-normality of the s.p. states implied in Φ.
Linearization then means to approximate

$$\Phi(t) \approx \left(1 + \mathrm{i}\hat{G}(t)\right)\Phi_0, \qquad (2.43)$$

i.e. to consider only the immediate neighborhood of $1ph$ excitations. The linearized TDHF/TDKS equations are obtained from inserting the expansion
everywhere into the equations and collect all terms linear in $\hat{G}$. The result is
a matrix equation in $1ph$ space. The procedure is straightforward but tedious.

For a detailed derivation of the linearized TDKS equations, we refer the reader to [RG92, RGB96, MRS10].

2.2.4.2 TDCIS

As seen above, TDDFT is widely used to describe the dynamics of electronic systems. The analogous dynamical theory at the Hartree-Fock level is TDHF (section 2.2.2) which in its full form has found much less attention. More often used is a simplified version thereof called Time-Dependent Configuration-Interaction Singles (TDCIS) which we review in this section. The CI method at its full extent aims to approximate successively the fully correlated many-body state of a system. It belongs clearly to the methods beyond mean-field and will be discussed in section 2.3. But TD-CIS, staying at the lowest level of expansion of the many-body state, still belongs to theories of one-body dynamics and so fits into the present section.

The idea behind TDCIS is conceptually simple. The time-dependent state of the system $|\Psi(t)\rangle$ is expanded as

$$|\Psi(t)\rangle = c_0(t)|\Phi_0\rangle + \sum_{ph} c_{ph}(t)\hat{a}_p^\dagger\hat{a}_h|\Phi_0\rangle \tag{2.44}$$

where $|\Phi_0\rangle$ is the stationary HF ground state and $|\Phi_{ph}\rangle = \hat{a}_p^\dagger\hat{a}_h|\Phi_0\rangle$ is a particle-hole excitation about $|\Phi_0\rangle$ in which one particle is removed from an occupied s.p. state h and placed into an unoccupied state p. This ansatz can be viewed as lowest-order CI expansion or as lowest-order expansion of a TDHF state in the spirit of the Thouless theorem [Tho60, FW71]. The dynamical degrees of freedom in the TDCIS ansatz are the expansion coefficients c_0 and c_{ph}. Their equations of motion can be derived from the time-dependent variation principle $\delta_{c^*}\langle\Psi|i\hbar\partial_t - \hat{H} - \hat{U}_{ext}(t)|\Psi\rangle$ where $\hat{H}$ is the Hamiltonian of the system embracing kinetic energy, two-body Coulomb interaction, and ionic background potential while $\hat{U}_{ext}(t)$ is a one-body operator representing the external field (section 2.1.1).

TDCIS belongs, like HF, to the class of ab initio methods because it works with the given microscopic Hamiltonian. Similar to TDHF, TDCIS does not include correlations. It has, nonetheless, its range of applications because in many situations correlations are less important for excitation properties. The expansion (2.44) makes the method particularly attractive for systems with high symmetry in the ground state (atoms, dimer molecules) because the basis of HF s.p. wavefunctions can exploit the symmetry of the system. Think, e.g., of an atom where everything can be reduced to the very efficient spherically symmetric wavefunctions. The method has thus found widespread applications in atomic and molecular physics, e.g [SSL07, GHP+10, IGRS19, BKS21].

A word of caution is in order, though. The restriction to a $1ph$ basis limits excitations to small amplitude. Thus it has some similarities with the Random Phase Approximation (RPA) which can be derived as a small amplitude limit of TDLDA, see section 2.2.4.1. RPA is the more consistent approach as it includes also the backward-going amplitudes in terms of $1hp$ excitations [RS80]. Thus TDCIS has problems at low energies and overestimates hyper-polarizabilities. But it could be shown that TDCIS can perform equally well for high excitations in strongly bound systems [GHP+10] in which case one is happy to take advantage of the simplifications brought with TDCIS.

2.2.4.3　Time-Dependent DFT Tight Binding (TD-DFTB)

DFT and wavefunction methods scale poorly with the number of atoms, thus preventing from studying the dynamics of large, chemically complex systems like proteins. Even the simplest DFT approximation, the LDA, cannot afford more than a few thousand atoms, while *actin*, for example, has 13002 atoms, without counting the environment. A similar situation occurs when attempting to study dynamical phenomena in materials that involve multiple bond breaks e.g., the propagation of fractures. It is therefore desirable to develop simplified methods for electron transport in large systems at low energies. One line of development is orbital-free density functional theory inspired by the Thomas-Fermi approach, which avoids orbitals and diagonalization, but has been successful only in some cases and not in general [WW13].

A more promising line exploits the fact that many systems consist of considerably localized electrons. The Starting point is the basis representation Eq. (3.25) in terms of the atomic orbitals of its constituents, the widely used Linear Combination of Atomic Orbitals (LCAO), which is still an exact expansion. The basic approximation consists of taking only as many orbitals per atom as there are electrons at this site. Further approximations at different levels are invoked to simplify the evaluation of the Hamiltonian matrix elements. In that spirit, semi-empirical methods have been used in quantum chemistry for long. Based on Hartree-Fock, extended Hückel, CNDO, MINDO, and PF7 [Ste13] approaches among others, are implemented, e.g., in code MOPAC [Ste16], and have shown their usefulness by a systematic increase in the quality of the predictions. These methods reduce significantly the cost of the calculations by neglecting expensive terms in the Hamiltonian matrix and parametrizing the remaining ones in a way that partially compensates for the neglected terms.

Along similar lines, the Tight-Binding TB method was pioneered by Slater and Koster [SK54] as an approximation to the band structure of solids using a minimal LCAO basis. Still, the Coulomb and exchange terms of the Hamiltonian matrix contain two-, three-, and four-center integrals, according to which atoms are the basis orbitals centered on. Three and four-center integrals are expensive to calculate and there are too many of them. The TB Hamiltonian reduces that to only one- and two-center integrals, and is written

$$H_{\alpha I,\beta J} = \varepsilon_{\alpha I}\, \delta_{IJ}\delta_{\alpha\beta} + t_{\alpha I,\beta J}\,, \qquad (2.45)$$

where we use a composed index with I, J labeling the atoms and α, β the orbitals (s, p, etc). The first term $\varepsilon_{\alpha I}$ represents the energy of orbital α in the isolated atom I, and it is called the on-site energy. The second one, $t_{\alpha I, \beta J}$ represents the sharing of electrons between orbital α in atom I and orbital β in atom J, and receives the name of hopping or resonance integral. If we consider a set of basis orbitals $B_{\alpha I}(\mathbf{r})$, the two-center hopping integrals are given by

$$t_{\alpha I, \beta J} = \int d^3r\, B_{\alpha I}(\mathbf{r}) \left[-\frac{\hbar^2}{2m_e} \nabla^2 + U_I(\mathbf{r} - \mathbf{R}_I) + U_J(\mathbf{r} - \mathbf{R}_J) \right] B_{\beta J}(\mathbf{r}),$$
(2.46)

with $U_I(\mathbf{r} - \mathbf{R}_I)$ the potential felt by electrons due to atom I. In practice, hopping integrals are parameterized as a function of the distance between atoms I and J. Since the interatomic forces arising from the TB Hamiltonian are purely attractive, a repulsive two-body potential is added, and fitted to reproduce geometric and elastic properties. With this addition, interatomic forces enable the possibility of performing TB MD simulations.

A step forward to a more robust derivation of the TB Hamiltonian has been achieved with Density-Functional Tight Binding (DFTB), proposed in [PFK+95], and reviewed in [FSE+02]. In that case, the Kohn-Sham potential $U^{KS}(\mathbf{r})$, Eq. (2.33) is decomposed approximately as a sum of atom-centered potentials $U_I^{KS}(\mathbf{r})$, so that the hopping integrals retain the form of Eq. (2.46), but replace $U_I(\mathbf{r})$ with $U_I^{KS}(\mathbf{r})$. Careful calibration of the repulsive two-body contributions with respect to full DFT calculations allows to pack a great deal of DFT quality into the TB Hamiltonian. This approach has been expanded into the general-purpose and widely used package DFTB+, the details of which can be found in [HAB+20]. The extension to Time-Dependent DFTB (TD-DFTB) was proposed in [NHTF05], based on earlier work by Todorov for the semi-empirical TB model [Tod01]. A recent implementation that allows not only for Ehrenfest dynamics but also for trajectory surface hopping simulations (see sections 2.5.3 and 2.5.4.2), can be found in [BAH+20]. TD-DFTB has been used in the example in section 4.2.1.5.

2.2.4.4 Single Electron Approximation

The power of TDDFT lies in the dynamical rearrangement effects embodied in the time-dependent KS Hamiltonian $\hat{h}_{KS}(t)$. These may be neglected for very fast and energetic processes. This allows to use of the static KS Hamiltonian instead which saves the costly re-evaluation of the dynamical mean field which reads

$$\hat{h}_{KS,0}\varphi_i \;\; = \;\; i\hbar\partial_t\varphi_i \quad, \quad \hat{h}_{KS,0} = \hat{h}_{KS}(t{=}0). \qquad (2.47)$$

This leads to the Single-Active Electron approximation (SAE) [Kul88] and its improved version d-SAE [RGS06], in which the propagation of only one electron is considered explicitly. The same is also often presented under the name

Time-Dependent Schrödinger Equation (TDSE). We will use both names depending on the work reported.

2.3   BEYOND MEAN FIELD

2.3.1   WHAT ARE CORRELATIONS

The use of the term "correlations" is abundant in many-body physics. A trivial understanding would be that all particles in an interacting many-body system are correlated because all are connected by interactions. But the notion is usually meant more specifically: *Correlation is everything beyond a mean-field description.* In wavefunction representation, a mean-field model is distinguished by the fact that the state of the system can be described completely in terms of a set of s.p. wavefunctions $\{\varphi_i, i = 1...\Omega\}$ where $\Omega = N_e$ for zero-temperature situations (only fully occupied s.p. states). States at non-zero temperatures have to allow for $\Omega > N_e$ and add occupation numbers $n_i \in [0,1]$ to the description. We start the discussion here with the zero-temperature case.

As outlined in section 2.2.1, a mean-field state, or independent-particle state respectively, in the wavefunction picture is a Slater determinant $|\Phi\rangle$ as specified in Eq. (2.17a). A complete, ortho-normal set of s.p. wavefunctions spans the s.p. space. If $i \equiv h = 1, ..., N_e$ represent the occupied states, then the expansion into $1ph$, $2ph$, ... excitations, or as they are usually called, single, double, ... excitations, reads

$$|\Psi\rangle = \left(c_0 + \hat{S}_1 + \hat{S}_2 + ...\right)|\Phi\rangle, \tag{2.48a}$$

$$\hat{S}_1 = \sum_{ph} c_{ph}\hat{a}_p^\dagger\hat{a}_h, \tag{2.48b}$$

$$\hat{S}_2 = \sum_{pp'hh'} c_{pp'hh'}\hat{a}_p^\dagger\hat{a}_{p'}^\dagger\hat{a}_h\hat{a}_{h'}, \;\; ... \;, \tag{2.48c}$$

with arbitrary coefficients $c_{p...h...}$ spanning a complete set of wavefunctions in many-body space. Recall that $p > N_e$ stands for initially unoccupied states which are coined particle states because they are filled with a *particle* in an excited state while a *hole* is cut into the initially occupied state $h \leq N_e$. The basis of the expansion is the Slater state $|\Phi\rangle$ and it is an uncorrelated state by definition. One is tempted then to call everything with a sprinkle of $c \neq 0$ a correlation. This is, however, a premature conclusion.

The $1ph$ part $\hat{S}_1$ of the expansion makes an exception because of the Thouless theorem Eq. (2.42) which states that every two Slater determinants $|\Phi\rangle$ and $|\Phi'\rangle$ are connected uniquely by a transformation operator as exponential of a $1ph$ operator, written here as

$$|\Phi'\rangle = \exp\left(\hat{S}_1\right)|\Phi\rangle \tag{2.49}$$

with an appropriate choice of coefficients c_{ph} [Tho60]. This means that $1ph$ excitations $\hat{S}_1$ merely change the reference Slater determinant. This is the reason why we placed section 2.2.4.2 about TDCIS in the mean-field part of that theory chapter. Truly correlated states start from $\hat{S}_2$ on. The expansion Eq. (2.48) is the starting point for the Time-Dependent Configuration Interaction method (TDCI) which will be discussed in section 2.3.4.1.

The Taylor expansion of the Thouless transformation Eq. (2.42),

$$|\Phi'\rangle = \left(1 + \hat{S}_1 + \frac{1}{2}\hat{S}_1^2 + ...\right)|\Phi\rangle,$$

contains any order nph excitations in the form of $\hat{S}_1^n$. This shows that the expansion Eq. (2.48) mixes correlated and uncorrelated components in the higher terms $\hat{S}_n$ with $n > 1$. The uncorrelated contributions $\hat{S}_1^n$ becomes the larger the larger the amplitude of the Thouless transformation. A better discrimination between the mean-field part and correlations is achieved by exponentiating all contributions in the expansion leading to the ansatz

$$|\Psi\rangle = \exp\left(1 + \hat{S}_1 + \hat{S}_2 + ...\right)|\Phi\rangle \qquad (2.50)$$

with the $\hat{S}_n$ defined as in Eq. (2.48). This is the starting point of the coupled cluster method (and of Time Dependent Coupled Cluster, TDCC), also named $\exp(S)$ method, which will be discussed in section 2.3.4.2.

Both TDCI and TDCC expand about a time-independent reference state Φ, which means that the time dependence of the process is only and fully contained in the $c_m(t)$ coefficients. An obvious modification consists in expanding with respect to a time-dependent reference, usually a TDDFT or TDHF trajectory $|\Phi(t)\rangle$. This allows for the incorporation of a great deal of time-dependent interaction effects into the time-dependent mean field basis state $\Phi(t)$ and leaves less to do for the $\hat{S}_n(t)$. Once looking at the expansion this way, suggests a most general ansatz as

$$|\Psi\rangle = \sum_m c_m(t)|\Phi_m(t)\rangle \qquad (2.51)$$

where the set $\{|\Phi_m(t)\rangle\}$ is a set of TDHF trajectories and $c_m(t)$ expansion coefficients. Note that the set can be any choice and need not to be orthogonal. However, many implementations employ an orthogonal basis, which simplifies the formalism, but lets MCTDHF stay close to a sort of generalized TDCI. The MCTDHF method is exemplified in section 2.3.4.3.

An alternative to the wavefunction picture is the description of the state of a system in terms of density matrices. A mean-field model carries all information about the state of a many-body system in the one-body density matrix $\rho(\mathbf{x}, \mathbf{x}', t)$. The mean field many-body density matrices are generated from $\rho(\mathbf{x}, \mathbf{x}', t)$ as anti-symmetrized product Eq. (2.18b) of desired order, see section 2.2.1. Correlations appear if an m-body density matrix deviates from the product form. For example, this reads, for the two-body density matrix (skipping here the time label),

$$\rho(\mathbf{x}_1, \mathbf{x}_2, \mathbf{x}_1', \mathbf{x}_2') = \mathcal{A}\left\{\rho(\mathbf{x}_1, \mathbf{x}_1')\rho(\mathbf{x}_2, \mathbf{x}_2')\right\} + \delta\rho_2(\mathbf{x}_1, \mathbf{x}_2, \mathbf{x}_1', \mathbf{x}_2'), \qquad (2.52)$$

where the deviation $\delta\rho_2$ stands for the correlations beyond the mean-field. Strategies for handling time-dependent correlations within the density matrix formalism based on the BBGKY hierarchy will be addressed in the section 2.3.5.

A similar anti-symmetrized product form applies to the independent-particle states in terms of Green's functions (Eq. (2.21) in section 2.2.1) and similarly, one expands correlations about the product state. The expansion is here at first glance formally more involved, but technically more versatile and allows to derive in systematic steps models at different levels of correlations which can be tailored for any given case as a compromise between need and feasibility. Along that line, we will present in section 2.3.6 the still rather general GW extension of DFT, from which one can derive the quantum Boltzmann equation as outlined in section 2.3.7.1. The latter can be numerically realized by stochastic techniques as in Stochastic TDLDA (STDLDA) discussed in section 2.3.8 together with a brief summary of emerging time-dependent quantum Monte-Carlo approaches. Another simplification of the quantum Boltzmann equation is the relaxation-time approximation explained in section 2.3.7.2.

Thus far for a brief first glance at the various approaches to deal with many-body correlations. Whatever method, it is obvious that the treatment of correlations will be much more demanding than treating mean-field theories. Take, e.g., the CI expansion Eq. (2.48). The magnitude of $1ph$ space in $\hat{S}_1$ is $N_p N_h$ where N_p and N_h are numbers of particle or hole states. Let us assume that this characterizes the expense of a mean-field calculation. The expense of treating two-body correlations with $\hat{S}_2$ is then $\propto N_p^2 N_h^2$, not to dream of even higher correlations. In view of these enormous efforts, the step beyond the mean field needs good reasons. A couple of them will be discussed in the next subsection.

2.3.2 WHY DO WE NEED CORRELATIONS

Mean-field theories are designed for the description of energy and one-body observables. Genuine two-body observables may be computed but without guarantee for a pertinent description. A clear case of failure of DFT is for example double ionization of atoms which can only be overcome by an exact calculation, see e.g., [LvL98, LJY$^+$19]. Other examples are electron transfer in atomic/molecular collisions and multi-fragmentation of highly excited molecules. In both examples, one deals with several different reaction channels which deviate largely in the course of time while DFT can at best describe an average over the channels. Think, e.g., of a charge transfer reaction where two neutral atoms enter and either two neutral atoms or anion-cation come out. It makes no sense to average over these two outgoing channels at late stages with large atomic distances.

One qualitative change introduced by TDHF and TDDFT is that the resulting equations of motion become highly non-linear due to embodying the interactions into the self-consistent feedback while the original exact many-body Schrödinger equation was linear. One of the consequences of non-linearity is spontaneous symmetry breaking, such as the pairing phase transition in BCS theory [Hak76] or shape transitions in nuclei [RS80]. The symmetry-broken state carries still useful information if interpreted with care. In critical cases, one can produce a better approximation of the exact state through symmetry restoration by projection [RS80]. Take as an example H_2 at very large molecular distance. The exact state is a spin singulet which is a superposition of two Slater states, thus not a mean-field state. There are two degenerate mean-field solutions one with spin-up left and spin-down right and the other vice versa. Both states violate reflection symmetry. All physical properties except spin (and related observables) are still fine. The exact state can be restricted by projection, i.e. building a reflection symmetric superposition.

LDA is, as the name says, a local approximation which can raise also problems in dynamical applications. For example, the LDA for exchange leads to a lack of electrostatic attraction between the excited electron and hole (excitonic effects) which, e.g., produces incorrect peak shifting and charge transfer dynamics [LGI+20]. Some of these problems in dynamics are resolved by the SIC, see section 2.2.3.5. TDLDA usually makes the additional approximation of locality in time, or adiabatic approximation, in which memory effects are either ignored or not treated properly. Memory times shrink rapidly with increasing excitation energy due to increasing level density. This validates locality in time (Markovian approximation) for energetic processes. Theories for dynamical correlations within Markovian approximation will be addressed in sections 2.3.7.1, 2.3.7.2, and 2.3.8. There remains a regime of lower excitations where memory effects may be required. Not much is available so far for finite Fermion systems. One proposal is a TDDFT extended by a dependence on currents (Current DFT, section 2.1.5) [KB04]. More developments are still needed.

A major drawback of mean-field approaches in highly excited systems is the lack of dynamical correlations which account for direct electron-electron collisions driving dissipation in electronic dynamics toward thermalization of the electron cloud. In pure mean-field theory, the system retains too much (coherent) memory of the past and never reaches an equilibrium situation like, e.g., a Fermi-Dirac distribution for the occupation numbers of the electronic states. This absence of equilibration can be demonstrated in the density-matrix formulation. The time-evolution of the one-body density in mean-field theory is given by the Liouville-van-Neumann equation $i\hbar\partial_t\hat{\rho}(t) = \left[\hat{h}_{\mathrm{mf}}, \hat{\rho}(t)\right]$ (where

$\hat{h}_{\mathrm{mf}}$ is the one-body mean field Hamiltonian), which was given for example for TDHF in Eq. (2.25c), but holds similarly also for any other time-dependent mean-field theory. Let us assume that we start with a one-body density matrix in natural orbital representation Eq. (2.20) and map the Liouville-van-Neumann equation to an equation for the matrix elements ρ_{ij}. Then one can see easily for the diagonal elements $n_i = \rho_{ii}$ that $i\hbar\partial_t n_i = [h_{ii}n_i - n_i h_{ii}] = 0$ so that the occupation numbers n_i remain constant at their original values all along the time evolution, while the one-electron orbitals evolve in time. If an orbital was initially occupied, it will remain so forever, which is un-physical even for closed quantum systems as it inhibits any dissipation and thermalization.

2.3.3　EXACT SOLUTION OF THE MANY-BODY SCHRODINGER EQUATION

The dream is to produce an exact solution of the TD Schrödinger equation $i\hbar\partial_t\Psi = \hat{H}\Psi$ given in Eq. (2.6a) with the many-body Hamiltonian $\hat{H}$ as given in Eq. (2.3a). Analytical solutions exist for one-electron systems like the hydrogen atom and the H_2^+ molecule, and for model systems such as the 1D-Hubbard, low-dimension Ising, and Lipkin model. Exact numerical solutions can be thought of within different numerical schemes, see section 3.5. Basis set expansions amount to TDCI (section 2.3.4.1) with sufficiently large, actually gigantic, expansion spaces. Coordinate-space representations are for-mally even more straightforward, but very soon they hit a hard computational wall.

For present-day computing facilities, an exact numerical treatment of the time-evolution of a two-electron system, i.e. seven coordinates, is at the limit of capabilities. Indeed, there are a couple of exact numerical calculations around. To mention a few: Calculations in reduced symmetry are well feasible and often used for principle studies of mechanisms and approximations as, e.g., in [BYL+11]. Meanwhile, there are also principle investigations on the basis of a fully three-dimensional treatment of the two-electron Schrödinger equation [LDSO19].

Realistic applications to laser-driven ionization dynamics of the He atom are feasible when exploiting the axial symmetry of the process and by careful tailoring of the spatial representation [PDMT03, FNP+08]. These examples put into evidence the high computational demand of describing electron corre-lations already at the level of two-electron systems. There is a world of exciting experiments measuring two electrons from double ionization in coincidence, which delivers unambiguous signals of correlations e.g., [WGW+00, UMD+03]. The two-body Schrödinger equation can give here answers without doubts. But the time-evolution of two-electron systems is presently at the edge of our capabilities for an exact numerical solution. Anything beyond this requires physical approximations that will be discussed in what follows.

2.3.4 CORRELATIONS CONSTRUCTED FROM WAVEFUNCTIONS

2.3.4.1 Time-Dependent Configuration Interaction (TDCI)

The correlated state $|\Psi\rangle$ in TDCI is given in the form Eq. (2.48) with a constant basis state in a Slater determinant form, $|\Phi\rangle$, and time-dependent expansion coefficients $c_{ph}(t)$, $c_{pp'hh'}(t)$, ... The equations of motion for the coefficients can be derived from the time-dependent variational principle Eq. (2.7a) by variation with respect to the coefficients c^*. Expanding up to $2ph$ terms, the equations read

$$i\hbar\partial_t c_0 = \langle\Phi|\hat{H}|\Psi\rangle, \tag{2.53a}$$

$$i\hbar\partial_t c_{ph} = \langle\Phi|\hat{a}_h^\dagger \hat{a}_p \hat{H}|\Psi\rangle, \tag{2.53b}$$

$$i\hbar\partial_t c_{pp'hh'} = \langle\Phi|\hat{a}_{h'}^\dagger \hat{a}_h^\dagger \hat{a}_p \hat{a}_{p'} \hat{H}|\Psi\rangle. \tag{2.53c}$$

The matrix elements with the microscopic many-body Hamiltonian $\hat{H}$ are formally straightforward, but tedious, to evaluate. The initial state is an appropriate stationary CI solution. The propagation of these TDCI equations can proceed with standard solvers for one-dimensional ordinary differential equations.

Stationary CI is a standard method for static ab-initio calculations in quantum chemistry [HOJ13]. It can achieve considerable precision with a sufficiently large expansion basis. Modern diagonalization schemes allow for a large number of expansion states. However, the time-dependent extension to TDCI is more demanding because real-time simulations at high energy or over long times drive the system far off the ground state, thus requiring the enlargement of the expansion considerably for keeping up with the steadily increasing active Hilbert space. Practical TDCI calculations are thus limited to small amplitude dynamics. This allows for a number of useful applications, e.g., ab initio calculations of optical absorption spectra, [BKP97, BKVM01]. On the other hand, TDCI is bound to remain within the framework of ab-initio wavefunction calculations because it requires a microscopic Hamiltonian $\hat{H}$. The equations of motion require nph matrix elements with $\hat{H}$, which are not easily extended to the DFT formulation.

In Section 2.2.4.2 the basis set was restricted to determinants including only single electron-hole excitations (TDCIS) which, similarly to TDHF, still lacks correlation effects. The successes and failures of TDCIS have already been discussed in section 2.2.4.2. In short, TDCIS is useful when chemical processes are mostly driven by single-electron dynamics, e.g., linear optical response and photo-ionization, but also High Harmonic Generation (HHG).

2.3.4.2 Time-Dependent Coupled Cluster Method (TDCC)

The exponentiation of the excitation operators in the TDCC ansatz Eq. (2.50) promises to overcome the limitation of TDCI to small amplitudes. However, just this exponential form causes a considerable formal overhead. A direct application of the stationary or time-dependent variational principle is inhibited because it is technically impossible to evaluate the basic energy expression $\langle\Psi|\hat{H}|\Psi\rangle/\langle\Psi|\Psi\rangle$ exactly, and approximations have to stop too early to be meaningful. The remedy is to forget the safer variational formulation and to take a risk by applying approximations at the level of the equations of motion. The principle strategy is the same in static and in dynamic applications. Thus a detailed reference for the basics can be found in the review article [KLZ78] which deals with static $\exp(S)$, or coupled-cluster, formalism. The time-dependent case requires an intricate recoupling of dynamical variables which is motivated and explained, e.g., in [SRT90]. A nice short summary can be found in [SBK20b]. We repeat here briefly the final procedure.

For a compact formulation, we summarize the basic nph excitation operators as $\hat{A}_\mu^\dagger \in \{\hat{a}_p^\dagger \hat{a}_h, \hat{a}_p^\dagger \hat{a}_{p'}^\dagger \hat{a}_h \hat{a}_{h'}, ...\}$ with a super-label μ counting the various configurations. The time-dependent CC ansatz then reads

$$|\Psi(t)\rangle = \exp(\hat{S}(t))|\Phi\rangle e^{-i\epsilon t}, \quad \hat{S}(t) = \sum_\mu c_\mu(t)\hat{A}_\mu^\dagger, \qquad (2.54)$$

where ϵ follows Eq. (2.55c). The idea is to exploit the fact that the Taylor expansion of the combination $e^{-\hat{S}}\hat{H}e^{\hat{S}}$ yields a series of multiple commutators

$$e^{-\hat{S}}\hat{H}e^{\hat{S}} = \hat{H} + [\hat{H}, \hat{S}] + \frac{1}{2}[[\hat{H}, \hat{S}], \hat{S}] + ...$$

which is known to terminate at order $n + 1$, where n is the order of the expansion in $\hat{S}$ (Eq. (2.50)) and the $+1$ applies if $\hat{H}$ contains at most a two-body interaction. The practical evaluation of the multiple commutators may be hard. But it is relieving to know that the efforts remain finite.

The evaluation of the equations of motion then does not use projection into the mere excitation basis $\langle\Phi|\hat{A}_\mu$ as in TDCI. In order to get the $e^{-\hat{S}}$ into play, one projects with the transformed conjugated states $\langle\Phi|\hat{A}_\mu e^{-\hat{S}}$. This strategy provides a consistent scheme for stationary CC theory. However, it is not that trivial in TDCC because the conjugated states become time dependent with $e^{-\hat{S}(t)}$. The solution is to double the set of coefficient by introducing a dual state

$$\langle\tilde{\Phi}(t)| = \langle\Phi|(1 + \hat{\Lambda}(t))e^{-\hat{S}(t)}, \quad \hat{\Lambda}(t) = \sum_\mu \tilde{c}(t)\hat{A}_\mu,$$

with a complementing set of dual expansion coefficients $\tilde{c}_\mu$. The equations of motion for the dynamical coefficients c_μ, $\tilde{c}_\mu$ and ϵ are finally

$$i\hbar\partial_t c_\mu = \langle\Phi|\hat{A}_\mu e^{-\hat{S}}\hat{H}e^{\hat{S}}|\Phi\rangle, \qquad (2.55a)$$

$$i\hbar\partial_t \tilde{c}_\mu = \langle\Phi|(1 + \hat{\Lambda}(t))e^{-\hat{S}}[\hat{H}, \hat{A}_\mu]e^{\hat{S}}|\Phi\rangle, \qquad (2.55b)$$

$$\hbar\partial_t \epsilon = \langle\Phi|e^{-\hat{S}}\hat{H}e^{\hat{S}}|\Phi\rangle. \qquad (2.55c)$$

Although cumbersome to handle, the TDCC equations are manageable. Compared to TDCI, they have the advantage for covering also high excitations and large-amplitude evolution. However, TDCC propagation is plagued by instabilities that render it inapplicable to ab-initio dynamics in nuclear physics, with its extremely strong short-range interaction [SRT90]. The stability problem persists at lesser degree for the comparatively weaker Coulomb interaction such that calculations can be stabilized for a certain time span [KSKP20] and promising applications to strong-field dynamics exist [SBK20b]. As with TDCI, TDCC is designed for ab initio wavefunction calculations because the equation of motion involve a microscopic Hamiltonian $\hat{H}$. Commutators with $1ph$ operators $\hat{S}_1$ could still be evaluated with DFT techniques [RGB96], but higher order operators leave the realm of DFT.

2.3.4.3 Multi-Configuration TDHF (MCTDHF)

Multi Configuration Hartree Fock (MCHF) and MCTDHF start from the ansatz Eq. (2.51). Their static or dynamic equations can be derived variationally with the energy written simply as $\langle \Psi | \hat{H} | \Psi \rangle$. Much less simple is the outcome if the $c_m(t)$ and $|\Phi_m(t)\rangle$ are varied simultaneously. The equations of motion are strongly inter-related and, worse, are prone to redundancies which cause, in turn, singularities. The typical treatment takes the time evolution of the basis determinants $|\Phi_m(t)\rangle$ from elsewhere, e.g., as TDHF trajectories, while the variation of the expansion coefficients c_m is given by

$$\mathrm{i} \sum_{m'} \langle \Phi_m | \Phi_{m'} \rangle \hbar \partial_t c_{m'} = \sum_{m'} \langle \Phi_m | \hat{H} | \Phi_{m'} \rangle \hbar \partial_t c_{m'} - \mathrm{i} \sum_{m'} \langle \Phi_m | \hbar \partial_t \Phi_{m'} \rangle c_{m'}.$$

$$(2.56)$$

The norm kernel $\langle \Phi_m | \Phi_{m'} \rangle$ appears because the set $\{|\Phi_m\rangle\}$ is not necessarily ortho-normal. Imagine, e.g., that each $|\Phi_m(t)\rangle$ follows its own TDHF trajectory. Then non-orthogonality will develop because each state evolves in a different mean field. One often avoids that difficulty by using a time-dependent ortho-normal basis generated from the same mean field. This comes already close to TDCI but with time dependent expansion basis $|\Phi(t)\rangle$ and $\hat{a}_i^\dagger(t)$. This has the advantage that large amplitude motion is embraced by the underlying time-dependent mean field. On the other hand, the larger expense of the MCTDHF equations reduces the affordable size of expansion which, in turn, makes it harder to cover many-body correlations sufficiently.

All considered, MCTDHF is promising as it allows in its full formulation Eq. (2.56) to include much different mean fields thus being capable of describing, e.g., electron transfer or multi-fragmentation. The problem is that MCTDHF is an ab-initio theory dealing with a microscopic Hamiltonian $\hat{H}$ thus requiring a sufficiently large expansion basis to account for correlations. But the technical overhead as compared to TDCI or TDCC limits the size of the expansion and even bulkier expansions than in TDCI are needed to cover many different mean fields emerging along the dynamical evolution. A detailed discussion of technicalities as well as strengths and limits of MCTDHF can be found, e.g., in [ZKBS04, NKS05].

The ideal compromise would be to combine the multi-configuration ansatz with TDDFT, i.e. a Multi Configuration TDDFT scheme. However, this requires to evaluate non-diagonal elements of energy $\langle \Phi_m | \hat{H} | \Phi_{m'} \rangle$ for which DFT does not have an unambiguous answer. Moreover, the method is applicable if one expects only very few reaction channels known ahead of time. Such an example was also discussed in a nuclear context in [RGC83] where one even found an approximate way to deal with a DFT basis. But even there, the technical expense has hindered large-scale applications so far. Notice that there are schemes that go under the name of multi-configuration time-dependent DFT, but what they do is to combine long-range MCTDHF with short-range DFT correlations, thus combining static and dynamic correlations [FKJ13], but this is different from using a manifold of DFT Slater determinants as a basis.

The MCTDHF approach (sometimes called MCTDSCF) includes and optimizes all the determinants of the CI expansion. If the core electrons are not involved in the process, e.g., for valence-shell ionization, one can restrict the number of determinants to the so-called active space, i.e., all the states that are actively involved. The static version is called Complete Active Space SCF (CASSCF), and its time-dependent version is TD-CASSCF [SI13]. It is a less costly variant that makes larger systems affordable while not compromising accuracy provided that the active space is properly chosen. A further reduction in computational cost is obtained by considering fixed determinants, as in a regular CI expansion, but restricting them to the active space. This method is called CASCI, and its time-dependent version is TD-CASCI [PFL18]. If, instead of using SCF determinants, the basis functions of the expansion are obtained from second-order perturbation theory, thus including dynamical correlation, this becomes the CASPT2 (TD-CASPT2) approach. CASCI has been used to compute the APES in the QM/MM example of section 4.2.3.2.

2.3.5 DYNAMICS IN TERMS OF DENSITY MATRICES

The dynamics of density matrices are given by the quantum mechanical version of the BogoliubovBornGreenKirkwoodYvon (BBGKY) hierarchy [HM76]. We discuss it here in compact operator form where we abbreviate the one-body density matrix as $\rho(\mathbf{x}, \mathbf{x}') \to \hat{\rho}_1$ for particle 1, $\rho(\mathbf{x}, \mathbf{x}') \to \hat{\rho}_2$ for particle 2..., the two-body density matrix as $\rho(\mathbf{x}_1, \mathbf{x}_2, \mathbf{x}_1', \mathbf{x}_2') \to \hat{\rho}_{12}$, etc. We assume a Hamiltonian consisting of a one-body operator $\hat{T}$ (kinetic energy and external fields) and a two-body interaction $\hat{V}$. The coupled hierarchical equations of motion then read

$$i\hbar\partial_t\hat{\rho}_1 = [\hat{T}_1, \hat{\rho}_1] + \mathrm{tr}_2\left([\hat{V}_{12}, \hat{\rho}_{12}]\right) \tag{2.57a}$$

$$i\hbar\partial_t\hat{\rho}_{12} = [\hat{T}_1 + \hat{T}_2, \hat{\rho}_{12}] + [\hat{V}_{12}, \hat{\rho}_{12}] + \mathrm{tr}_3\left([\hat{V}_{13} + \hat{V}_{23}, \hat{\rho}_{123}]\right) \tag{2.57b}$$

$$\ldots \quad \ldots \quad \ldots$$

The hierarchy couples ever higher reduced density matrices. The dynamics of $\hat{\rho}_1$ calls for the knowledge of $\hat{\rho}_{12}$ which, in turn, requires $\hat{\rho}_{123}$ and so forth.

To terminate at a given order, one has to make a model for the next higher density matrix. To make that more transparent, we separate independent-particle content from correlations similarly as in Eq. (2.52) which gives

$$\hat{\rho}_1 = \hat{\rho}_1 \tag{2.58a}$$

$$\hat{\rho}_{12} = \mathcal{A}\{\hat{\rho}_1\hat{\rho}_2\} + \hat{c}_{12} \tag{2.58b}$$

$$\hat{\rho}_{123} = \mathcal{A}\{\hat{\rho}_1\hat{\rho}_2\hat{\rho}_3\} + \mathcal{A}\{\hat{\rho}_1\hat{c}_{23}\} + \mathcal{A}\{\hat{\rho}_2\hat{c}_{13}\} + \mathcal{A}\{\hat{\rho}_3\hat{c}_{12}\} + \hat{c}_{123} \tag{2.58c}$$

$$\ldots \quad \ldots \quad \ldots \tag{2.58d}$$

For example, we can generate a one-body theory by ignoring two-body correlations, namely taking $\hat{c}_{12} = 0$, which delivers exactly the TDHF equation (2.25b) with $\hat{h}_{HF} = \mathrm{tr}_2\{\hat{V}_{12}\hat{\rho}_2\}$. It sounds obvious that a theory at the two-body level can be obtained by setting $\hat{c}_{123} = 0$ which yields (after singling out the mean-field part in the second order equation)

$$i\hbar\partial_t\hat{\rho}_1 = [\hat{T}_1 + \hat{h}_{\mathrm{HF}}, \hat{\rho}_1] + \mathrm{tr}_2\left([\hat{V}_{12}, \hat{c}_{12}]\right), \tag{2.59a}$$

$$i\hbar\partial_t\hat{c}_{12} = [\hat{T}_1 + \hat{T}_2 + \hat{V}_{12}, \hat{c}_{12}] \tag{2.59b}$$

$$+\mathrm{tr}_3\left([\hat{V}_{13} + \hat{V}_{23}, \mathcal{A}\{\hat{\rho}_1\hat{c}_{23}\} + \mathcal{A}\{\hat{\rho}_2\hat{c}_{13}\} + \mathcal{A}\{\hat{\rho}_3\hat{c}_{12}\}]\right).$$

The second equation can and should be reduced further. But even here we realize that the scheme gets rather involved.

Worse than that, it has intrinsic consistency problems: the crucial trace relation $\mathrm{tr}_3\{\hat{\rho}_{123}\} = (N_e - 2)\hat{\rho}_{12}$ becomes increasingly violated over time evolution [SRT90]. One may run the scheme over a certain period, the length of which depends on the strength of the correlations. One may dream of varied cutoff schemes which overcome the problem. But those have not yet been found. A preliminary solution in the meantime is to correct the mismatch "by hand" along the propagation. With that strategy, there have been, indeed, a couple of successful applications of Two-body density matrix propagation to strong field dynamics [LBS+15, LBS+17].

2.3.6   THE GW EXTENSION OF DFT

The GW approach is formulated within the Green's function description. To make the formalism more transparent, we abbreviate the Green's function by combined indices similarly as we did above for the density matrices. This means for the one-body Green's function $G(x_1, t_1; x_1', t_1') \rightarrow \hat{G}_1$ and similarly for higher Greens functions and interactions. The time evolution of Green's functions follows the Martin-Schwinger hierarchy [vLS13] which is the Green's function analogue of the quantum BBGKY hierarchy Eq. (2.57) for density matrices. The lowest-order reads

$$(i\hbar\partial_t - \hat{T}_1)\hat{G}_1 = \hat{\delta}_1 - \text{itr}_2\{\hat{V}_{12}\hat{G}_{12}\} = \hat{\delta}_1 - \text{itr}_2\{\hat{\Gamma}_{12}\mathcal{A}\{\hat{G}_1\hat{G}_2\}\}, \quad (2.60\text{a})$$

$$\hat{\delta}_1 = \delta^3(x_1 - x_{1'})\delta(t_1 - t_{1'}), \quad (2.60\text{b})$$

$$\hat{V}_{12}\hat{G}_{12} = \hat{\Gamma}_{12}\mathcal{A}\{\hat{G}_1\hat{G}_2\}, \quad (2.60\text{c})$$

where the one-body Green's function $\hat{G}_1$ couples to the two-body Green's function $\hat{G}_{12}$ the same way as the one-body density matrix coupled to the two-body one in the BBGKY hierarchy Eq. (2.57).

The first-order equation can be reshaped into the form of a TDHF equation by shifting the correlations from $\hat{G}_{12}$ into the four-point vertex function $\hat{\Gamma}_{12}$ which takes the place of an effective two-body interaction. Note that this is not a simple local and instantaneous interaction. It embodies an intricate non-local structure and memory effects. So far, all correlations are plugged into the vertex function which is, of course, intractable. We would otherwise have an exact solution of the whole many-body problem. It is, however, feasible to deal with subsets of correlations. In electronic systems, the leading correlation stems from the polarization of the electron cloud which is produced by the long-range Coulomb force and which is described diagrammatically as the "bubble series" [Mah93, RT94]. This amounts to replacing the bulky full $\hat{\Gamma}_{12}$ with the screened Coulomb interaction $\hat{W}_{12}$. The emerging coupled equations of motion are then

$$(i\hbar\partial_t - \hat{T}_1)\hat{G}_1 = \hat{\delta}_1 - \text{itr}_2\{\hat{W}_{12}\mathcal{A}\{\hat{G}_1\hat{G}_2\}\}, \quad (2.61\text{a})$$

$$\hat{W}_{12} = \hat{V}_{12} + \hat{V}_{12}\hat{G}_1^r\hat{G}_2^a\hat{W}_{12}, \quad (2.61\text{b})$$

where the second line is the summed Dyson series for the screened Coulomb interaction [FW71]. The product $\hat{G}_1^r\hat{G}_2^a$ is the free polarization propagator combined from the retarded (or particle) Green's function $\hat{G}_1^r$ and the advanced (or hole) Green's function $\hat{G}_2^a$ (section 2.1.4). These two coupled equations constitute the much celebrated GW approximation, also introduced by Hedin [Hed65].

In the GW approximation, the self-energy, which includes exchange and correlation effects, is calculated as the product of Green's function with the screened Coulomb interaction $\hat{W}_{12}$. The latter is computed in the Random Phase Approximation (RPA), see section 2.2.4.1, which includes only the screening by independent electron-hole pairs. The equation can be solved nearly analytically for the homogeneous electron gas and this was the basis for one of the first practicable electronic energy density functionals [GL76].

The early applications of the GW equations dealt with static properties for which Green's function formalism along the normal time arrow, from past to future, applies. Dynamical applications of the GW equations lead to non-equilibrium stages which require the Keldysh time path for a proper description [KB62, Kel65]. This path is composed of a forward part running from $t = -\infty$ up to $t = +\infty$ followed by a backward part in the reverse direction. This is necessary to resolve properly degenerated stationary states as they occur, e.g., at finite temperatures [KB62]. The formal structure of the above GW equations is not changed by that. The extension becomes only visible if one writes out the time integrations explicitly.

The GW equations are well manageable in symmetry-restricted systems or reduced dimensions. For periodic systems, they have been implemented into the YAMBO code [AGM11], and used to compute the nonlinear optical response of carbon nanotubes and nanoribbons [ACG17], amongst other systems. Another implementation has been used to study ultrafast carrier and exciton dynamics in 2-D materials [PPS22]. With considerable effort, one can also treat finite electronic systems. To this end, Perfetto and Stefanucci have developed a non-equilibrium Green's function Kadanoff-Baym approach in the code CHEERS [PS18], and used it, among other things, to study time-resolved, transient optical phenomena, and transport phenomena in carbon-based molecular junctions [TvLPS21]. To achieve that goal, they use a representation in terms of localized basis functions (see section 3.5.1). With a sparse choice of basis states and exploiting zero entries in the Coulomb matrix, one can evaluate the eight-fold summations involved in the GW equations. A similar application in coordinate-space representation allowing for large excursions and emission of electrons is beyond feasibility. This requires further approximations which will be addressed in the subsequent sections. So far, that looks all very promising. However, as many time-dependent approaches to correlations (see, e.g., section 2.3.5), time-dependent GW can raise problems in actual numerical realizations [VRBL18]. Applications thus always require careful testing.

An important feature of the GW approximation in a dynamical context is that it allows for the description of dissipation and thermalization. To understand that, consider the second order of the polarization equation (2.61b), $\hat{W}_{12} = \hat{V}_{12}\hat{G}_1^>\hat{G}_2^<\hat{W}_{12}$, and insert it into the effective TDHF equation (2.61a). You will see that together with the $1ph$ pair $\hat{G}_1^>\hat{G}_2^<$ another one-body Green's function is active in the same time span. This is of the same type as the incoming Green's function. Let us assume that this is a $1h$ Green's function $\hat{G}^<$. Thus we access during dynamical evolution an intermediate $1p2h$ space which may have a high spectral density of real $1p2h$ excitations at the given excitation energy, the more so the higher the energy. This opens a decay channel in which the energy from the incoming $1h$ state is distributed over energetically matching $1p2h$ states, thus being lost for further propagation in the one-body channel. This describes the dissipation of coherently propagating excitation energy into intrinsic channels, eventually leading to the thermalization of the electron cloud.

The GW approximation is, again, an ab initio theory as the other models for correlations above. In fact, GW in bulk matter had been used to derive an electronic energy-density functional [GL76]. The corresponding correlations are thus imprinted into the functional in a local-instantaneous approximation. Adding GW simply on top is prone to double counting. The situation looks different in dynamical calculations where GW propagation allows for dissipation which is impossible to incorporate into a local-instantaneous energy-density functional. Such dissipation is automatically included in a full, dynamical GW treatment, but at the gigantic expense of creating all correlations from scratch (as typical for an ab-initio calculation). One can combine the advantages of DFT (efficient parametrization of static correlations) with the power of GW to treat dynamical correlation in GW. To this end, however, one must remove the static part from the screened interaction which is already contained in DFT. This can be done by mapping $\hat{W}_{12}$ from time to frequency domain as $\tilde{W}(\omega)$. The static part is obviously given by $\tilde{W}(\omega = 0)$. The exclusively dynamical part remains as $\tilde{W}^{(\mathrm{dyn})}(\omega) = \tilde{W}(\omega) - \tilde{W}(\omega = 0)$, for examples from nuclear DFT see [TR88, Tse13]. This subtraction helps against double counting. But the expense of a full dynamical GW calculations remains and it may be overkill. Having embodied most of the many-body correlations into an energy density functional, one can expect that the dynamical corrections from GW are small which allows a perturbative treatment. This leads to the quantum Boltzmann equation which will be addressed in the next section.

2.3.7 QUANTUM KINETIC EQUATIONS AND RELATED METHODS

2.3.7.1 Quantum Boltzmann Equation (QBE)

Isolated quantum systems evolve in time without energy loss. This, however, does not mean that there is no energy flow between the various degrees of freedom. If the electronic subsystem is driven away from equilibrium, then dissipation from electron-electron collisions should bring it back to equilibrium, characterized by a Fermi-Dirac distribution at some finite temperature. We have argued in the previous section that a proper treatment of dynamical correlations implies, indeed, dissipation. This happens independent of the approach, BBGKY hierarchy of density matrices, the GW approximation to the Martin-Schwinger hierarchy, or any other comparable many-body theory. All derivations lead to the same final outcome, the Quantum Boltzmann Equation (QBE).

We motivate it here briefly by continuing with the GW approximation of the previous section. The second-order polarization equation (2.61b) shrinks to $\hat{W}_{12} = \hat{V}_{12} + \hat{V}_{12}\hat{G}_1^{>}\hat{G}_2^{<}\hat{V}_{12}$. We insert that explicitly into the effective TDHF equation (2.61a) and write the result now with explicit time indices to visualize in a somewhat sloppy notation the crucial time structure:

$$i\hbar\partial_t \hat{G}_1(t',t) = \hat{h}_{mf}\hat{G}_1(t',t) + \hat{\delta}_1(t'-t)$$

$$- i\int dt''\mathrm{tr}_2\left\{\hat{V}_{12}\hat{G}_1^>(t',t'')\hat{G}_2^<(t',t'')\hat{V}_{12}\mathcal{A}\{\hat{G}_1(t'',t)\hat{G}_2(t',t'')\}\right\},$$

where the standard mean-field Hamiltonian $\hat{h}_{mf}$ includes kinetic energy and the lowest order term, the trace with the mere two-body interaction $\hat{V}_{12}$. One sees here more clearly that three one-body Green's functions propagate simultaneously along the same time span $[t'',t']$ which opens the dissipation channels as discussed above. As the correlations are treated here only in second order perturbation theory the basic static correlations are grossly underestimated. Thus one employs TDDFT for one-body propagation which means to inserting for $\hat{h}_{mf} = \hat{h}_{KS}$ (Eq. (2.33)) the Kohn-Sham Hamiltonian and for the interaction in the second order term the static screened Coulomb interaction $\hat{W}_{12}^{(0)}$, see e.g., Eq. (2.61b). This combination of effective interactions can be justified by elaborate considerations of many-body theory [SR83].

Equation (2.62) with its fully detailed time structure and some double counting in the principle-value part of the r.h.s. is still hard to handle in practice. Further reductions are commonly done. We sketch briefly the steps following [GRR86]. One first expresses the Green's functions by density matrices time one-body evolution operators. Then one assumes that the phase oscillations in the evolution operators are much faster than the temporal change of the density matrices and evaluates the time integral for the oscillatory factors only. This eliminates memory effects and treats the electron-electron collisions as instantaneous process (Markovian approximation). The final Quantum Boltzmann Equation (QBE) then reads in detailed matrix notation

$$i\hbar\partial_t\hat{\rho}_{nn'} = \left[\hat{h}_{KS},\hat{\rho}\right]_{nn'} + \hat{I}_{nn'}, \tag{2.62a}$$

$$I_{nn'} = \pi \sum_{ij...k'l'} \delta\left(\varepsilon_i + \varepsilon_j - \varepsilon_k - \varepsilon_l\right)$$

$$\{\delta_{kn'}W_{nlij}\widetilde{\rho_{ii'}\rho_{jj'}}W_{i'j'k'l'}\widetilde{\bar{\rho}_{k'k}\bar{\rho}_{l'l}} - \delta_{kn'}W_{nlij}\widetilde{\bar{\rho}_{ii'}\bar{\rho}_{jj'}}W_{i'j'k'l'}\widetilde{\rho_{k'k}\rho_{l'l}} -$$

$$\delta_{in}W_{kln'j}\widetilde{\rho_{ii'}\rho_{jj'}}W_{i'j'k'l'}\widetilde{\bar{\rho}_{k'k}\bar{\rho}_{l'l}} + \delta_{in}W_{kln'j}\widetilde{\bar{\rho}_{ii'}\bar{\rho}_{jj'}}W_{i'j'k'l'}\widetilde{\rho_{k'k}\rho_{l'l}}\} \tag{2.62b}$$

$$\text{with } \hat{h}_{KS}\varphi_n = \varepsilon_n\varphi_n, \tag{2.62c}$$

and where $\widetilde{\rho_{ii'}\rho_{jj'}} = \rho_{ii'}\rho_{jj'} - \rho_{ij'}\rho_{ji'}$, $\bar{\rho}_{ii'} = \delta_{ii'} - \rho_{ii'}$. Note that $\widetilde{\rho_{ii'}\rho_{jj'}}$ is an abbreviation for an anti-symmetrized product and the diagonal elements of the density matrix become the occupation numbers $\rho_{ii} = n_i$. The collision term $\hat{I}$ does not change the mean field nor the mean-field propagation. By virtue of the Markovian approximation, it describes exclusively dissipation and it thus free of double counting.

This quantum Boltzmann equation constitutes an enormous simplification as compared to a coherent treatment of dynamical correlations. And yet, it involves eight-fold summations over the s.p. basis which is affordable only in small basis sets (as in the code CHEERS [PS18]), with high symmetries (e.g., homogeneity), or in reduced dimensions. The hindrance comes from the non-diagonality of the one-body density matrices. We cannot avoid this because the collision is formulated in the instantaneous eigen-basis Eq. (2.62c) to recover properly the energy matching δ function. In truly dynamical systems, $\hat{h}_{KS}$ and $\hat{\rho}$ will not be simultaneously diagonal.

To reduce the collision term further, we make the additional assumption of near diagonality

$$\rho_{ij} \approx n_i \tilde{\delta}_{ij}$$

which yields the collision term in the QBE resembling that in a master equation

$$I_{nn'} = 2\pi\delta_{nn'} \sum_{jkl} \delta\left(\varepsilon_n + \varepsilon_j - \varepsilon_k - \varepsilon_l\right)$$

$$\left|v_{klnj}\right|^2 \left[n_n n_j (1 - n_k)(1 - n_l) - (1 - n_n)(1 - n_j) n_k n_l\right]. \quad (2.62d)$$

One can see here, immediately, that this collision term has only non-vanishing diagonal elements $I_{nn} \neq 0$ which drives a change of the occupation numbers n_n in the one-body density matrix. This is one effect of dissipation, the other one being a reduction of the oscillations of the non-diagonal elements of ρ_{ij}. The KS Liouville equation (2.62a) together with this much-reduced form Eq. (2.62d) of the collision term is often coined Extended TDHF, or TDLDA respectively. It can describe dissipative dynamics for not too high excitations [LSRD19]. The limitation in energy is due to the assumption of diagonal $\hat{\rho}$. Higher energies validate often semi-classical approaches (section 2.4). If quantum effects remain, nonetheless, one can tackle the problem with a stochastic treatment of correlations (section 2.3.8) or by modeling the collision term through relaxation times (section 2.3.7.2).

2.3.7.2 The Relaxation Time Approximation

The Boltzmann equation in the classical domain as well as the semi-classical generalization in the form of the Vlasov Uehling Ulhenbeck (VUU) equation (section 2.4) have proven over decades to be a very versatile and powerful approach to describe dissipative dynamics of all sorts of many-body systems from molecules to bulk matter [BB97]. And even there, the evaluation of the (semi-)classical collision term was often found burdensome. The observation that dissipation is often dominated by one leading relaxation time leads to the formulation of the Relaxation-Time Approximation (RTA) which is since long well established in homogeneous systems [BGK54, PN66]. This suggests applying the RTA to the QBE Eq. (2.62) with its much more laborious collision term.

The RTA applied to the quantum Boltzmann equation, built formally analogous to semi-classical RTA, reads [RS15]

$$i\hbar\partial_t\hat{\rho} = \left[\hat{h}_{\mathrm{KS}}[\rho], \hat{\rho}\right] - i\hat{\Gamma}[\rho] \quad , \quad \hat{\Gamma} = \frac{1}{\tau_{\mathrm{relax}}}\left(\hat{\rho} - \hat{\rho}_{\mathrm{eq}}[\varrho, \mathbf{j}, E]\right), \qquad (2.63\mathrm{a})$$

where $\varrho\,(\mathbf{r}, t)$ is the actual density, $\mathbf{j}(\mathbf{r}, t)$ the actual current, and E the energy. Key piece is $\hat{\rho}_{\mathrm{eq}}$ which is the local-instantaneous equilibrium state for local density $\varrho\,(\mathbf{r}, t)$, local current distribution $\mathbf{j}(\mathbf{r}, t)$, and energy E, all three entries computed from the actual state $\hat{\rho}(t)$. The RTA collision term thus drives relaxation toward local-instantaneous equilibrium in accordance with the experience that local equilibration is fast while global equilibration involving collective flow over the whole system (i.e. transport by the KS equation) can take a long time. The situation is comparable to a sound wave which evolves along local equilibrium states [Tho61].

The key task is to determine the instantaneous equilibrium density-operator $\hat{\rho}_{\mathrm{eq}}, [\varrho, \mathbf{j}, E]$. It is the mean-field state of minimum energy under the constraints of given local density ϱ, current $\mathbf{j}$, and energy E. The density constrained mean-field Hamiltonian then reads

$$\hat{h}_{\mathrm{dens.co.}}[\varrho] = \hat{h}_{KS}[\varrho] - \int d^3r\,\lambda_\varrho(\mathbf{r})\hat{\varrho}(\mathbf{r}) - \int d^3r\,\boldsymbol{\lambda}_j(\mathbf{r})\hat{\mathbf{j}}(\mathbf{r}) \quad (2.63\mathrm{b})$$

where $\hat{h}_{KS}[\varrho]$ is the standard KS Hamiltonian for the given local density $\varrho\,(\mathbf{r})$, $\hat{\varrho}(\mathbf{r})$ is the operator of the local density, $\hat{\mathbf{j}}(\mathbf{r})$ the operator of local current, while λ_ϱ and $\boldsymbol{\lambda}_j$ stand for the associated Lagrange parameters. These are determined iteratively such that the solution of the corresponding Kohn-Sham equations yields the wanted density $\varrho\,(\mathbf{r})$ and current $\mathbf{j}(\mathbf{r})$ following a scheme which was developed in [CRM+85].

As a further constraint, we need to adjust the equilibrium state to given energy $E_{\mathrm{goal}} = E_{\mathrm{mf}} = E[\hat{\rho}]$ and actual electron number $N_{\mathrm{goal}} = N_{\mathrm{e}}(t)$ (mind that electron loss renders the electron number time dependent). This is achieved by regulating temperature T and chemical potential μ in the Fermi equilibrium distribution associated to $\hat{\rho}_{\mathrm{eq}}$

$$n_i^{(\mathrm{eq})} = \frac{1}{1 + \exp\left((\varepsilon_i - \mu)/T\right)}. \qquad (2.63\mathrm{c})$$

The other crucial entry in the RTA collision term is the relaxation time τ_{relax}. Here one borrows well-settled experience from Fermi liquid theory [PN66] which gives

$$\frac{\hbar}{\tau_{\mathrm{relax}}} = 0.40\frac{\sigma_{ee}}{r_s^2}\frac{E_{\mathrm{intr}}^*}{N} \quad , \qquad (2.63\mathrm{d})$$

where E_{intr}^* is the intrinsic (thermal) energy of the system, $N(t)$ the actual number of electrons, σ_{ee} the in-medium electron-electron cross-section, and $r_s = (3/(4\pi\overline{\varrho}))^{1/3}$ is the Wigner-Seitz radius of the electron cloud [RS15]. It employs an average density $\overline{\varrho}$ because τ_{relax} is a global parameter. Examples of application will be given in Chapter 4. A very detailed description of RTA together with a ready-to-use code can be found in [DVC$^+$22].

It is to be noted that a similar scheme as RTA is also used in bulk matter [GZ15] with a slightly different relaxation operator which reads in matrix notation

$$\Gamma_{nn'} = \begin{cases} \frac{1}{\tau_{\text{diag}}} \left(\rho_{nn} - \rho_{\text{eq},nn}\right) & \text{if } n = m \\ \frac{1}{\tau_{\text{off}}} \rho_{nm} & \text{if } n \neq m \end{cases}, \qquad (2.64)$$

with τ_{diag} and τ_{off} being two phenomenological relaxation times for the populations and coherence, and $\hat{\rho}_{\text{eq}}$ the target equilibrium density matrix. This relaxation operator is a bit more general as it allows for different speed of the relaxation for diagonal and off diagonal elements with respect to the local equilibrium basis Eq. (2.63b).

2.3.8　STOCHASTIC APPROACHES

The ground state of interacting electron systems can be obtained using stochastic methods that sample the many-body wave function either in real space or in determinant space (for fermions). There are several techniques based on minimizing the energy or the variance of the wave function, beginning with the variational Monte Carlo's approach that optimizes the parameters of a trial wave function. A special, widely used, case is Diffusion Monte Carlo (DMC) which employs Green's functions and solves on the imaginary-time axis a diffusion equation [RTG90]. A slightly different variant is path integral Monte Carlo, which is based on Feynman's approach [KDL97]. In imaginary time, this allows us compute finite temperature properties. In the case of fermions, notably electrons, Quantum Monte Carlo (QMC) methods suffer from a severe problem associated with sign changes of the wave function upon particle permutation. Elaborate techniques have been developed over the past decades to cure this problem. These imaginary-time methods also allow for the computation of a few of the lowest excited states [HWAL94].

When moving to real-time QMC, the sign problem becomes a phase problem, and it is even more severe because now the wave function has to be phase-averaged. Nonetheless, quantum dynamics using real-time path-integral Monte Carlo has been successfully used for model systems or systems of reduced dimensionality [Mak98]. There have been a few attempts at developing a real-time DMC method. One such proposal exploits the similarities of the walkers used in DMC with the de Broglie-Bohm (dBB) approach to quantum mechanics, in which the wave function is represented as an ensemble of classical trajectories that follow a generalized Hamilton-Jacobi equation [Chr07]. This TD-DMC method has been used, again, only in very small model systems, with little impact. A more recent attempt uses a variant of QMC based on quantum field theory, in which the stochastic sampling is performed on a set of auxiliary classical variables introduced via a Hubbard-Stratonovich transformation, which recasts the two-body propagator into a sum of one-body ones [CR21]. This promising method has been used to study the dynamics of the Hubbard model with encouraging results, while more realistic systems have not yet been addressed.

A direct attack to the quantum mechanical many-body problem with stochastic methods is probably too demanding for present-days tools. More promising is to look for stochastic methods to deal with existing approaches. That strategy worked with success in the semi-classical domain as the example of VUU solvers shows, for a review in nuclear dynamics see for example [AARS96]. It is likely to help in the quantum domain as well. Within static calculations, a stochastic full CI scheme has been realized by stochastically sampling the space of excited Slater determinants, with considerable success [Gre95, BTA09]. A similar, more affordable approach based on MCHF has also been proposed [TSAB15]. A time-dependent working scheme along these lines is the Stochastic TDLDA (STDLDA). This stochastic approach starts from MCTDHF presented in section 2.3.4.3. The MCTDHF ansatz Eq. (2.51) consists of a coherent superposition of Slater states. Experience tells that coherence is rapidly lost for sufficiently large excitation energies. The idea is thus to replace the coherent superposition by an incoherent ensemble of Slater states expressed in terms of N-body density matrices Eq. (2.9) as

$$\hat{D} = \sum_{n=1}^{\mathcal{N}} |\Phi^{(n)}\rangle w_n \langle \Phi^{(n)}| \tag{2.65}$$

where $\mathcal{N}$ is the size of the ensemble, $|\Phi^{(n)}\rangle$ a Slater state, and W_n its weight in the ensemble. In a formally simpler version, one can omit the w_n (i.e. set them all to 1) and let the distribution of $|\Phi^{(n)}\rangle$ do the job. STDLDA had been proposed first in [RS92] and worked out to practicability in [SR14]. We sketch it here briefly.

Let us assume that we stay at a certain stage with the ensemble of Slater states $\{|\Phi^{(n)}(t)\rangle, n = 1...\mathcal{N}\}$ and want to do the next step. Each Slater state is defined by a set of occupied s.p. states $\varphi_i^{(n)}, i \leq N_\mathrm{e}$. It is important to carry a sufficient amount of unoccupied (particle) states $i > N_\mathrm{e}$ to supply space for the stochastic jumps to come. The enlarged space (size Ω) allows to unfolding the hierarchy of n-particle-n-hole (nph) excitations. The $1ph$ excitations need not be taken care of, as they are already accounted for in the TDHF/TDLDA propagation of the Slater state. The first true excitations beyond the mean field come along with $2ph$ states

$$|\Phi_{pp'hh'}^{(n)}\rangle = \hat{a}_p^{(n)\dagger} \hat{a}_{p'}^{(n)\dagger} \hat{a}_{h'}^{(n)} \hat{a}_h^{(n)} |\Phi^{(n)}\rangle \tag{2.66}$$

which as such are, again, Slater states. For the derivation of the stepping scheme, we propagate the coherent MCTDHF state Eq. (2.51) with TDHF/TDLDA for the $|\Phi^{(n)}(t)\rangle$ and with Eq. (2.56) for the $c_{pp'hh'}$ with initial condition $c_{pp'hh'} = 0$. This is done for a short time step δt after which we assume that coherence should be lost and reduce the coherent state to the incoherent ensemble Eq. (2.65) with probability $w_{pp'hh'} = |c_{pp'hh'}(\delta t)|^2$ and the complementing probability for the case of no transition ($0ph$ state). This yields with standard steps of statistical mechanics Fermi's golden rule [RS92]. In the $2ph$ picture it reads in detail

$$w^{(n)}_{pp'hh'} = \delta t \left|\langle \Phi^{(n)}_{pp',hh'}|\hat{W}|\Phi^{(n)}\rangle\right|^2 \delta_\Gamma(\varepsilon^{(n)}_p + \varepsilon^{(n)}_{p'} - \varepsilon^{(n)}_h - \varepsilon^{(n)}_{h'} + E^{(\text{rearr})}_{pp'hh'})$$
$$\delta_\Gamma(\varepsilon) = \Theta(\varepsilon - \Gamma/2)\Theta(\Gamma/2 - \varepsilon) \qquad (2.67)$$

where $\hat{W}$ is the screened interaction, as in the QBE Eq. (2.62), complementing the KS Hamiltonian $\hat{h}_{\text{KS}}$. The rearrangement energy $E^{(\text{rearr})}_{pp'hh'}$ corrects for the change in the mean field due to the actual $2ph$ transition. It is negligible in large systems because two changing states out of hundredth of particles will not matter much, but may play a role in small systems. The Θ in the $\delta_\Gamma(\varepsilon) = \Theta(\varepsilon - \Gamma/2)\Theta(\Gamma/2 - \varepsilon)$ of Eq. (2.67) is the Heaviside function. The finite width is realistic as it can be related to a finite sampling time and, even for long sampling times, to the fluctuations of the s.p. energies in a mean-field evolution.

Jumping to each accessible $2ph$ state in each time step would blast the ensemble in a few cycles. And here is where the stochastic aspect comes into play. One chooses for each sample n one state out of the $0ph$ plus the $2ph$ states with probability $w^{(n)}_0$, or $w^{(n)}_{pp'hh'}$ respectively. We identify the chosen state as the new representative $|\Phi^{(n)}(t + \delta t)\rangle$ of sample n in the ensemble. Doing that for every n yields finally the full ensemble $\{|\Phi^{(n)}(t)\rangle, n = 1...\mathcal{N}\}$ at the next stage. First practical applications of STDLDA (often also coined STDHF) and critical tests can be found in the series of publications [SR14, SRS15, LRD16, LSRD19].

2.3.9 COVERING STATISTICAL FLUCTUATIONS

Several approaches for dynamical correlations, as GW (section 2.3.6), QBE (section 2.3.7.1), RTA (section 2.3.7.2), and later on VUU (section 2.4.1), manage very well to describe dissipation. But they stay bound to one common mean-field trajectory. In practice, dissipation comes along together with fluctuations of the mean field. This is appropriately covered in genuinely stochastic approaches as, e.g., STDLDA (section 2.3.8) which generates an ensemble of trajectories and, of course, in many-body approaches as MCTDHF (section 2.3.4.3) or TDCI (section 2.3.4.1) which are not tied at all to one leading mean-field (however, at the price of limited applicability).

Mean-field theories of dissipation can be extended to cover fluctuations by adding a Langevin term which acts as stochastic force. This is known as Boltzmann-Langevin equation extending the Boltzmann equation or VUU [AARS96]. Similar extensions of QBE or RTA are conceivable, but have not been done so far. A Langevin term is also used in the Lindblad equation which is commonly used in spin-dynamics and quantum optics [Man20] and to describe stochastically the coupling to an environment, see section 2.5.5.1. The alternative to a statistical ensemble is to describe the fluctuations explicitly in terms of distribution functions with the Fokker-Planck equation [Rei16, AARS96]. However, this is manageable only if few degrees-of-freedom are involved. A distribution of mean-field trajectories is intractable that way.

2.3.10 PRELIMINARY CONCLUSIONS ON THEORIES OF CORRELATIONS

Looking back at all the approaches to quantum mechanical many-body dynamics, we see that the case is everything else than settled. We dispose of a couple of approaches all of them more or less demanding, and yet, tuned to special ranges of applications. Due to the urgency of the problem, we will see steady progress of these methods in the future. Let us here briefly summarize what each one of the methods can presently do for us.

Coherent two-body correlations are properly treated in TDCI (section 2.3.4.1), TDCC (section 2.3.4.2), MCTDHF (section 2.3.4.3), density-matrix dynamics (section 2.1.3), and the GW approach (section 2.3.6). This would allow, e.g., to evaluate genuine two-body observables as they could be measured experimentally by coincidence experiments. Except for MCTDHF, all the approaches are limited to small amplitudes in some way. TDCI is tied to a stationary reference state and requires a small amplitude even for the mean-field. TDCC, density matrices, and GW allow for larger excursions of the mean field while the two-body correlations are still tied to the leading mean-field trajectory.

One of the problems of mean-field models is that they often produce sharp phase transitions related to spontaneous symmetry breaking as, e.g., the transition to superfluidity in BCS (see also the discussion in section 2.3.2). This is unphysical in finite systems where the exact solution produces always a smooth transition (for an example see section 7.6.3. of [MRS10]). Already including two-body correlations in any one of the coherent approaches suffices to soften the transitions.

Dissipation is implicitly or explicitly contained in all approaches described above. One will probably not recover it in TDCI because it is limited to low excitations due to the explosive growth of expansion space with energy. The same holds true for MCTDHF which is also limited to rather small basis sets. Dissipation is a possibility in GW which becomes even more formally obvious in the QBE (section 2.3.7.1) derived therefrom. And dissipation is a center issue in RTA (section 2.3.7.2) and STDLDA (section 2.3.8).

Large deviations of different mean fields as they are necessary to describe electron transfer of molecular (multi-)fragmentation are provided only in MCTDHF and its stochastic realization in STDLDA, the latter having great potential for future applications with available computing facilities.

2.4 SEMICLASSICAL AND MACROSCOPIC APPROXIMATIONS

Situations with large particle numbers N and/or excitation energies E^* drive fully quantum mechanical treatments quickly beyond the bounds of feasibility. Fortunately, large N and E^* are exactly those regimes where quantum effects loose importance which allows to employ (semi-)classical approximations. This section will summarize three levels of approximations: first, a semi-classical phase-space picture which dismisses quantum shell effects but still maintains the Pauli principle among the electrons; second, a classical molecular dynamics with effective electron-electron interactions; and finally a discussion of macroscopic approaches used for describing plasmon response in very large systems (plasmonics).

2.4.1 VLASOV- AND VUU DYNAMICS

The quantum mechanical one-body density matrix $\rho(\mathbf{r}, \mathbf{r}')$ can be mapped to a phase-space distribution $f(\mathbf{r}, \mathbf{p})$ by virtue of the Wigner transformation and with the help of the Wigner-Kirkwood expansion thereof, an expansion in order of $\hbar$, one can derive on lowest order $\hbar$ a semi-classical equation of motion for $f(\mathbf{r}, \mathbf{p})$ [BB97]. This is the much celebrated Vlasov equation

$$\frac{\partial}{\partial t} f = -\frac{\mathbf{p}}{m} \cdot \nabla_\mathbf{r} f + \nabla_\mathbf{r} V_{\text{eff}}(\mathbf{r}, t) \cdot \nabla_\mathbf{p} f \qquad (2.68)$$

which is widely used in plasma physics [Vla50]. The old applications just contained the Coulomb mean-field for V_{eff} which is, indeed, the by far dominating agent in plasma physics. In cluster physics, the Vlasov equation is considered as the semi-classical limit of the KS equation and thus it is natural to use the KS mean field, see Eq. (2.33). This combination leads to the Vlasov-LDA approximation. Although looking straightforward, the derivation of the Vlasov-LDA equation via the Wigner functions runs into difficulties in connection with the Coulomb field. The more robust semi-classical limit is derived via the Husimi transformation [Pru78, LSR95] which also works well for Coulomb interactions [DLRS97].

The Vlasov equation in terms of the phase space distribution f as such is not much of a simplification as compared to a description in terms of the one-body density matrix Eq. (2.18b). In both cases, we deal with functions of seven variables. However, the robustness of the semi-classical Vlasov propagation allows a representation in terms of test particles as $f(\mathbf{r}, \mathbf{p}, t) = \sum_{n=1}^{\Omega} g(\mathbf{r} - \mathbf{R}_n(t), \mathbf{p} - \mathbf{P}_n(t))$ where g is a well concentrated basis function. The $\mathbf{R}_n, \mathbf{P}_n$ are the coordinates of the nth test particle which obey classical equations of motion in the KS mean-field [HM76, AT87, BD88, GRS02, FBMB04]. This, however, raises problems when propagating the pure Vlasov-LDA equation: the Pauli principle, still present in the semi-classical picture, becomes violated with time going on [RS95]. The problem can be cured, however at some computational expense [DRS97]. The mere Vlasov equation as such is thus of limited use for simulating the dynamics of finite quantum systems.

The strength of the semi-classical phase space picture with test-particle representation is unleashed when including collisional correlations. Starting point is the particle-particle collision term in the classical Boltzmann equation which can be augmented by the quantum features of the Pauli blocking yielding the Uehling-Uhlenbeck collision term. Together with the Vlasov equation, this leads to the Vlasov-Uehling-Uhlenbck (VUU) equation [UU32]

$$\frac{\partial}{\partial t} f = -\frac{\mathbf{p}}{m} \cdot \nabla_{\mathbf{r}} f + \cdot \nabla_{\mathbf{r}} V_{\text{eff}}(\mathbf{r}, t) \cdot \nabla_{\mathbf{p}} f + I_{\text{UU}}, \tag{2.69}$$

$$I_{\text{UU}}(\mathbf{r}, \mathbf{p}) = \int d\Omega \, d\mathbf{p}_1 \frac{|\mathbf{p} - \mathbf{p}_1|}{m} \frac{d\sigma(\theta, |\mathbf{p} - \mathbf{p}_1|)}{d\Omega}$$
$$\times \left[f_{\mathbf{p}'} f_{\mathbf{p}_1'} (1 - \tilde{f}_{\mathbf{p}})(1 - \tilde{f}_{\mathbf{p}_1}) - f_{\mathbf{p}} f_{\mathbf{p}_1} (1 - \tilde{f}_{\mathbf{p}'})(1 - \tilde{f}_{\mathbf{p}_1'}) \right].$$

The collision term embodies a local gain-loss balance for elastic electron-electron scattering $(\mathbf{p}, \mathbf{p}_1) \leftrightarrow (\mathbf{p}', \mathbf{p}_1')$ determined by the differential cross-section $d\sigma(\theta, |\mathbf{p}_{\text{rel}}|)/d\Omega$, the local phase-space density $f_{\mathbf{p}} = f(\mathbf{r}, \mathbf{p})$, and the Pauli blocking factors in parenthesis as functions of the relative phase-space occupation for paired spins $\tilde{f}_{\mathbf{p}} = (2\pi\hbar)^3 f_{\mathbf{p}}/2$. The velocity-dependent scattering cross-section has to be calculated for a screened electron-electron potential using standard quantum scattering theory [DRS00, KRF12]. It is to be noted that the VUU equation can also be derived as semi-classical limit of the QBE Eq. (2.62) the same way as the Vlasov equation can be derived from the Liouville equation for the one-body density matrix.

The full VUU equation involving a five-fold integration of seven-dimensional phase space functions is even less manageable than the Vlasov equations alone. Again, a robust and affordable solution scheme is offered by the test-particle method. Unlike the case of the pure Vlasov equation, the Pauli principle raises no problems here because the Pauli blocking in the VUU collision term automatically takes care of that [BD88, GRS02, FBMB04]. After all, whenever quantum shell effects can be ignored, the VUU equation is the approach of choice for dynamical simulations. This is typically the case for sufficiently excited Fermi fluids as in nuclei [BD88] or metal clusters [GRS02, FBMB04]. The Boltzmann equation has also been used to model X-ray created warm dense matter and plasma, for a recent review see [ZBM+23].

There remains the problem to determine the initial distribution $f(\mathbf{r}, \mathbf{p}, t = 0)$. Setting $\partial_t f = 0$ in the Vlasov or VUU equation defines just a stationary state, but not necessarily the ground state. We need to begin with a semi-classical description of the ground state and that is found in Thomas-Fermi theory. This is found from DFT at the level of the Hohenberg-Kohn theorem. The starting point is the energy in LDA with the kinetic energy also as functional of local density in the Thomas-Fermi approximation Eq. (2.28). Variation of this energy with respect to the local density $\varrho(\mathbf{r})$ with a constraint on the wanted electron number yields the Thomas-Fermi equation for the ground state distribution [Tho27, Fer28]

$$f_0(\mathbf{r}, \mathbf{p}) = \frac{2}{(2\pi\hbar)^3} \Theta(p_\mathrm{F}(\mathbf{r}) - |\mathbf{p}|), \ p_\mathrm{F}(\mathbf{r}) = \sqrt{2m[\mu - V_{\mathrm{eff}}(\mathbf{r})]}, \qquad (2.70)$$

where Θ is the Heaviside function, p_F the local Fermi momentum, and μ the chemical potential, the latter to be tuned such that the wanted electron number is obtained as $N_e = \int d\mathbf{r} d\mathbf{p}/(2\pi\hbar)^3 f_0(\mathbf{r}, \mathbf{p})$. This delivers a fully six-dimensional phase-space distribution. The initial state for a Vlasov dynamics in test-particle representation requires a sampling of the Thomas-Fermi distribution or a modification of the Thomas-Fermi Eq. (2.70) directly in terms of test particles, for details, see e.g., [FBMB04]. Note that the ground state of metal clusters can be rather well described with Vlasov-LDA [LSR06] because electrons are sufficiently delocalized in such systems. The case of covalent binding is more problematic and is usually not treated with Vlasov-LDA.

The Thomas-Fermi distribution Eq. (2.70) is a sphere in momentum space around momentum zero. The momentum sphere is the signature of an equilibrium state. Dynamical excitation produces states with non-zero, position-dependent average momentum $\overline{\mathbf{p}}(\mathbf{r}, t) = \int d\mathbf{p} \, f(\mathbf{r}, \mathbf{p}, t)$. Still, there are many dynamical regimes where the momentum distribution about the average momentum remains close to spherical. In fact, most excitation mechanisms first deform the momentum distribution. But the collision term drives the system rather quickly to the local instantaneous equilibrium with its spherical momentum distribution [Bal75, Rei16]. This allows to reduce the description to the time evolution of local density $\varrho\,(\mathbf{r}, t)$ and local momentum $\overline{\mathbf{p}}(\mathbf{r}, t)$, or local velocity $\mathbf{v}(\mathbf{r}, t)$ respectively, which turns out to be simply a classical hydrodynamic description [HM76]. When derived from VUU based on Vlasov LDA, it contains all information on the exchange correlation effectively imprinted via the density functional used and all information on dissipation via the Uehling-Uhlendbeck collision term.

There is an alternative road to the hydrodynamic description if dissipation can be ignored. This is the time-dependent Thomas-Fermi theory which is derived directly from the given density functional by modeling the wavefunctions in terms of density and momentum distribution [DRS98]. Hydrodynamical approaches are suited for Fermi fluids in dynamical regimes which are dominated by the collective flow. As such they have been widely employed in the nuclear dynamics of heavy-ion collisions [SG85]. The applications in electronic dynamics are more restricted, basically to metal clusters where collective flow often plays a leading role as long as electron emission remains small.

2.4.2 MD APPROXIMATIONS FOR ELECTRONS

For even larger excitations, one reaches a regime where a purely classical treatment of electrons is applicable. This allows a description in terms of classical Molecular Dynamics (MD) for the electrons. The equations of motion are trivially Hamiltonian equations

$$\partial_t \mathbf{r}_n = \frac{\mathbf{p}_n}{m_e} \quad , \quad \partial_t \mathbf{p}_n = -\nabla_n \sum_{n' \neq n} V(|\mathbf{r}_n - \mathbf{r}_{n'}|) \, . \qquad (2.71)$$

The interaction V is simply the Coulomb interaction to begin with. However, it is often appropriate to cut off the Coulomb singularity in accordance with the spatial dimensions one aims to describe. In a way this step toward electronic MD covers an extremely broad range of phenomena, from bulk motion over dynamical correlations up to large fluctuations. With specifically tuned effective interactions it even allows to deal with the behaviour of deep core and valence electrons. Not surprising then that classical MD models for highly excited electron dynamics had found great interest in the past, see e.g., [RPSWB97, IB00, Bau04, JRZB05, SSR06, BMR+06, FRB07, SI09]. Electronic MD thus renders the description of many processes feasible which would have been far out of reach of the more refined techniques. And yet, MD has its limits too. The limitation comes up for very large systems because the expense of two-body forces grows quadratically with particle number. Huge systems as they occur, e.g., in plasmas of astrophysical scenarios are handled with tree strategies mixing a global treatment for remote particles and a detailed MD for close ones [PG96].

The other limitation of pure MD approaches is encountered if very different length and/or energy scales are involved when considering deep core and valence electrons simultaneously. The deep core electrons reside still in the quantum regime and their field-induced tunneling from the inner core to the valence cloud (called inner ionization, section 1.1.1) cannot be described by classical MD. This process can be included in the classical MD simulations by rate equations generating new electrons for explicit dynamics stochastically, according to the given tunneling probability [SSR06, AF10]. For an application example see Figure 4.18. The modeling of such mixed models (rate equations together with electronic MD) is very specific to a system and needs to be looked up for each case anew.

2.4.3 PLASMONS IN LARGE SYSTEMS

As the number of atoms in a metallic cluster increases, the atomistic details of the ionic potential become less relevant and the positive ionic background can be profitably approximated as homogeneous. This is the so-called jellium approximation [KV95, Bra93, Bec87] (section 2.1.1). Apart from simplifying the electronic structure calculations, the jellium approximation provides a logical link between atomistic modeling based on DFT and hydrodynamic modeling based on the macroscopic "optical" approach itself based on the solution of the Maxwell equations [Mor21, ZFR+14, SZGV+13]. In fact, hydrodynamic models (see also section 2.4.1) can be related to an orbital-free formulation of TDDFT [Cir17, Yan15, BH00], as already introduced in [RG84, GDP96].

2.4.3.1 The Case of Bulk Metals

In the upper limit of large metal clusters, the bulk properties are retrieved. Translational invariance, or at least lattice symmetry, simplifies the modeling such that many results can be summarized nearly analytically. It is thus enlightening to discuss the optical response in bulk metals, and in particular the dielectric function (or relative permittivity), $\varepsilon(\omega, q)$. It is also desirable to compute the dieletric function from first principles because $\varepsilon(\omega, q)$ is used to model the plasmonics response of nanoparticles and nanostructures from the solution of the Maxwell equations [Mor21].

A simple picture of the response of a bulk metal is based on the Random Phase Approximation (RPA). Note, however, that RPA in solid state theory means a different level of approximation. It includes only the Coulomb residual interaction while the RPA as linearized TDLDA (see section 2.2.4.1) takes also into account the exchange-correlation contributions from the onset. Treating RPA with Coulomb interaction only and using the jellium approximation yields the well-known Lindhard's model of the macroscopic dielectric function [Woo72, Lin54]:

$$\varepsilon_{\text{RPA}}(\omega, q) = 1 - \frac{4\pi e^2}{q^2} \chi_0(\omega, q) \tag{2.72a}$$

$$\chi_0(\omega, q) = \lim_{\eta \to 0} \frac{2}{V_c} \sum_{\mathbf{k}} \frac{n_T(\varepsilon(\mathbf{k}+\mathbf{q}), \varepsilon_F) - n_T(\varepsilon(\mathbf{k}), \varepsilon_F)}{\varepsilon(\mathbf{k}+\mathbf{q}) - \varepsilon(\mathbf{k}) - \hbar\omega - i\eta} \tag{2.72b}$$

where χ_0 is the Lindhard response function, V_c the volume of the unit cell, $\hbar\mathbf{k}$ the electron momentum, and $n_T(\varepsilon(\mathbf{k}), \varepsilon_F)$ the Fermi-Dirac distribution for the single particle energy ε of momentum $\hbar\mathbf{k}$ and Fermi energy ε_F (Eq. (2.63c)). The choice of the unit cell is somehow arbitrary in the jellium approximation, e.g., it can be taken as a cube of side L, with a large value of L. The summation in Eq. (2.72) is over the first Brillouin zone and can be approximated by an integral if L is large.

The dielectric function in the limit $q \to 0$ can be expressed by the Drude formula

$$\varepsilon(\omega) = 1 - \frac{\omega_p^2}{\omega(\omega + i\Gamma)}, \tag{2.73}$$

where ω_p is the bulk plasmon frequency and Γ is a scattering rate which gives a finite spectral width to the plasmon. In the simple Drude theory, the bulk plasmon frequency is given by $\omega_p = \sqrt{4\pi e^2 \varrho/m^\star}$ where ϱ is the ground-state carrier density (density of dynamically active electrons) and $m^\star$ is the effective mass of the charge carriers. According to RPA, the bulk plasmon dispersion for finite, but small, q and small Γ is given approximately by

$$\Omega_p^2(q) \approx \omega_p^2 + \frac{3}{5} v_F^2 q^2, \tag{2.74}$$

where v_F is the Fermi velocity. The plasmon dispersion curve crosses a continuum of electron-hole excitations at a critical value of the momentum approximately given by the plasmon cut-off momentum, $|\boldsymbol{q}_c| \approx \hbar\omega_p/v_F$ [Woo72]. Above this critical momentum the plasmon gets damped by a process analog to the "Landau damping" of plasma physics [LP88].

The simple jellium-RPA picture is not a bad initial approximation for metals [KE99], but it neglects a few details: First, the residual interaction from exchange and correlations, second, band-structure effects, since the jellium approximation ignores the atomistic nature of the positive ion background, and third, crystal local field effects. We come to these corrections in the following.

2.4.3.2   Including Exchange and Correlations

The exchange and correlation interaction can be incorporated by summing the Dyson series for the exchange-correlation diagrams which yields the macroscopic dielectric function within the jellium approximation as [Stu93]

$$\varepsilon(\omega, \boldsymbol{q}) = 1 - \frac{4\pi e^2}{q^2} \frac{\chi_0(\omega, \boldsymbol{q})}{1 - f_{xc}(\omega, \boldsymbol{q})\chi_0(\omega, \boldsymbol{q})} \qquad (2.75)$$

where $f_{xc}(\omega, \boldsymbol{q})$ is the Fourier transform of the exchange-correlation kernel, namely the functional derivative $\delta U_{xc}/\delta\varrho$ of the time-dependent exchange-correlation potential U_{xc} with respect to the time-dependent electronic density, $\varrho(\mathbf{r}, t)$, evaluated at the ground-state density, $\varrho(\mathbf{r})$ [UY14].

The exchange-correlation kernel accounts for the physics neglected in the RPA. For instance, the plasmon relation Eq. (2.74) is complemented by a term $\varrho f_{xc}(\omega_p, \mathbf{0})q^2/m^\star$ [UY14]. In the most common approximation where f_{xc} does *not* depend on frequency, its imaginary part vanishes and $U_{xc}(\mathbf{r}, t)$ just depends on the instantaneous value of the density. This approximation is usually referred to as adiabatic. If working at the LDA level (section 2.2.3) one then recovers the ALDA (section 2.2.3.4). For then, f_{xc}^{ALDA} reduces to $f_{xc}^{ALDA}(t - t', \mathbf{r} - \mathbf{r}') = \delta(\mathbf{r} - \mathbf{r}')\delta(t - t')d^2 E_{xc}^{(LDA)}/d\varrho^2$, evaluated at the ground-state density.

The ALDA provides an accurate description of the long-wave response of monovalent metals, for which the uniform electron gas model already gives a fair approximation [UY14, Lie97]. Discrepancies may occur otherwise [CWH+11]. Any adiabatic functional neglects the plasmon decay through multiple electron-hole pairs [LST94], for which a frequency-dependent exchange-correlation kernel is needed [BSSR07, TSS01].

2.4.3.3   "Beyond" Jellium

The Lindhard equation Eq. (2.72) can be generalized to include band-structure as

$$\chi_0(\omega, \boldsymbol{q}) = \lim_{\eta \to 0} \frac{2}{\mathcal{V}_c} \sum_{\mathbf{k},n,n'} |\mathcal{M}_{\mathbf{k}+\mathbf{q},\mathbf{k}}^{n',n}|^2 \frac{n_T(\varepsilon(\mathbf{k}+\mathbf{q}, n'), \varepsilon_F) - n_T(\varepsilon(\mathbf{k}, n), \varepsilon_F)}{\varepsilon(\mathbf{k}+\mathbf{q}, n') - \varepsilon(\mathbf{k}, n) - \hbar\omega - i\eta}$$

$$(2.76)$$

where $\mathcal{M}_{\mathbf{k}+\mathbf{q},\mathbf{k}}^{n',n} = \langle u_{\mathbf{k}+\mathbf{q},n'} | u_{\mathbf{k},n} \rangle$ is the superposition integral over the unit cell between the periodic parts of the Bloch functions $\psi_{\mathbf{k},n}(\mathbf{x}) = e^{i\mathbf{k}\cdot\mathbf{x}} u_{\mathbf{k},n}(\mathbf{x})$. The wavevector $\mathbf{k}$ is summed over the first Brillouin zone. For the sake of simplicity, crystal local field effects are still neglected here. The $\varepsilon(\mathbf{k},n)$ is the energy of an electron of momentum $\hbar\mathbf{k}$ and depends on the band index, n, as well.

Equation (2.76) can be simplified by taking the nearly-free-electron approximation for the Bloch function, $\psi_{\mathbf{k},n}(\mathbf{x}) \approx e^{i\mathbf{k}\cdot\mathbf{x}}/\sqrt{V_c}$. There, one can distinguish between the *intra*band contributions to the dielectric function obtained by setting $n' = n$ into the summation and the *inter*band contributions obtained by setting $n' \neq n$, instead.

The intraband contributions yield a Drude-like form in the limit of vanishing $\mathbf{q}$,

$$\varepsilon_{\mathrm{RPA}}^{\mathrm{inter}}(\omega,\mathbf{0}) \approx 1 - \frac{\omega_p^2}{\omega^2} + i\pi\frac{\omega_p^2}{\omega}\delta(\omega) + \delta\varepsilon_{\mathrm{RPA}}^{\mathrm{inter}}(\omega,\mathbf{0}), \qquad (2.77)$$

where $\delta\varepsilon_{\mathrm{RPA}}^{\mathrm{inter}}(\omega,\mathbf{q})$ gives the interband contributions, $\delta(\omega)$ is a Dirac delta function centered at $\omega = 0$, and the plasmon frequency reads now

$$\omega_p^2 \approx -e^2 \sum_n \int d\mathbf{k} \, \frac{1}{3} v_{\mathbf{k},n}^2 \frac{\partial n_T}{\partial\varepsilon}\bigg|_{\varepsilon=\varepsilon(\mathbf{k},n)}, \qquad (2.78)$$

where the semi-classical electron velocity is $\hbar v_{\mathbf{k},n} = (\nabla_{\mathbf{k}} E_{\mathbf{k},n})$. In fact, Eq. (2.78) holds for crystals with cubic symmetry. Otherwise, the plasmon frequency depends on the direction along which $\mathbf{q}$ is brought to zero.

Since at low temperature, the factor $\partial n_T/\partial\varepsilon|_{\varepsilon=\varepsilon(\mathbf{k},n)}$ is sharply peaked at the Fermi energy, the contribution to Eq. (2.78) comes from the bands that cross the Fermi energy weighted by the square of the semi-classical electron velocity. It can be shown that in the case of a monovalent metal with a spherical Fermi surface, Eq. (2.78) gives the same result as the simple Drude's theory, see section 2.4.3.1.

The Full Width at Half Maximum (FWHM) of the plasmon resonance can be approximated as [Stu82]

$$\mathrm{Im}\{\varepsilon_{\mathrm{RPA}}(\omega_p,\mathbf{q})\} \left(\frac{\partial\mathrm{Re}\{\varepsilon_{\mathrm{RPA}}(\omega,\mathbf{q})\}}{\partial\omega}\bigg|_{\omega=\omega_p}\right)^{-1}. \qquad (2.79)$$

According to Eq. (2.77), the imaginary part of the dielectric function in the limit of vanishing $\mathbf{q}$ only comes from $\varepsilon_{\mathrm{RPA}}^{\mathrm{inter}}(\omega,\mathbf{0})$, if $\omega \neq 0$. This result is analogue to the RPA picture for the jellium approximation: there is no "Landau damping" for $|\mathbf{q}| < |\mathbf{q}_c|$ because no particle-hole excitations with $\omega = \omega_p(\mathbf{q})$ are available.

Band-structure effects can cause "Landau damping" even in the limit of vanishing $\mathbf{q}$ if $\varepsilon_{\mathrm{RPA}}^{\mathrm{inter}}(\omega_p,\mathbf{0}) \neq 0$. This is the case in particular for noble metals, but band-structure effects also have a strong influence on the plasmon dispersion of simple alkali metals, like potassium [CDRE00, EKS00, KE99, SO81].

Finally, note that "Landau damping" does not affect the $\omega \to 0$ limit of the dielectric function, for which we expect that $\varepsilon_{\text{RPA}}^{\text{inter}}(\omega, \mathbf{0}) \to 0$. In the absence of any proper scattering process, e.g., due to carrier-carrier or carrier-lattice interactions, the real part of $\varepsilon_{\text{RPA}}(\omega, \mathbf{q})$ still diverges in the limit of $\omega \to 0$. This is the behavior expected from an ideal (i.e., defect-free) normal (i.e., not superconducting) metal at zero temperature. To this extent, "Landau damping" is a collisionless process in metals, as it is in plasmas.

2.4.3.4 "Beyond" RPA + TDDFT

Time-Dependent Current Density Functional Theory (TDCDFT, section 2.2.3) [Vig12, Vig04, GD88] provides an alternative route to model plasmon damping by electron-electron interaction beyond RPA and ALDA. It allows to modify Eq. (2.77) by including electron-electron scattering equivalent to an effective electron-electron scattering rate $\Gamma_{e-e}(\omega)$ [BRvLdB06, RdB05].

The presence of disorder (e.g., points or extended defects), surfaces, and a finite temperature introduce new scattering mechanisms. Assuming the mechanisms independent of each other (Matthiessen's rule), the total scattering rate reads $\Gamma_{tot} = \Gamma_{e-e} + \Gamma_{imp} + \Gamma_{surf} + \Gamma_{e-ph}$, where Γ_{imp} is the scattering rate due to non-magnetic atomic defect (or impurities), Γ_{surf}, is the surface scattering rate, and Γ_{e-ph} is the electron-phonon scattering rate. Both Γ_{e-e} and Γ_{e-ph} can be evaluated from Many-Body Perturbation Theory (MBPT) [Ber16, CBG+06] which allows computing generation and relaxation of hot electrons and holes following plasmon decay [BSN+16a, MBNSGL16, BSN+16b, BMNL15].

The surface-assisted decay of the plasmon is also referred to as "Landau damping" in the plasmonics literature [Khu19, KS17]. The observed size-effect (D) on the localized plasmon FWHM can be modeled by setting $\Gamma_{tot} = \Gamma_{bulk} + \Gamma_{surf}(D)$, where the bulk value Γ_{bulk} is retrieved in the limit of $D \to \infty$ as $\Gamma_{surf}(D) \propto v_F/D$[KV95]. This size-effect is already accurately described by TDLDA when the dielectric properties of the environment are considered [CTC+19, Ler11, LBB+10].

2.5 COUPLING OF ELECTRONS TO NUCLEAR MOTION AND MOLECULAR ENVIRONMENTS

So far, we have considered electronic dynamics without atomic displacement. A common justification for neglecting atomic motion, at least in the first approximation, is based on the large ratio between the electron and nuclear masses. Even for the lightest atom, this ratio is about $1/2000$. For instance, if we equate the average electronic kinetic energy, K_e, to the average nuclear kinetic energy, K_n, we find that the nuclear velocity is at least one order of magnitude smaller than the electron velocity. Hence, it makes sense to neglect the atomic displacement during an ultrashort (≈ 1 fs) laser pulse when the typical timescale of molecular oscillations (e.g., bond stretching) is of order of

100 fs. On the other hand, molecular motion during the long evolution *after* an ultra-short laser pulse can involve non-negligible atomic displacements and need to be taken into account at these longer time scales.

Another separation of scales happens if only a small fraction of the system is optically active (often coined hot spot or chromophore) and the matter around (environment) reacts only indirectly and weakly. This allows to treat the environment at a lower level of approximation, mostly in terms of classical MD possibly augmented by polarization interactions. In this section, we will present various approaches to handle ionic motion and/or environment together with electronic dynamics. The coupling with ionic dynamics in the regime of low excitations is treated in adiabatic approximation where quasi-static electronic configurations are prepared ahead of time. More violent processes require to treat electronic and ionic propagation simultaneously which is much more demanding and presently not always handled to full satisfaction. The coupling to environment is even more complex because approximations have to be fine-tuned anew for every material and configuration. Here, we will present only a few typical strategies to give an impression of what to take into account and how to deal with it.

To simplify notation, we will summarize all ionic positions as $\mathbf{R} \equiv \{\mathbf{R}_I, I = 1...N_{ion}\}$ and all electronic positions as $\mathbf{r} \equiv \{\mathbf{r}_i, i = 1...N_e\}$. Only if inevitably needed, we will make the I and i explicit. To keep notations simple we use the same ionic mass M_{ion} for all ions.

2.5.1 ADIABATIC AND NON-ADIABATIC APPROXIMATION

For low energies, ions move slowly such that electrons can adjust almost immediately to the actual ionic configuration. Thus we take the basis for the electronic states from the adiabatic electron spectrum

$$\hat{H}\phi_{n,\mathbf{R}}(\mathbf{r}) = E_n(\mathbf{R})\phi_{n,\mathbf{R}}(\mathbf{r}) \tag{2.80a}$$

where $\hat{H}$ is constituted of $\hat{H}_{el} + \hat{H}_{coupl}$ plus the static part of $\hat{U}_{ext}$ (Eq. (2.1)) and ϕ_n are the full many-body wavefunctions. In practice, we use independent-particle states Eq. (2.17a) for the wavefunctions and the KS Hamiltonian $\hat{h}_{KS}$ in place of the Hamiltonian $\hat{H}_{el} + \hat{H}_{coupl}$. It is important to note that the eigenvalues E_n and the electronic states $\phi_{n,\mathbf{R}}(\mathbf{r})$ depend parametrically on the given ionic positions $\mathbf{R}$. A key for the subsequent evaluation of ionic dynamics is the total energy of the adiabatic states $\phi_{n,\mathbf{R}}$, which delivers the Adiabatic Potential Energy Surfaces (APES), the term "potential" refereeing here to the forthcoming motion of the system on such surfaces during the dynamical evolution.

$$V_{APES,n}(\mathbf{R}) = E_{many-body}(\phi_{n,\mathbf{R}}, \mathbf{R}) \xrightarrow{\text{LDA}} E_{DFT,el}(\varrho_{n,\mathbf{R}}, \mathbf{R}) + E_{ion}(\mathbf{R}) \tag{2.80b}$$

where $E_{many-body}$ stands for a full many-body treatment, E_{DFT} for DFT Eq. (2.26), and $E_{ion}(\mathbf{R})$ for the purely ionic configuration energy. The

$V_{\text{APES},n}(\mathbf{R})$ are the APES for ionic motion in the fixed electronic state n, for a simple example see Figure (4.20). The eigenvectors $\phi_{n,\mathbf{R}}(\mathbf{r})$ form a complete ortho-normal basis set for each value of $\mathbf{R}$, and they can be used as expansion. We denote with $n = 0$ the electronic ground state.

For sufficiently slow motion, i.e. sufficiently low energies, the system creeps along the electronic ground state ($n = 0$) which suggests the adiabatic ansatz for the total wavefunction

$$\Psi_0(\mathbf{r}, \mathbf{R}, t) \approx \chi_0(\mathbf{R}(t))\phi_0(\mathbf{r}, \mathbf{R}(t)) \tag{2.81}$$

which is also known as Born-Oppenheimer approximation [BH54]. The nuclear ground state wavefunction $\chi_0(\mathbf{R}, t)$ is determined by the nuclear time-dependent Schrödinger equation

$$i\hbar\partial_t\chi_0 = \left(-\frac{\hbar^2}{2M_{\text{ion}}}\nabla_{\mathbf{R}}^2 + V_{\text{BO}}(\mathbf{R})\right)\chi_0 \tag{2.82}$$

where $V_{\text{BO}} = V_{\text{APES},0}$ is often also called the Born-Oppenheimer (BO) surface. Equation (2.82) is a well manageable, widely used approach to describe the quantum states of low-energy electronic-ionic many-body systems (section 2.5.2.1).

The adiabatic basis $\{\phi_{n,\mathbf{R}}(\mathbf{r}), n = 1, ...\}$ Eq. (2.80a) is a complete and ortho-normal set of electronic wavefunctions. Thus any electronic-ionic wavefunction can be represented in the form

$$\Psi(\mathbf{r}, \mathbf{R}, t) = \sum_n \chi_n(\mathbf{R}, t)\phi_{n,\mathbf{R}}(\mathbf{r}) \tag{2.83}$$

called adiabatic, or Born-Huang expansion [BH54]. In general, $\chi_n(\mathbf{R}, t)$ is the probability amplitude to find electronic state n at ionic configuration $\mathbf{R}$. The spatial ionic density distribution can then be written as $\mathcal{P}(\mathbf{r}) \equiv \int d\mathbf{R}\,|\Psi(\mathbf{r}, \mathbf{R}, t)|^2 = \sum_n |\chi_n(\mathbf{r})|^2$.

The equations of motion for the ionic wavefunctions $\{\chi_n\}$ are

$$i\hbar\partial_t\chi_n(\mathbf{R}, t) = \sum_m \hat{\mathcal{H}}_{nm}\chi_m(\mathbf{R}) \tag{2.84a}$$

with

$$\hat{\mathcal{H}}_{mn} = \frac{1}{2M_{\text{ion}}}\sum_k \Pi_{mk}\Pi_{kn} + V_{\text{APES},n}(\mathbf{R})\delta_{mn} + \mathbf{D}_{mn}(\mathbf{R}), \tag{2.84b}$$

where the kinematic momentum is $\Pi_{mk} = (-i\hbar\nabla_{\mathbf{R}}\delta_{mk} - \mathbf{A}_{mk}(\mathbf{R}))$. It involves the non-adiabatic couplings [BMK$^+$03]

$$\mathbf{A}_{mk}(\mathbf{R}) = i\hbar \int d^3r\, \phi_m^*(\mathbf{r}, \mathbf{R})\nabla_{\mathbf{R}}\phi_k(\mathbf{r}, \mathbf{R}). \tag{2.84c}$$

These matrix elements can be computed from linear response, when one of the two states (m or k) is the electronic ground-state. In practice, the calculations are also extended to the non-adiabatic couplings between excited electronic states, although the linear response is not strictly applicable in this case [Mai16]. The last term $\mathbf{D}_{mn}(\mathbf{R})$ in Eq. (2.84b) occurs in the presence of an external electric field $\mathcal{E}$ and is proportional to the dipole moment $i\hbar \int d^3r \phi_m^*(\mathbf{r}, \mathbf{R})\mathbf{r}\phi_k(\mathbf{r}, \mathbf{R})$.

The ionic motion causes non-adiabatic couplings between the electronic states which add a generalized vector potential $\mathbf{A}_{mk}(\mathbf{R})$ to the kinetic energy. It looks at first glance similar to the kinetic energy of a charged particle in a magnetic field. However, there is a basic difference in the structure of the vector potential. The integral $\oint d\mathbf{r}' \, \mathbf{A}_{mn}(\mathbf{r}')$ over each closed path is not necessarily zero but can have a finite value if the path embodies a diabolic point. The value of the closed-path integral is called the Berry phase [BMK+03].

Looking at the effective ionic Hamiltonian, Eq. (2.84b), we see that electronic transitions among adiabatic states are promoted by two terms: the non-adiabatic coupling $\mathbf{A}_{mk}(\mathbf{R})$ appearing in the kinetic term and the dipolar coupling to the external field, $\mathcal{E} \cdot \mathbf{D}_{mn}(\mathbf{R})$. Both couplings depend parametrically on the ionic position $\mathbf{R}$. Rewriting Eq. (2.84c) one obtains

$$\nabla_{\mathbf{R}} V_{\text{APES},m} = \int d^3r \phi_m^*(\mathbf{r}, \mathbf{R}) \left(\nabla_{\mathbf{R}} \hat{\mathcal{H}}_{mm}\right) \phi_m(\mathbf{r}, \mathbf{R}) \quad \text{for } m = k, \quad (2.85a)$$

$$\mathbf{A}_{mk}(\mathbf{R}) = -i\hbar \frac{\int d^3r \phi_m^*(\mathbf{r}, \mathbf{R}) \left(\nabla_{\mathbf{R}} \hat{\mathcal{H}}_{mk}\right) \phi_k(\mathbf{r}, \mathbf{R})}{\varepsilon_k(\mathbf{R}) - \varepsilon_m(\mathbf{R})} \quad \text{for } m \neq k. \quad (2.85b)$$

Equation (2.85a) is the well-known Hellman-Feynman theorem and Eq. (2.85b) provides a compact expression for the vector potential driving diabatic processes. The latter allow to deduce a quantitative criterion for the validity of the Born-Oppenheimer approximation. For that, diabatic processes should be small, i.e. $\mathbf{A}_{mk}(\mathbf{R})$ should be negligible, which yields $|\int d^3r \phi_m^*(\mathbf{r}, \mathbf{R}) \left(\nabla_{\mathbf{R}} \hat{\mathcal{H}}_{mk}\right) \phi_k(\mathbf{r}, \mathbf{R})| \ll |\varepsilon_m(\mathbf{R}) - \varepsilon_k(\mathbf{R})|$. This is the same condition of the adiabatic theorem which states that for an infinitely small perturbation the "quantum numbers", i.e. the indices of the adiabatic states, are conserved. The criterion makes it also clear that non-adiabatic effects get unavoidably relevant close to an electronic level crossing or avoided crossing. In these cases, even an extremely slow ionic motion can promote an electronic transition.

2.5.2 ADIABATIC MOLECULAR DYNAMICS

Adiabatic molecular dynamics refers to a class of methods in which the nuclei move according to Newton's equations of motion while the electronic component (density or wavefunction) follows the nuclei while remaining always in the same stationary state, typically the ground state. Over the years this approach has been called in various, equivalent ways: Ab Initio MD (AIMD), First-Principles MD (FPMD), Density Functional MD (DFMD) when electrons were described by DFT, and sometimes even with the misleading term Quantum MD (QMD), which introduces confusion with methods treating nuclei quantum-mechanically. Below we describe the two main approaches within the AIMD class, i.e. Born-Oppenheimer and Car-Parrinello MD.

Until the introduction of AIMD by Car and Parrinello [CP85], the dynamics of systems of atoms, i.e., solids, liquids, molecules, etc., was studied using classical mechanics interacting via inter-atomic potentials, or classical Force Fields (FF), designed by selecting suitable functional forms and fitting their parameters to experimental data and/or static ab initio calculations. A FF generally comprises three types of interactions between atoms: bonded interactions such as bond stretching (harmonic or anharmonic like Morse potentials), bond bending for angular forces, and 4-body torsional cosine-like terms. Beyond 4-body, the interactions are considered non-bonding, and are generally described via repulsive-dispersive interactions such as Lennard-Jones or Buckingham potentials. The third type is electrostatic interactions, modeled by fitting the electrostatic potential computed quantum-mechanically with the one generated by a collection of fixed point-like partial charges located in the atoms (ESP methods). A more detailed account of FF can be found in [Koh06].

Force Fields have their limitations arising from a variety of sources, i.e., the choice of functional form of the potentials, the fitting procedure, the choice of target properties to fit either to experiment or to calculations, the choice of configurations to fit the forces, and the theory level used to calculate these forces. In recent years, the fitting of FF has been taken up by machine learning procedures, which do away with some of the above limitations at the expenses of introducing many more fitting parameters and increasing the computational cost by one order of magnitude. This, however, is still much more affordable than 1000-fold more expensive quantum methods [DCC19].

2.5.2.1   Born-Oppenheimer Molecular Dynamics

The simplest way to introduce the electronic component along with the atomic motion is to consider that the electrons follow instantaneously the motion of the nuclei (or ionic cores), adjusting their wave function (Hartree-Fock) or density (DFT) so that it corresponds to the electronic ground state at all times along a molecular dynamics trajectory. This approach treats the nuclei as classical particles. If this electronic structure optimization is carried out at every step [BZ83], the scheme is called Born-Oppenheimer Molecular Dynamics (BO-MD), because the electronic component moves strictly on the Born-Oppenheimer surface, excluding any kind of transitions between electronic states. This scheme requires an excellent convergence of the electronic component at each MD step. The equations of motion for the (classical) nuclei are the Newtonian equations along the BOS

$$M_I\ddot{\mathbf{R}}_I(t) = -\nabla_I V_{\mathrm{BO}}(\mathbf{R}). \tag{2.86}$$

Within BO-MD the quantum treatment of the nuclei is also possible, and indeed necessary, e.g., in photo-chemistry. This will be briefly discussed later.

2.5.2.2 Car-Parrinello Molecular Dynamics

A scheme related to BO-MD was proposed in 1985 by Car and Parrinello [CP85], within the context of DFT electronic structure. Here, the equation of motion for the nuclei is the same as for BO-MD, but the electronic Kohn-Sham orbitals are propagated with a fictitious electron mass μ which lies closer to the ionic masses. This saves computing expenses as the time step for electronic evolution can be taken much larger and less electronic steps are needed relative to the ionic steps. The larger electron mass degrades the adiabatic decoupling from the ions and drives the electrons a bit farther off the BO path. This mismatch is reduced by employing a second-order time derivative in the electronic equation of motion [CP85]

$$\mu\frac{d^2\varphi_i}{dt^2} = \hat{h}_{KS}\varphi_i + \sum_j \Lambda_{ij}\varphi_j. \tag{2.87}$$

The last term represents the ortho-normality constraint with Λ_{ij} being a set of Lagrange multipliers that ensure the preservation of the norm and orthogonality of the Kohn-Sham dynamical orbitals. The Car-Parrinello equations of motion can be derived from an extended Lagrangian formalism, and their properties have been thoroughly studied in [PSB91]. This scheme is called Car-Parrinello Molecular Dynamics (CPMD).

2.5.3 EHRENFEST DYNAMICS

2.5.3.1 Ehrenfest Dynamics and TDLDA-MD

The Born-Huang expansion with respect to the adiabatic basis Eq. (2.80a) becomes extremely inefficient for highly excited dynamics. This calls for a new starting point. Inspired by the product ansatz for the electronic many-body state, one takes a product ansatz for the combined electronic-ionic wavefunction as $\Psi(\mathbf{r}, \mathbf{R}, t) = \chi(\mathbf{R}, t)\phi(\mathbf{r}, t)$. At the typical high energies under consideration, the ions are well described by classical motion. Thus one performs the classical limit for the ions and starts from the total energy in terms of classical ionic coordinates $\mathbf{R}(t)$ and quantum mechanical electronic wavefunctions, in the case of DFT reduced to the local electron density only. Variation with respect to the electronic wavefunction and to the ionic coordinates yields the coupled equations of motion, here written immediately for the case of DFT

$$M_I \frac{d^2 \mathbf{R}_I(t)}{dt^2} = -\nabla_I E_{\text{tot}}, \tag{2.88a}$$

$$i\hbar \partial_t \varphi_i = \hat{h}_{KS}(\text{R}) \varphi_i, \tag{2.88b}$$

where $E_{\text{tot}} = E_{\text{DFT,el}} + E_{\text{ion}}$ is the total energy corresponding to the nuclear configuration $\mathbf{R}^{cl}$. This approach receives the name Ehrenfest Dynamics (ED).

The Ehrenfest equations apply to all sorts of approximations for the electronic dynamic. This could be the simplified one-electron schemes like semi-empirical or tight-binding, like in the Time-Dependent Tight-Binding approach [Tod01] (see also section 2.2.4.3) or any of the methods described above, mean-field and beyond. In case of a many-body treatment one replaces $E_{\text{tot}} \longrightarrow \langle \hat{h}_{\text{tot}} \rangle$, $\hat{h}_{\text{KS}} \longrightarrow \hat{H}_{\text{tot}}$, and $\varphi_i \longrightarrow \phi$. Due to the optimal trade-off between accuracy and computational cost, the most popular implementation of ED remains the one based on TDDFT. This is why we have provided the Eqs. (2.88b) already for DFT. Because of this overwhelming use of TDDFT for the electronic part, Ehrenfest Dynamics has become practically a synonym of TDDFT-MD (also coined rt-TDDFT-MD to emphasize the solution in real-time propagation), where MD stands for molecular dynamics for the nuclear degrees of freedom. We will use both terms throughout the book depending on the original work we discuss.

Ehrenfest Dynamics can be traced back to the early times of quantum mechanics [Fre34], later reformulated by McLachlan [McL64]. A number of works appeared during the early 80's, based on TDHF [Mic83, KGNR84]. One of the first implementations of rt-TDDFT-MD was that of Theilhaber [The92]. He developed a scheme to integrate the Time-Dependent Kohn-Sham (TDKS) equations using the Suzuki-Trotter split operator method, for simulating liquid metals and degenerate plasmas. The formal basis of the coupling of rt-TDDFT and ED was analyzed in [AAE$^+$08, ACZ$^+$09, ACER12]. Nowadays rt-TDDFT-MD has become practically mainstream to study a variety of problems that involve both electronic excitation and nuclear motion, as will be illustrated in section 4.2.1.

Recently, the advent of intense ultrashort (attosecond) X-ray pulses generated by free electron lasers (XFEL), prompted the need to consider non-linear effects in light-matter interactions at the molecular scale [FR11]. Here, the electromagnetic field, the electronic density and nuclear motion are all coupled, and one then needs to solve in parallel the usual electronic and ionic equations of motion together with Maxwell equations. This approach, originally proposed within the framework of single-molecule spectroscopy with XFEL and applied to the HNCO molecule, can also be used to describe the coupled dynamics of the near field in finite systems, not necessarily arising from XFEL sources. Moreover, it has been extended to describe light propagation with a multi-scale set of equations combining Maxwell's equations with TDDFT for crystalline solids, and applied to silicon [YSS$^+$12], and more recently extended to lattice vibrations via Ehrenfest Dynamics, and applied to coherent phonons generated in a stimulated Raman experiment in diamond [YY19].

2.5.3.2 Time Steps and Time Scales: CPMD, BOMD, and ED

Typical integration time steps in CPMD are of the order of 0.1 fs, i.e. an order of magnitude smaller than in BO-MD (section 2.5.2.1), to ensure accurate integration of the electronic variables. While the CPMD scheme (section 2.5.2.2) involves the dynamical evolution of the KS orbitals, this is not the actual electron dynamics of ED, because the orbitals do not follow the TDKS equations. Moreover, the CPMD approach requires the strict ortho-normalization of the KS orbitals at every MD step for stability, something that is unnecessary in the case of ED, as the correct propagation of the KS orbitals is unitary. The price to pay in ED is that the time step is of the order of attoseconds, i.e., about 100 times smaller than in CPMD. While AIMD simulations (section 2.5.2) can reach total times of hundreds of ps, in ED the ps timescale is only achievable for fairly small molecular systems, e.g., the photo-isomerization of small molecules [MTOO16]. Conversely, since the implementation of ED does not require the orthogonalization of the orbitals, with an appropriate rescaling of the orbital time-derivative it can provide an alternative to CPMD [AAE$^+$08].

2.5.4 NON-ADIABATIC DYNAMICS BEYOND EHRENFEST

In AIMD (section 2.5.2) there is a strong correlation between electronic and nuclear motion, as electrons are slaves of the nuclei. AIMD simulations are thus useful and reliable when the time evolution of the system is adiabatic, i.e. there are no transitions between electronic states. Therefore, within the irradiation context, these are only useful once the excitation has decayed completely into atomic motion or when there is a constant and homogeneous finite electronic temperature leading to a thermally averaged electronic density. In this latter case, the Kohn-Sham orbitals are weighted with the Fermi-Dirac distribution.

In ED there is also a strong correlation between electronic and nuclear motion. On the one hand, TDDFT remembers *too much* of the previous evolution due to the lack of electron-electron collisions. Hence, the electron dynamics is fully coherent and electrons never equilibrate to the Fermi-Dirac distribution. Possible ways to overcome this problem have been discussed in section 2.3.2.

On the other hand, in ED the nuclear dynamics correlates directly to the instantaneous electronic density, ignoring fluctuations on both sides. As a consequence, electronic excitation processes like those occurring due to irradiation with ions are captured efficiently by ED. However, the characteristics of energy transfer from electrons to phonons are distorted, in particular failing to properly describe some very ubiquitous phenomena such as Joule heating

[HBF+05] or the thermalization between electronic and nuclear degrees of freedom [RTKC16]. In fact, an important caveat of ED is that it does not satisfy the principle of detailed balance [Tul98, PT06, ABMC+21]. Therefore, spontaneous phonon emission, which is crucial for these phenomena, is suppressed. This can be attributed to the lack of incoherent scattering of electrons by vibrations (phonons). Part of the blame for this can be put into the classical description of the nuclear degrees of freedom, but this is not the whole story. Progress beyond ED requires also improved approximations for the electron-nuclear correlation.

2.5.4.1 Direct Quantum Dynamics vs Trajectory-Based Methods

The mainstream approach in quantum nuclear dynamics with electronic transitions has been, for a long time, to map the relevant adiabatic (or diabatic) potential energy surfaces and then run the nuclear quantum dynamics on them. This is an essentially exact methodology but with two main bottlenecks.

Firstly, it requires an efficient representation of the many-body nuclear wave function. One of the most sophisticated approaches in this respect is the Multi-Configurational Time-Dependent Hartree (MCTDH) method [MMC90, BJWM00], nowadays available in the general purpose package Quantics [Wor20], amongst others. In the MCTDH the nuclear wavefunction is expanded in a basis,

$$\chi(\mathbf{q},t) = \sum_{j_1}^{n_1} \cdots \sum_{j_f}^{n_f} C_{j_1\cdots j_f}(t) \, \xi_{j_1}^{(1)}(q_1,t) \cdots \xi_{j_f}^{(f)}(q_f,t) \,, \qquad (2.89)$$

where $\mathbf{q} = (q_1,\cdots,q_f)$ is a set of f nuclear coordinates, generally normal modes. Typical choices for the basis functions are either harmonic oscillator solutions, or a Discrete Variable Representation (DVR) in terms of orthogonal polynomials and associated Gaussian quadratures to calculate the integrals, e.g., Lagrange, Hermite, or Chebyshev polynomials. The wavefunction for coordinate q_κ is expanded in n_κ basis functions. All the basis functions of the f coordinates are coupled through the time-dependent coefficients $C_{j_1\cdots j_f}(t)$. The MCTDH Equations Of Motion (EOM) read:

$$i\hbar \dot{C}_J(t) = \sum_L \langle \Xi_J \mid H \mid \Xi_L \rangle C_L(t) \qquad (2.90)$$

$$i\hbar \, \dot{\xi}_j^{(\kappa)}(t) = \left(1 - P^{(\kappa)}\right) \sum_{k,l=1}^{n_\kappa} \left(\hat{\rho}^{(\kappa)^{-1}}\right)_{jl} \langle H \rangle_{lk}^{(\kappa)} \, \xi_k^{(\kappa)}(t) \qquad (2.91)$$

where $J = (j_1,\cdots,j_f)$, $\Xi_J = \xi_{j_1}^{(1)} \cdots \xi_{j_f}^{(f)}$, $\hat{\rho}$ is the density matrix, and $1 - P^{(\kappa)}$ projects out degree of freedom κ.

In Eq. (2.90) $H = \sum_{\kappa=1}^{f} h^{(\kappa)} + H_R$ is the nuclear Hamiltonian that sometimes is split into terms that operate on a single degree of freedom, $h^{(\kappa)}$, and a residual part, H_R, that contains the correlations between all degrees of freedom. The quantity $\langle H \rangle^{(\kappa)}$ is the mean-field average of the Hamiltonian over degrees of freedom κ, and $\hat{\rho}^{(\kappa)}$ is the density matrix κ. Equations (2.90) and (2.91) incorporate the constraint that single-particle orbitals and their time derivatives are orthogonal. This choice, however, is not mandatory. For example, the scheme can be readily reformulated in terms of H_R, which can be more convenient in terms of efficiency and/or stability.

The second problem is to choose an efficient way to represent the multidimensional potential energy surface, with an affordable number of quantum-mechanical calculations, especially if a high-level theory is used. The most widely used method, POTFIT, uses a linear combination of products of one-dimensional functions [JM96, Ott14]. The most recent advances use a canonical polyadic decomposition combined with a Monte Carlo estimation of numerical integrals that reduce the exponential scaling to almost linear [Sch20]. However, even with these state-of-the-art methodologies, the number of dimensions accessible with this approach is quite limited, e.g., 21 dimensions corresponding to 7 atoms in a recent study of CO scattering from a Cu(100) surface [MSM21]. This is a huge field that has produced a wealth of interesting results in photo-chemistry and related areas. Here, however, we are focusing on calculating forces on nuclei on-the-fly through the time-evolving electronic wave function or density. An excellent account of the various methodologies of both types available can be found in a recent book [GL20].

In section 2.5.3, we have seen that the simplest approximation is to represent the nuclei as classical particles following a single average trajectory, i.e., Ehrenfest Dynamics. To go beyond one has to consider more than one trajectory to take into account possible electronic transitions. If on top of this one wants to introduce the quantum character of the nuclei, one would need to know the time-dependent electronic density everywhere where the nuclear wave function is non-negligible, which has practically the same complexity as mapping the APES as discussed above. A possible solution is to represent the nuclear wave function in terms of an ensemble of classical trajectories, for which it is possible to compute forces on-the-fly. Another possibility is to expand the nuclear wave function in the momenta of the distribution These methods will be discussed in the following sections, see also the comprehensive review [COB18].

2.5.4.2   Independent Classical Trajectories: Surface Hopping

While ED is straightforward to implement and computationally efficient, going beyond is far from trivial. To improve the electron-nuclear correlation, a possibility is the Trajectory Surface Hopping (TSH) method, whose most widespread version, the Fewest Switches algorithm (FSSH), was proposed by Tully in 1990 [Tul90]. In TSH the electronic component evolves in the same way as in ED, i.e. according to Eq. (2.88b) , while the forces on the nuclei are determined from one single electronic APES at a time (Eq. (2.88a)). The difference with ED resides in that hops between different APES are allowed in order to include non-adiabatic effects.

TSH can be considered the classical limit of the Born-Huang propagation Eq. (2.84a). A system propagates usually along one APES $V_{\mathrm{APES},n}$ until another surface $V_{\mathrm{APES},n}$ comes energetically close. For a short time interval, the quantum mechanical coupling dynamics between the crossing surfaces is evaluated and eventually reduced to a hopping probability. The FSSH hopping probability between electronic states n and m is proportional to the non-adiabatic coupling $\mathbf{A}_{mk}(\mathbf{R})$ (Eq. (2.84c)). It becomes very large when APES are close together, leading to a vanishingly small denominator in $\mathbf{A}_{mk}(\mathbf{R})$ (Eq. (2.85b)). A simpler approximation to this probability is given by the Landau-Zener expression, derived within the context of a single avoided crossing between two APES. Essentially, the dynamics run on a single APES until a second APES gets close enough so that the probability becomes sizeable enough. At that point, a jump to the second APES is allowed with this probability, and the nuclear dynamics continues on the new APES until another avoided cross occurs, and so forth. This procedure is repeated for many independent trajectories, and the results are then averaged.

Trajectory Surface Hopping introduces the quantum character of the nuclei in a minimalistic way, by rescaling their kinetic energy to ensure energy conservation every time there is a discrete electronic transition. The velocity is rescaled in the direction of the non-adiabatic coupling vector $\mathbf{A}_{mk}(\mathbf{R})$ in an amount corresponding to the energy difference between the two APES [FKT08]. While TSH is not derived directly from first-principles, it has been shown that is it connected to the quantum-classical Liouville equation by dropping out terms related to decoherence [SOL13, EVR+15, Kap16]. At variance with ED, TSH does preserve the detailed balance condition. TSH works reasonably well when non-adiabatic transitions occur between a small number of APES, but not for a dense manifold of excited states [SOL13]. It has been combined with Linear Response TDDFT to calculate the forces on excited electronic states [CDP05, TTR07] and has evolved into widely available codes, e.g., [AP13, AP14]. This approach has enabled the calculation of photo-induced dynamical phenomena, e.g., the degradation of materials under light irradiation [HFLP20].

The main limitation of surface hopping is that measurable quantities are computed as averages of a collection of independent classical trajectories. This means, from a density matrix point of view, that the off-diagonal elements do not vanish during the dynamics, hence leading to a fully coherent propagation. This happens in ED, and also in the FSSH method, despite the stochastic nature of the hopping between APES. One of the consequences of this is that the internal consistency of the algorithm is generally not preserved, in the sense that the average of electronic populations over trajectories differs from the number of individual trajectories in each state. Therefore, decoherence corrections are required to make TSH reliable [SOL13]. Decoherence can be introduced at the level of individual trajectories or by coupling them [PA21].

For a summary and discussion of Decoherence Corrected (Trajectory) Surface Hopping (DCSH) see [COB18]. See also the example discussed in the section 4.2.2.2.

2.5.4.3 Gaussian Trajectories: Multiple Spawning, Multi-configurational Enrenfest, and Multiple Cloning

The main reason for overcoherence in TSH is the localization of the nuclear wave function at the classical coordinates. When a wave packet splits into two different states due to strong non-adiabatic coupling, the off-diagonal elements of the density matrix should decay rapidly to zero, but they do not. This is because in TSH the amplitude of the unoccupied states is propagated with the trajectory corresponding to the occupied one. A simple solution is the instantaneous decoherence approach, in which all but one of the amplitudes are set to zero after a hop, but this method is not internally consistent. Several other methods have been proposed under the generic name of Decoherence Corrections [COB18]. An alternative route is to introduce some level of quantum mechanics in the description of the nuclear wave function that allows for a smooth transition between electronic states, as in the Multiple Spawning (MS) methodology. The advantage of this kind of approaches is that they also incorporate tunneling.

In MS the nuclear wave function is expanded in Gaussian functions propagated as classical trajectories. The number of Gaussian functions $N_n(t)$ is allowed to change via *spawning* events to represent bifurcations of the wave function when the non-adiabatic coupling is important. The derivation starts from the Born-Huang expression Eq. (2.83), with the nuclear wave function written as a linear combination of multidimensional frozen Gaussians [Tan07]

$$\chi_n(\mathbf{R}, t) = \sum_{m=1}^{N_n(t)} \Lambda_n^m(t)\, g_n^m(\mathbf{R}, \mathbf{R}_n^m, \mathbf{P}_n^m, \gamma_n^m, \alpha_n^m) \tag{2.92a}$$

$$g_n^m(\mathbf{R}) = \frac{1}{\mathcal{N}_n^m} \exp\left(-\alpha_n^m (\mathbf{R} - \mathbf{R}_n^m)^2 - \frac{1}{i\hbar} \mathbf{P}_n^m \cdot (\mathbf{R} - \mathbf{R}_n^m) - \frac{1}{i\hbar} \gamma_n^m \right) \tag{2.92b}$$

where the APES involved are indicated with the index n and the normalized ($\mathcal{N}_n^m$) Gaussians g_n^m are given by Eq. (2.92b). The parameters $\mathbf{R}_n^m$ and $\mathbf{P}_n^m$ set the average position and momentum of the Gaussian wave packet g_n^m. These Gaussians are usually called Trajectory Basis Functions (TBF). Since these are frozen Gaussians, the coefficients α_n^m are kept constant. Position and momentum Gaussian centers follow "classical" equations of motion: $d\mathbf{R}_n^m/dt = \mathbf{P}_n^m/M_I$ and $d\mathbf{P}_n^m/dt = -\nabla_I E_{\text{tot}}$.

The phase factors γ_n^m are propagated as

$$\frac{d\gamma_n^m}{dt} = \frac{1}{2M_I} \langle \mathbf{P_I} \rangle^2 - \nabla_I E_{\text{tot}} - \frac{\hbar^2 \alpha_n^m}{M_I}. \tag{2.93}$$

for the case of DFT. For the general case, one replaces E_{tot} by $\langle \hat{\mathcal{H}}_e(\mathbf{R}) \rangle_e$ where $\hat{\mathcal{H}}_e(\mathbf{R})$ is the electronic Hamiltonian at ionic configuration $\mathbf{R}$ and the suffix e at the ket indicates a partial average with respect to the electronic degrees of freedom only.

The coefficients $\Lambda_n^m(t)$ of the linear combinations in Eq. (2.92a) are subject to complex electronic equations of motion involving overlap matrices between the TBFs and the Hamiltonian matrix which generates a correlation between trajectories. Details of the MS algorithm can be found in [BNM02]. In a nutshell, it goes like this: non-adiabatic couplings $\mathbf{A}_{mk}$ Eq. (2.84c) are calculated at each time step for each nuclear basis function, and a specific number of new Gaussians are evenly spawned in regions with large effective coupling, in general on a different APES. However, they can be spawned in the same APES in order to describe tunneling. When MS is combined with a first-principles electronic description, the method receives the name of Ab Initio Multiple Spawning (AIMS) [CM18, LHC22].

The Multi-Configurational Ehrenfest (MCE) approach, proposed and developed in [Sha09, SS12] is similar in spirit to MS, but instead of the standard Ehrenfest guiding trajectories the forces that determine the Ehrenfest trajectories are obtained by averaging the Hamiltonian over two or more electronic APES. At variance with MS, MCE does not expand the basis set by spawning new TBFs. A scheme that combines the advantages of both approaches, i.e. MCE and spawning, has been proposed under the name of multiple cloning algorithm [MGMS14].

2.5.4.4 Correlated Electron-Ion Dynamics (CEID)

So far, nuclear dynamics has been described via classical mechanics. Introducing the quantum nuclear dynamics on-the-fly is not trivial at all. The Correlated Electron-Ion Dynamics approach (CEID) [HBF+04, HBF+05] is an example of a methodology that does this in an approximate, yet systematic and controlled manner. CEID starts from the bare electron-nuclear Hamiltonian and solves the Louville-von Neumann (LvN) equation by separating the electronic and nuclear density matrices and treating the nuclear part approximately by a perturbative expansion in momenta of the fluctuations about the mean-field trajectory. By means of this expansion, supplementary time-dependent operators are introduced, e.g., $\hat{\mu}_I = \left\langle \Delta\hat{\mathbf{R}}_I \right\rangle_I$ and $\hat{\lambda}_I = \left\langle \Delta\hat{\mathbf{P}}_I \right\rangle_I$, where $\Delta\hat{\mathbf{R}}_I = \hat{\mathbf{R}}_I - \langle \mathbf{R}_I \rangle$ and $\Delta\hat{\mathbf{P}}_I = \hat{\mathbf{P}}_I - \langle \mathbf{P}_I \rangle$. Averages indicated with angle brackets with no subscript, $\langle \cdots \rangle$, are taken with respect to both electronic and ionic degrees of freedom. This is the same as tracing the product with the total electron-ion density matrix, $\langle \cdots \rangle = \text{Tr}\left\{ \cdots \hat{\rho}_{\text{tot}} \right\}$. A subscript will be used to indicate a partial average, e.g., the subscript I indicates a partial average with respect to the ionic degrees of freedom, only: $\langle \cdots \rangle_I = \text{Tr}_I\left\{ \cdots \hat{\rho}_I \right\}$. The operators $\hat{\mu}_I$ and $\hat{\lambda}_I$ give the first moments of the ionic positions and momenta, respectively. Since the $\hat{\mu}_I$ and $\hat{\lambda}_I$ operators are obtained by means of a partial trace with respect to the ionic degrees of freedom, only, they are still operators which act on the electronic Hilbert space, i.e., on the electronic wave functions. CEID supplements a new equation of motion for each of these supplementary electronic operators. Alternative versions of CEID differ in the way this hierarchy of equations of motion is truncated.

In the original mean-field second-order scheme [HBF$^+$04, HBF$^+$05], the expansion is terminated at the second order by imposing a "mean-field" approximation to the second moment of the fluctuations, e.g., $\left\langle \Delta\hat{\mathbf{R}}_I \Delta\hat{\mathbf{R}}_J \right\rangle_I \approx$ $\mathrm{Tr}_I \left\{ \Delta\hat{\mathbf{R}}_I \Delta\hat{\mathbf{R}}_J \hat{\rho}_I \right\} \hat{\rho}_e$, where $\hat{\rho}_I$ and $\hat{\rho}_e$ are the partial traces of the total density matrix $\hat{\rho}_{\mathrm{tot}}$ with respect to the electronic and ionic degrees of freedom, respectively. In this "mean-field" approximation, the total density matrix is taken factorized, i.e., , $\hat{\rho}_{\mathrm{tot}} \approx \hat{\rho}_e \otimes \hat{\rho}_I$, in the second moments to derive a closed set of equations of motion. Accordingly, in the mean-field second-order CEID scheme, the ED equations of motion are supplemented by two additional sets of equations for $\hat{\mu}_I$ and $\hat{\lambda}_I$, along with the classical equations of motion for the nuclear position and second momentum correlations.

Although a relatively low order expansion, the original mean-field second-order CEID is already effective in modeling the Joule heating in models of nano-wires and molecular junctions [MWD$^+$10]. This CEID scheme is indeed well-suited for systems with weak electron-phonon interaction, but fails to reproduce strong coupling behaviors like coherent oscillations between electron and phonon degrees of freedom. An alternative CEID scheme better suited for systematic expansion has then been developed [SMFH07] and shown to converge to the exact solution of the electron-ion Schrödinger equation. This alternative scheme has been successfully used to describe the analogue of Rabi oscillations in a model of strongly coupled electron-ion system [SMHF11] and the decay to the non-radiative relaxation to the lowest singlet exciton in a model [SSH79] of excited conjugated polymers [MWD$^+$10].

CEID has emerged as a powerful tool for problems in which the transfer of energy between electrons and nuclei is crucial. A cost-effective alternative that limits the nuclear motion to harmonic vibrations, named Effective CEID (ECEID), has been proposed recently [RTKC16], and applied to inelastic electron transport in water chains [RTK17] and thermoelectric phenomena [Chu20]. In ECEID the evolution of the phonon system is described through the dynamics of phonon occupation numbers, thus simplifying enormously the original CEID formulation in terms of momenta.

2.5.4.5 Exact Factorization

The Born-Huang expansion (2.83), although fairly general, is still tied to a previously prepared stationary adiabatic basis. An obvious extension toward more energetic processes is to allow the electronic basis states ϕ to become time-dependent. Doing that, it suffices to use only one electronic state leading to the ansatz

$$\Psi\left(\mathbf{r},\mathbf{R},t\right) = \chi\left(\mathbf{R},t\right)\phi_{\mathbf{R}}\left(\mathbf{r},t\right), \qquad (2.94)$$

where the electronic part is subject to the partial normalization condition $\int d\mathbf{r}\,|\phi_{\mathbf{R}}\left(\mathbf{r},t\right)|^2 = 1$, which must be valid for any nuclear configuration $\mathbf{R}$. The eXact Factorization (XF) approach has been shown to be exact and completely general [AMG10, AMG12]. For simplicity of notation, we present here the formalism in terms of Hamiltonian formulation, rather than for DFT.

Inserting Eq. (2.94) into the time-dependent Schrödinger equation for Ψ leads to the following equations of motion for the electronic and nuclear wave functions

$$i\hbar\partial_t\phi_{\mathbf{R}}\left(\mathbf{r},t\right) = \left(\hat{\mathcal{H}}_{el}^{XF} - \varepsilon(\mathbf{R},t)\right)\phi_{\mathbf{R}}\left(\mathbf{r},t\right) \qquad (2.95\mathrm{a})$$

$$i\hbar\partial_t\chi\left(\mathbf{R},t\right) = \left(\sum_{I=1}^{N_{\mathrm{ion}}}\frac{1}{2M_I}\left(-i\nabla_I + \mathbf{A}_I(\mathbf{R},t)\right)^2 + U_{\mathrm{ext},e}(\mathbf{R},t) + \varepsilon(\mathbf{R},t)\right)\chi \qquad (2.95\mathrm{b})$$

where the electronic Hamiltonian is given by

$$\hat{\mathcal{H}}_{el}^{XF}(\mathbf{r},\mathbf{R},t) = \hat{\mathcal{H}}_{el}^{BO}(\mathbf{r},\mathbf{R},t) + U_{\mathrm{el,ion}}[\phi_{\mathbf{R}},\chi] + U_{\mathrm{ext},e}(\mathbf{r},t), \qquad (2.95\mathrm{c})$$

with $\hat{\mathcal{H}}_{el}^{BO}$ the usual Born-Oppenheimer electronic Hamiltonian, including the electronic kinetic energy and the electron-electron, electron-nuclear, and nuclear-nuclear Coulomb interactions. The second term,

$$U_{\mathrm{el,ion}}[\phi_{\mathbf{R}},\chi] = \sum_{I=1}^{N_{\mathrm{ion}}}\left[\frac{1}{M_I}\frac{(-i\hbar\nabla_I - \mathbf{A}_I(\mathbf{R},t))^2}{2} + \right.$$
$$\left.\left(\frac{-i\hbar\nabla_I\chi}{\chi} + \mathbf{A}_I(\mathbf{R},t)\right)(-i\hbar\nabla_\nu - \mathbf{A}_I(\mathbf{R},t))\right], \qquad (2.95\mathrm{d})$$

with $\nabla_I = \nabla_{\mathbf{R}_I}$, is a new electron-nuclear coupling term beyond Born-Oppenheimer. The potentials V_n^{ext} and V_e^{ext}, represent external sources like laser fields and act on nuclei and electrons, respectively. The index I labels the N_{ion} nuclei. The time-dependent scalar and vector potentials are given by

$$\varepsilon(\mathbf{r},t) = \left\langle\phi_{\mathbf{R}}(t)\left|\hat{\mathcal{H}}_{el}^{XF}(\mathbf{R},t) - i\hbar\partial_t\right|\phi_{\mathbf{R}}(t)\right\rangle_e \qquad (2.95\mathrm{e})$$

$$\mathbf{A}_I(\mathbf{R},t) = \left\langle\phi_{\mathbf{R}}(t)\left|-i\hbar\nabla_I\right|\phi_{\mathbf{R}}(t)\right\rangle_e, \qquad (2.95\mathrm{f})$$

where the subindex e indicates integration over the electronic variables. The former can be identified as a time-dependent potential energy surface. The definition of $\mathbf{A}_I(\mathbf{R},t)$ is similar to that of Eq. (2.84c), although in this context $\phi_{\mathbf{R}}(t)$ is not necessarily an adiabatic electronic state.

The full set of XF Eqs. (2.95) is not simpler to solve than the direct Schrödinger equation for the full electron+ion many-body problem. In fact, it is even more involved. The merit of the XF methodology is that it constitutes a framework from which approximate methods can be derived in a systematic manner. This helps interpretation as it reveals the forces between ions and electrons. For example, one can make a classical approximation for the nuclei and obtain a mixed quantum-classical method that has the correct electron-nuclear correlation, thus allowing for the assessment and improvement over the standard Ehrenfest dynamics [AAG14]. Moreover, this quantum-classical approach also allows for a formulation in terms of coupled trajectories that, unlike the case of independent trajectories as in TSH, introduces decoherence in a natural way [AMAG16, VZMM22]. The first XF calculation of a realistic system, the ring-opening in the oxirane molecule after photoexcitation, was presented in [MATG17], together with an assessment of the quality of the FSSH approach (section 2.5.4.2). Coupled quantum XF trajectories were also used to study a model of photo-isomerization of a retinal, and used to validate coupled classical trajectory methods [TLA22].

An interesting aspect of XF is that, within this framework, the role of ions and electrons can be swapped, depending on whether one is interested in the electronic effects on nuclear dynamics or vice versa. In summary, the XF method appears to be getting traction and might become mainstream in the future.

2.5.5 THE ROLE OF THE ENVIRONMENT: DISSIPATIVE DYNAMICS

All methods described above, namely Ehrenfest Dynamics, surface hopping, CEID, and even the exact factorization approach, treat all ions and all electrons explicitly. This rapidly limits the size of the manageable systems. There are many situations where the key electron dynamics remain within a comparatively small spatial hot spot. This offers a natural hierarchy of relevance in space where electrons and ions far from the hot spot serve merely as "environment" and can be treated at a lower level. The simplest is to consider the environment as a heat bath with thermal coupling to the active region. Examples of this approach are given in sections 2.5.5.1 and 2.5.5.2.

A more elaborate approach is to describe the ions and electrons in the environment still explicitly, however, simplified as classical particles with appropriate effective forces and sometimes allowing for internal polarization. This will be discussed in section 2.5.5.3.

An alternative is to trace out the environmental degrees of freedom, entirely or partially, and replace them with an effective bath acting on the electronic subsystem, or on a reduced model including electronic and some ionic degrees of freedom, thus leading to an open quantum system interacting with the bath.

2.5.5.1 Open Quantum Systems: The Lindblad Master Equation

It is convenient here to move away from the traditional TDKS formulation in terms of the evolution of Kohn-Sham orbitals. This formulation does not render itself naturally to the inclusion of decoherence effects. Instead, this can be achieved starting from the equivalent Liouville-von Neumann (LvN) equation for the evolution of the density matrix. Indeed, an extension of the LvN equation in the form of a master equation leads to the Lindblad Master equation

$$\hbar\partial_t\hat{\rho}^{N_e} = -i[\hat{H},\hat{\rho}^{N_e}] + \sum_i \Gamma_i \left(L_i\rho^{N_e}L_i^\dagger - \frac{1}{2}\left\{L_iL_i^\dagger,\rho^{N_e}\right\}\right). \qquad (2.96)$$

Equation (2.96) describes the evolution of the electronic many-body density matrix $\hat{\rho}^{N_e}$, which can account for both mixed and pure quantum states. The second term in the right-hand-side incorporates the interaction between electrons and the environment within a Markovian approximation, and hence it is able to address decoherence. Here, the curly brackets indicate an anti-commutator or Poisson bracket, and L_i are called Lindblad (or jump) operators, which represent the effect that the interaction between system and environment has on the system. The sum runs over the various environmental modes that are coupled to the system, and can represent various forms of dephasing and vibrational relaxation. The factor Γ_i represents a decay rate associated to process i. If this second term is zero one recovers the LvN equation which is equivalent to the Schrödinger equation. The Lindblad master equation is a general Markovian framework for introducing the interaction of a quantum many-body open system with the environment, which is central for the success of quantum technologies like quantum computing [Man20].

2.5.5.2 Quantum Kinetics and the One–Body Reduced Density Matrix

An alternative to the Lindblad equation is offered by non-equilibrium MBPT, with a set of coupled equations for the electronic Green's function $G^>(t,t')$ on the forward part of the Keldysh path, see Figure 2.1, and the corresponding ionic propagator $D^\uparrow(t,t')$ which would capture both the coherent and non-coherent dynamics. The main advantage of such an approach is the use of reduced quantities in place of the many-body density matrix. A possible example of such an approach for small atomic displacements has been discussed in [KvLP+21], where the ionic propagator is simplified by replacing it with a bosonic phonon propagator, similarly to the ECEID approach [RTKC16] (section 2.5.4.4). This leads to a generalized version of the coupled evolution equations for the electronic, $\hat{\rho}^{N_e}$, and phononic, $\hat{\rho}_b$, density matrices

$$\begin{aligned}
\hbar\partial_t\hat{\rho}^{N_e}(t) &= -i[\hat{H}(t),\hat{\rho}^{N_e}(t)] - \left(I(t)+I^\dagger(t)\right)\\
\hbar\partial_t\hat{\rho}_b(t) &= -i\left[\boldsymbol{\alpha}\boldsymbol{\Omega}\hat{\rho}_b(t) - \hat{\rho}_b(t)\boldsymbol{\alpha}\boldsymbol{\Omega}\right] + \boldsymbol{I}_b(t)+\boldsymbol{I}_b^T(t) \qquad (2.97)
\end{aligned}$$

where $I(t)$ and $\boldsymbol{I}_b(t)$ are electronic and bosonic collision integrals, $\boldsymbol{\alpha}$ is proportional to a Pauli matrix $\boldsymbol{\sigma}_y$, and $\boldsymbol{\Omega}$ are bosonic frequencies.

This scheme is too demanding and impractical for ab initio applications beyond small atomic displacements. The couplings, i.e. the scattering mechanisms, originate from the non-adiabaticity of the many-body self-energy. An ab initio implementation of the coupled semi-classical Boltzmann equations, that ignores the coherent part of the propagation, i.e., the commutators, has been proposed recently [Car21]. There, phonon populations are updated similarly to ECEID (section 2.5.4.4). The full description of the coupled electron-phonon relaxation dynamics would also require the phonon-phonon matrix elements [TB21]. Notice that the evolution of phonon population is complementary to Ehrenfest Dynamics (ED), representing quantum harmonic motion around the classical positions. A practical scheme that combines ED with the evolution of a general nuclear wave function or density matrix is still in the wish list.

2.5.5.3 Hierarchical Approaches

It is frequently the case that the electron dynamics is confined to a limited region in space. For example, in the photo-excitation of a chromophore, a photon is absorbed locally by a molecule embedded in an environment, typically a solvent, or in a specific moiety belonging to a larger molecule. The energy absorbed is then transferred to nuclear motion that can lead to isomerization [SRF+11], bond breaks [NKR+18], or heating of the surrounding medium [VdMdC+22]. Since the motion of the electrons is restricted to a specific region, while the role of the surrounding medium is essentially structural, acting as a scaffold or enforcing steric constraints, then it is not required to treat the environment at a quantum-mechanical level. It is sufficient to use a classical description, which does not have an impact on the electron dynamics and is, computationally, much less onerous.

When the photo-excited molecule is immersed in a molecular or solid framework environment, the standard approach is to treat the environment as a classical Force Field (FF), frequently called also Molecular Mechanics (MM), while the whole molecule is treated quantum-mechanically, at the required theory level, i.e. TDDFT, TDCAS, or other. When the molecule is part of a larger entity, then the approach has been to identify the active region and separate it from the rest of the scaffold by cutting preferably homonuclear bonds, e.g., C-C, and replacing them with a pseudo-potential. Then, the rest of the interactions in the non-active region are described using MM, while the active, quantum-mechanical region is described by the usual electronic structure methods, either DFT or wavefunction-based. This quantum-classical approach is usually called QM-MM. An excellent account of QM-MM methods, including electron dynamics, is given in the review article [MAF+18].

In this approach the energy of the entire system is divided into three terms: the energies of the fragments treated classically, E_{MM}, and quantum-mechanically, E_{QM}, and the energy of the interaction between the two regions, E_{QM-MM}. Therefore,

$$E_{tot} = E_{QM} + E_{QM-MM} + E_{MM}, \tag{2.98}$$

The interactions included in the MM region, and represented by E_{MM}, have been briefly described in section 2.5.2. The most delicate part in the QM-MM approach is the coupling between the two regions. Here on has to take into account the electrostatic interactions between the MM partial charges and the ionic charges as well as the electronic distribution in the QM region. In addition, non-bonding repulsive-dispersive interactions between atoms in the MM and the QM region should be included to adjust structural properties at the interface, and especially to avoid the collapse of MM atoms with negative partial charges onto the (positively charged) nuclei of the QM region. This embedding energy is given by the following expression,

$$E_{QM-MM} = e^2 \sum_{I \in QM} \sum_{A \in MM} \frac{Z_I Q_A}{\mid \mathbf{R}_I - \mathbf{R}_A \mid}$$
$$- e^2 \sum_{A \in MM} Q_A \int \frac{\varrho(\mathbf{r})}{\mid \mathbf{r} - \mathbf{R}_A \mid} d\mathbf{r} + E_{RD}(\mid \mathbf{R}_I - \mathbf{R}_A \mid) \tag{2.99}$$

where eQ_A are the partial charges in the MM region, $\varrho(\mathbf{r})$ is the electronic density, and E_{RD} is a repulsive dispersive potential, e.g., Lennard-Jones. Sub-indices I and A refer to the QM and MM regions, respectively.

This approach has fixed charges, and therefore cannot account for polarization effects on the environment. It is possible to allow for variable partial charges that are either adjusted self-consistently or evolve dynamically at each MD step, in the same way as the electronic density is optimized at each step in ground state DFT, or evolves dynamically as in TDDFT (similarly in wavefunction approaches). In molecular systems inter-molecular charge transfer does not generally happen. Polarization is mostly due to a redistribution of charge within each molecule, to adjust the electrostatic multipole moments. Therefore, an alternative is to endow atoms in the MM region with a polarizability (and if required also higher-order multipoles), in addition to their partial charges. Details are described in the perspective article [BNC+20].

These approaches take into account both the effect of the QM part to fluctuations in the environment and the influence of the QM region on the environment. A simpler alternative is to replace the environment with a polarizable medium, or reaction field. The idea here is to define the surface of the QM region, e.g., through the van der Waals radii of the outermost atoms, and then treat the environment outside this surface as a polarizable continuum, characterized by a dielectric constant. This has led to the very successful Polarizable Continuum Model (PCM) [Men12]. Here, the environment is not modified by what happens in the QM region, but it does react to it by generating an induced electrostatic field that influences the behavior in the QM region. The PCM and its variants are frequently used in multi-layer embedding schemes, in which the QM part is embedded in an MM region, which in turn is embedded in a polarizable continuum.

QM-MM schemes can be used with the whole range of methods for electron-nuclear dynamics, from Born-Oppenheimer MD to Ehrenfest Dynamics, surface hopping, and multiple spawning, thus allowing for the inclusion of environmental effects on spectroscopic quantities and photo-chemical reactions. A demanding task is the extension of these methodologies to allow for the electrons to leave (or enter) the QM region as it occurs, e.g., in irradiation dynamics, where ionization and electron transport phenomena are central. A first step in that direction is found in [DRS10].

3 Physical Mechanisms and How to Access Them

We have now reached a stage at which we have gathered sufficient material, in terms of examples, physical mechanisms and available theories to develop a critical view of numerous irradiation scenarios. The aim of this chapter is to provide practical modeling tools to analyze irradiation scenarios, in complement to the formal developments of Chapter 2, but now with a practitioner's viewpoint. This will prepare Chapter 4 where various examples of applications will be given to illustrate both theories themselves and their use. But the aim is not only to reach dedicated applications. It is to give the reader the capability, once facing a given irradiation scenario, to be able to choose the most relevant theory and actual tools to apply it. By this we mean the best adapted-approach to a given physical situation. This will be widely illustrated in Chapter 4. In many cases one will be facing several options. They may provide different outputs and/or possibly be combined to cover a wider range of times and scales. The actual choice will of course depend on the situation but the more the readers will be able to critically analyze the situation and the theoretical tools they may rely on, the better they will be able to make a proper decision. This is the meaning of the critical dimension we advertise here as a central issue.

In Chapter 1 we have discovered the richness of the many possible dynamical scenarios one may be facing during irradiation. We have seen the major role played by the electrons in the early times of irradiation processes. On the longer term coupling to ionic degrees of freedom has also been identified as an important aspect, as well as the impact of a potential environment, as many realistic situations involve embedded or deposited systems. The few examples we have considered have furthermore pointed out the importance of key physical mechanisms such as eigenmodes and relaxation thereof. From these examples we figured out a general irradiation scenario covering the multiple spatial and time scales involved in irradiation processes. This also allowed to identify major observables (section 1.2.3) characterizing the response of target systems. Among the many possible such observables ionization, via electronic signals, represents a major one. It can be analyzed at various levels of sophistication from angle and energy-resolved versions up to global one and/or time-resolved versions, in particular in the case of pump and probe scenarios.

Chapter 2 was devoted to theoretical approaches available to address irradiation events. A large fraction of the chapter focused on electron dynamics which is central and which requires most of the efforts. Indeed in most situations, ions can be treated classically while electrons require a quantum

DOI: 10.1201/9781003127949-3

treatment. There are of course exceptions in both directions. The description of electron dynamics may rely on highly elaborate approaches usually limited in their field of applications. Most practical approaches are built around effective mean-field approaches such as in particular Density Functional Theory in its time-dependent versions (section 2.2.3). But the latter lack some correlations, which are precisely defined as what is beyond mean field. They nevertheless may play an important role in many situations such as in relaxation processes and energy redistribution within an irradiated system. This is a difficult problem which has been little attacked yet. The connection between electrons and ions is in a slightly better situation although formal and technical difficulties still remain. The huge span of spatial and time scales involved between electronic and ionic time scales represents a major hindrance for the development of unified theories. And the treatment of environment remains a practical challenge in particular, again, in view of the time and spatial scales involved.

The chapter is organized in five sections. Section 3.1 will first quickly summarize and reanalyze the examples extensively discussed in section 1.1. We will in particular discuss the capabilities of the theories used in these examples. Altogether this survey will provide us with a general view of physical mechanisms involved. We shall then briefly discuss theoretical tools used to simulate excitation processes themselves, mostly on the basis of lasers and charged projectiles (section 3.2). We shall then summarize, in section 3.3, the major physical mechanisms to be analyzed by theoretical approaches. The next section will discuss how to evaluate theoretically relevant experimental observables (section 3.4). Finally, section 3.5 will provide some technical details on numerical implementations.

3.1 AN AFTER DINNER REASSESSMENT OF APPETIZERS

The section provides a critical analysis of the irradiation examples (section 1.1) presented in Chapter 1. Here, we start from these examples and work out the key physical mechanisms involved in the various scenarios. These mechanisms will be further analyzed on other examples in Chapter 4, after a short summary and sorting of key mechanisms discussed in section 3.3. The second aspect concerns the theoretical tools used to analyze the examples of section 1.1. They were only superficially addressed and we shall here discuss them a bit more in detail, with a focus on their capabilities, the pros and cons, and their limitations. Note that we shall heavily refer here to section 1.1 and we advertise the reader to consider it accordingly.

3.1.1 PLASMONICS AND IONIC DYNAMICS

The first example (section 1.1.1) concerns a large Ag cluster (about 2500 atoms) irradiated by a dual pulse (pump and probe). While the first laser pulse (pump) deposits a sizable excitation in the system the net ionization (outer ionization, namely electrons stripped from the cluster itself) is not sufficient to immediately destroy it but suffices to provoke an ionic expansion. The system is metastable with the formation of a nanoplasma formed from the inner ionized electrons (electrons stripped from their parent atom but not from the cluster) and the ions. Via resonant coupling to the plasmon frequency of the excited system, the second laser pulse (probe) drives the final explosion of the cluster.

A key mechanism involved in this scenario is the appearance of a metastable state via inner and outer ionization. The pump laser strips electrons from their parent atom and from the valence cloud binding the cluster together. The fraction of the electrons which actually leave the system makes it highly charged which in turn binds the remaining stripped electrons into the cluster, at least temporarily. This creates a nanoplasma whose electron cloud remains temporarily bound to the ionic background but which tends to oscillate against it as a whole, whence the key role of the plasmon frequency in the whole irradiation scenario and the term plasmonics linked to it. Beyond this electron dominated mechanism, starting at very early times, lies the key role of ions and their coupling to electrons. As a consequence of outer ionization the cluster starts to grow and ions to separate from each other up to the final explosion. The ionic expansion drives the cluster plasmon frequency into the laser frequency, which at a certain delay between pump and pulse leads to resonant ionization and final explosion. The pace of the process is set by the ions thus extending the dynamics in the picosecond range.

Experiments access net ionization via Photo Electron Spectra (PES) and Photo Angular Distribution (PAD) namely in an energy and angle-resolved manner. The dual pulse setup with the two pump and probe laser pulses and variable delay between them adds a time component here making these observables time resolved, at least at the time resolution set by the delay. Remember that processes are rather slow due to the large mass of Ag atoms and thus lie in the ps range. The measurement of electrons indirectly provides information on the ionic dynamics via plasmonic coupling. The high energy deposited in the system, in turn, is not accessible directly. It is mostly read off from the modeling, as the actual ionic shape. The photo absorption signal is not recorded as well, at least directly. But again plasmonic coupling provides an indirect estimate of at least the major plasmon peak and the dynamics thereof due to ionic expansion.

The theory used here is classical Molecular Dynamics (MD, section 2.4.2) for both electrons and ions. The justification for the use of such a simple theory lies in the high excitation energies, the large size of the system and the long timescales involved. High excitation energies allow to overlook most of quantum details both at the side of the cluster structure and of the ensuing dynamics. The large size of the system with 2500 atoms furthermore makes quantum calculations computationally demanding, even more so on a ps time scale. The large size is not only computationally demanding but also justifying the fact that detailed quantization effects may become nondominant. The key role of the plasmon, also, is a crucial issue. Indeed the basic mechanism of plasmon oscillations is classical. Quantum effects may indeed be important in particular in small systems but tend to fade out for larger systems.

Beyond its computational interest, the major formal advantage of a MD treatment is the fact that it delivers by construction all many body correlations. The latter are restricted to the classical domain but are likely to dominate quantum ones because of the high excitation energy involved. Clearly an MD approach would become insufficient for gentler scenarios. Alternative approaches exist. The first step is to switch from full MD to VUU (section 2.4.1) for the electrons in order to restore minimal quantum features at the side of electrons. The price to pay is the loss of the classical-electron-electron correlations inherent to MD which are replaced by approximate ones at the level of a kinetic equation. The gain is an extension toward lower excitation energies where fluctuations, absent from an extended mean field treatment, are not expected to play a major role. The range of applicability of VUU, nevertheless, remains limited as true quantum effects easily pop up with decreasing excitation energies and/or in smaller systems. The sole computationally feasible option there could be real time TDLDA (section 2.2.3.4). Both, TDLDA and VUU, do not account for dynamical correlations and thus do not properly describe energy relaxation. This is an important issue to properly understand energy redistribution of the energy deposited by laser pulses in which relaxation effects compete with ionization. Dissipation and relaxation can be added on top of TDLDA by models treating dynamical correlations, see sections 2.3.7.1, 2.3.7.2, and 2.3.8. It is implicitly contained in MD and many-body quantum mechanics. Dissipation and relaxation toward equilibrium can be made visible by time averaging which replaces here the ensemble averaging in stochastic approaches [Rei16, SvL13].

3.1.2 ULTRA FAST ELECTRONIC DYNAMICS

The second example (section 1.1.2) concerns an Ultra Fast (UF) excitation of a rather small organic molecule, Naphthalene (Naph, $C_{10}H_8$). The scenario is again an IR pump laser excitation (Naph $\to$ Naph$^+$) followed by a XUV probe laser analysis (Naph$^+$ $\to$ Naph^{++}). Time scales now lie in the tens of fs range and the setup focuses on electronic response although coupling to vibrational motion is accessible. The setup explores hole migration within the molecule. Pump ionization may lead to various well-defined excited states (3 are analyzed in detail). The probe pulse allows to measure the lifetimes of these excited states via VMI (Velocity Map Imaging) which gives access to both PES and PAD, see section 1.2.3. Vibrational coupling can be identified from long-living oscillations in some of the electronic population signals which give access to lifetimes of excited states.

The interesting physical mechanisms at work here are many-body effects which strongly impact the response of the system to the pump excitation.

The whole scenario clearly gives access to detailed quantum features of the excited states within a manageable experimental setup. Because of the rather short time scales involved (several tens of fs) dynamics are dominated by electronic effects which are also the targets of the analysis. But interestingly enough there are signs of coupling to ionic degrees of freedom within the electronic measurement of excited state lifetimes. This is quite plausible as Naph contains numerous hydrogen atoms whose ionic motion, because of their small mass, takes place in the tens of fs range. Mind that the ground state vibration of H_2 has a period of 7.6 fs only. A particularly interesting aspect of this scenario is the fact that one measures life times which are typical dynamical quantities reflecting complex microscopic mechanisms.

A theoretical description of such a setup is a priori quite demanding as it requires a quantum analysis of the dynamics including correlations. Only very small systems (very few atoms/electrons) on very short time scales (some fs) can be treated at that level of sophistication. Even if Naph is rather small a proper treatment requires account of the 48 valence electrons on times of order several tens of fs to cover experimental widths of laser pulses and delays between pump and probe pulses. This is practically well beyond the scope of theoretical dynamical approaches including correlations and one is thus bound to make compromises.

The treatment adopted in the present example is a mixed approach associating several theoretical modelings. The ionization spectra of Naph have been calculated by Green's functions (section 2.3.6). The vibrational modes have been obtained by perturbation theory using quantum chemistry software. A dedicated vibronic coupling Hamiltonian has then been constructed to simulate the coupled nuclear and electronic dynamics (section 2.5).

Finally time propagation of this Hamiltonian has been done using the MCTDH method (section 2.5.4.1). Agreement between theory and experimental results is rather good. Simplifications fortunately popped up due to the rather limited number of states to be taken care of, due to the fact that only a few electronic states did contribute to the excited states. The MCTDH evolution had nevertheless to account for 23 electronic states and 25 relevant vibronic modes.

Without entering details one can see that even in a not too demanding case, the complexity of the process is enormous. As a consequence, one is bound to mix several theoretical approaches to access various facets of the process. This by itself is a nuisance but one might hope that the underlying theories are sufficiently robust, which is most probably the case. Still this mixing remains an annoying aspect as it is clearly specific to the actual scenario in the sense that changing experimental conditions would imply changing inputs of the calculations for example the number of states or modes to be included in the MCTDH dynamical evolution. The approach is thus still far away from being generic.

The use of other theoretical approaches is rather open. It is not clear at all that more robust approaches covering all aspects of the dynamics (electronic excitation, ionization, vibronic coupling) are available. We are not facing here insurmountable computational difficulties. They might pop up later on. We are already facing formal difficulties. A real-time treatment of a molecule like Naph on typically 100 fs can only be practically envisioned at TDDFT level. But standard TDDFT approaches may have difficulties in the evaluation of some electronic excited states (section 2.1.5). And at the Ehrenfest dynamics level coupling to ionic degrees of freedom is not perfect (section 2.5). There thus remains plenty of space for improvement here, both at the level of more elaborate effective mean-field (TDLDA) based approaches or at the level of alternative theories.

3.1.3 THE ROLE OF ENVIRONMENT

The last example discussed in section 1.1.3 deals with fragmentation of the Uracil molecule ($C_4H_4N_2O_2$) and of possibly hydrated Uracil clusters. Fragmentation corresponds to the exit channel of the irradiation scenario initiated via a collision of the target system with a high velocity $^{12}C^{4+}$ ion. The collision leads to moderate ionization (1 or 2 electrons) and energy deposition in the system. Rearrangements of both charge and energy lead to various fragmentation pattern depending on the nature of the target system (Uracil molecule, Uracil cluster, nano-hydrated Uracil cluster). The comparison between the various systems points out the impact of environment as a buffer and its consequences on fragmentation dynamics. Observation of the fragmentation pattern is made asymptotically in time but fragmentation itself is a slow process as compared to internal rearrangements, in particular at electronic level. Indeed fragmentation directly involves ionic motion and fragment separation which take even longer.

The major mechanisms pointed out in this example concern the sensitivity of fragmentation pattern to the environment. The notion of environment here might require some explanation, though. The interpretation is that in such an organic system electrons are rather localized so that the initial excitation and ionization take place in a rather localized region of the system, coined a chromophore, most probably in some sector of one Uracil molecule. Other Uracil molecules and/or water molecules thus constitute an environment for the initially excited molecule. They also provide a large reservoir for rearranging deposited energy and net charges. It is thus no big surprise that the final stages of these rearrangements will directly impact the actual fragmentation pattern. Stepping from Uracil molecule to Uracil clusters (hydrated or not) thus leads to the disappearance of some fragmentation channels and appearance of new ones. Observation of these fragmentation patterns is attained via Time-of-Flight measurements of the produced fragments. These are purely ionic quantities which integrate the whole history of the irradiation scenario.

From the theoretical side, an account of the whole irradiation process is quite challenging. In the present example, theory is reduced to structural considerations which at least allow to identify possible fragment formation and pathways to reach them. But this remains far away from a direct simulation of the dynamical evolution itself. There are two aspects of such dynamical simulations which need to be mentioned. The first one is technical. The fragmentation process is awfully slow as compared to electronic effects taking place at short times. This makes a direct simulation extremely time consuming to cover both short and long time scales. Furthermore, the size of the system when considering clusters rapidly becomes prohibitive for fully quantum mechanical approaches. A way out might be provided by Quantum Mechanics/Molecular Mechanics (QM/MM) approaches (section 2.5.5.3) where only a small part of the system is treated at full quantum mechanical level which makes sense in this case because of the assumed localized nature of the original excitation. Still, an extension of such approaches to time domain, especially with moving charges is far from being trivial as the boundary between QM active region and MM environments can change in time. This requires updates of the regions in the course of time for which strategies have yet to be developed. The presently working alternative is to take the QM region large enough from the onset to avoid later extensions of the region.

This computational issue is, after all, not new and as soon as one is dealing with long time propagation of an irradiated system, one is facing such difficulties. But there is another, more fundamental, aspect which is directly connected to the nature of the process itself (fragmentation) and the necessary level of a theory to account for such an effect. The process involves several competing channels leading, for a given fragmenting object, to several possible fragments. Such a multiple set of outputs cannot be described by a dynamical simulation following one single path, as would for example produce a time-dependent mean field or a kinetic equation description (section 2.3.7.1). The latter would lead to one set of fragments, not a distribution thereof and would not quantify how much such a set represents in terms of relative weight of the associated fragmentation channel. One is here facing a principle question on how to deal, in a dynamic approach, with such scenarios with multiple outputs, obviously involving bifurcations.

A way to solve such problems in quantum approaches is provided by Trajectory Surface Hopping methods (TSH), section 2.5.4.2, which can track bifurcations by sampling them stochastically to create statistical ensembles. Still, such approaches have been explored in rather simple situations involving only a very limited number of bifurcations which allows to evaluate the Adiabtaic Potential Energy Surfaces (APES) ahead of the dynamical evolution itself. The fragmentation scenario considered here involves more channels and pathways and lies probably outside the capability of such refined approaches as TSH. Other stochastic approaches might be conceivable (section 2.3.8) but require by construction ensembles of trajectories which becomes rapidly computationally prohibitive in view of the long time scales involved. Structural approaches avoid these problems by focusing on possible fragmentation patterns, at best with an estimate of the associated barriers to reach them so that one can then evaluate their relative weights as a function of the deposited excitation energy by simple statistical estimates.

3.2 EXCITATION MECHANISMS – FORMAL BACKGROUND

Simulations of electron dynamics usually start from the static ground state. Time evolution is triggered by external excitations. This is the purpose of this section to briefly discuss typical excitation mechanisms used in the analysis of irradiation dynamics. We shall restrict the discussion to theoretical aspects, namely explaining how an external excitation via for example a laser or a projectile, is actually delivered to the target system. We shall carefully avoid entering technical experimental issues which can be found, e.g., in [ME10]. Remind also that the target system may itself be in contact with an environment which may itself respond dynamically to an external excitation. This would then create an extra time-dependent field acting on the target system, which has then to be added to the original external excitation delivered to the system. We shall specifically discuss such situations when considering examples of systems in contact with a reactive environment (section 4.2). In the following, we thus first explain the most common excitation mechanisms in terms of pure external perturbation without specifying target system and/or environment.

3.2.1 LASER FIELDS

Coherent photon pulses from lasers or synchrotron radiation are the most widely used and flexible tool for excitation of electronic systems. Their electromagnetic field can be well approximated by a classical time-dependent electromagnetic field. Typical wavelengths are much larger than the spatial extension of atoms and molecules. Typical molecular and cluster sizes lie in the a_0 to nm range. A C_{60} has radius below 1 nm while the double helix in DNA has diameter of about 2 nm for a repetition rate of order 3–4 nm. A Na_{1000} cluster has radius of order 2 nm and one needs to reach systems in the million range to gain another order of magnitude in the radius. This still remains far below a standard IR laser of wavelength 800 nm. Only UV and XUV lasers with tens of nm wavelengths may thus compete with molecular sizes.

Most considered cases thereafter thus allow to ignore the spatial modulation of the photon field and to deal with a spatially homogeneous electrical field $\mathbf{E}$ and we can also neglect the effect of the magnetic field for the laser intensities of relevance here (up to typically 10^{14-16} W.cm^{-2}). There remains the freedom of gauge transformation [Jac62]. The external field operator in the velocity gauge reads:

$$U_{\text{ext}} = e\mathbf{E}_0 F(t) \cdot \hat{\mathbf{p}}, \quad F(t) = \int_0^t \mathrm{d}t'\, f(t') \exp\left(-\mathrm{i}\omega_{\text{las}}t'\right), \qquad (3.1)$$

where $f(t')$ is the envelop of the laser pulse. The same pulse in space gauge reads

$$U_{\text{ext}} = e\mathbf{E}_0 f(t)\cdot\hat{\mathbf{r}}\exp\left(-\mathrm{i}\omega_{\text{las}}t\right). \tag{3.2}$$

The latter form is numerically simpler to handle because it is a local operator. Transformation to velocity gauge, if needed, can be performed a posteriori by standard rules of gauge transformation [Jac62].

The photon pulse is parametrized by frequency ω_{las}, peak field strength $E_0 = |\mathbf{E}_0|$, polarization $\mathbf{E}_0/E_0$, and time profile $f(t)$. The field strength E_0 is usually given via the laser intensity I as

$$E_0 = c_{\text{EI}}I^{1/2}, \quad c_{\text{EI}} = 1.07\times 10^{-8}\,\frac{\text{eV}}{\text{a}_0}\left(\frac{\text{W}}{\text{cm}^2}\right)^{-1/2}. \tag{3.3}$$

Intensity provides energy per unit time and surface. Integrating intensity over pulse duration delivers the pulse energy per unit surface known as fluence which is an important quantity of a laser pulse to characterize energy deposit in a target. The latter strongly depends on target size and laser intensity but on pulse duration as well. For example, a 1 nm^2 target irradiated by a laser of intensity 10^{14} W/cm^2 and duration 10 fs is exposed to 60 keV energy input. It depends on the systems polarizability which fraction of that is soaked for excitation, typically 1% of that.

The experimental time profile of the laser pulse is rarely well known and usually taken as a Gaussian with a certain Full Width at Half Maximum (FWHM). However, Gaussians require long simulation times to cover the outer wings of the pulse. To avoid that, in numerical simulations it is thus preferable to use a pulse with finite support, such as a $\sin^2$ envelop:

$$f(t) = \begin{cases} \sin^2\left(\pi\dfrac{t}{2T_{\text{pulse}}}\right) & \text{for} \quad t \in \{0, 2T_{\text{pulse}}\} \\ 0 & \text{elsewhere} \end{cases}. \tag{3.4}$$

That is close to a Gaussian pulse in the vicinity of peak field strength and combines high spectral selectivity with finite bounds. Note that the form (3.4) renders the pulse parameter T_{pulse} comparable with the FWHM of a Gaussian pulse. Recall also that the FWHM of amplitude is not to be mixed with the FWHM of intensity as the latter scales with the square of E_0. In the case of a $\sin^2$ pulse for E_0 the FWHM of the amplitude is exactly $T_{\text{pulse}}/2$, while the FWHM of the intensity is slightly different, around 0.54 T_{pulse}.

A general order of magnitude estimate of laser pulse duration is, to some extent, irrelevant as different pulse durations lead to different scenarios and access to different properties. Here we shall mostly consider rather short pulses of duration below the 100 fs range where electronic effects play a leading role. One can nowadays reach very short pulses below the fs range down to the attosecond domain.

Furthermore, many experiments use a dedicated sequence of photon pulses as, e.g., in pump-and-probe setups [Zew94]. Well-chosen series allow to extract specific information on the system or to shape the pursuit of reactions [DDK$^+$02]. As stated by the acronym "pump-and-probe", the first laser pulse does excite the system (it pumps it into an excited) while the delayed second pulse gives access to de-excitation dynamics as a function of the delay (probe). An additional key parameter in such pump and probe setups is the tunable delay time between the two laser pulses. Each laser pulse (pump and probe separately) otherwise have the standard forms discussed in the present section with all characteristics chosen at will (duration, frequency, intensity...). We have already seen several such setups, as for example in sections 1.1.1 and 1.1.2, and we shall further illustrate pump and probe scenarios in several examples in Chapter 4.

3.2.2 CHARGED PROJECTILES

Remote collision with fast, highly charged projectiles is another often used tool to excite the electron cloud of atoms and molecules [BM92]. In this case, again, only the Coulomb field from the projectiles count. Charged ions are heavy and can be treated with classical trajectories $\mathbf{R}_{\mathrm{ext}}(t)$ delivering the time-dependent external field:

$$V_{\mathrm{ext}}(\mathbf{r}, t) \quad = \quad \frac{Z_{\mathrm{ext}} e^2}{|\mathbf{r} - \mathbf{R}_{\mathrm{ext}}(t)|}, \tag{3.5a}$$

$$\mathbf{R}_{\mathrm{ext}}(t) \quad = \quad \mathbf{R}_{\mathrm{ext}}(0) + \dot{\mathbf{R}}_{\mathrm{ext}}(0)t. \tag{3.5b}$$

For simplicity, we have inserted here a straight projectile trajectory which is an acceptable approximation for very fast and heavy projectiles. In the general case, one should insert the actual projectile trajectory computed along the whole process of interaction with the target. Magnetic effects are also neglected. They may play a role only for extremely fast ions in the relativistic domain [BJRT06].

The parameters of the collision are given by the charge of the projectile Z_{ext} which primarily governs the amplitude of the delivered electric field and thus the impact on the target and the resulting excitation mechanism. The other key quantities are the parameters of the trajectory which are determined by the impact parameter b and initial velocity $v_{\mathrm{ion,in}}$ [GPS01].

While the impact parameter is not directly accessible experimentally it is a trivial quantity from a theoretical viewpoint. Simulations of collision processes thus require an averaging over b to allow comparisons to experimentally relevant quantities. When dealing with charged particles one may have to scan numerous impact parameters because of the infinite range of the Coulomb potential. Collisions with neutral projectiles, which are also conceivable, are more forgiving in this respect.

The initial velocity is the other crucial quantity to characterize the interaction process. It governs the interaction "duration", energy deposition and

ionization. The range of projectile velocities is huge depending on the potential applications (fundamental physics, medical applications...). Without
entering such details it is interesting to recall a few typical values, as considered in the following. Inside a molecule or cluster a basic velocity unit is the
Bohr per femtosecond a_0/fs. A typical Fermi energy of valence electrons in a
metal is a few eV (2-5 eV), up to several of tens of eV in a covalent system.
This corresponds to a velocity of order 20 a_0/fs in a metal, and several times
more for covalent binding. Charged projectiles, like a proton with keV initial
energy has a quite comparable velocity of 8 a_0/fs. At 1 MeV the velocity goes
up to 260 a_0/fs for a proton. For heavier projectiles velocities simply scales
with square root of mass.

More involved projectiles come into play when considering atomic collisions.
Such scenarios could formally also be treated in TDDFT (section 2.2.3). However, it has to be kept in mind that TDDFT describes average trajectories.
As soon as electron transfer comes into play, the sub-states of target and projectile cover a mix of different charge states. The average description is still
possible in the reaction zone and reasonable results may be deduced short
after. But TDDFT is bound to fail at long times because trajectories for different charge states develop very differently in the course of time. MCTDHF,
see section 2.3.4.3, may be a solution in such situations as long as the number
of outgoing channels remains tractable.

3.2.3   INSTANTANEOUS BOOST

A simple limiting case emerges if the length of the laser pulse or the time
of impact of a charged projectile is shorter than the typical response time of
the dynamically active electron states. Then the Coulomb flash just instantaneously imprints a certain phase profile onto the s.p. wave functions which
then unfolds in the subsequent electron dynamics. This acts effectively as a
dipole boost

$$\varphi_i(\mathbf{r}, t=0) = e^{i\mathbf{p}_0 \cdot \mathbf{r}/\hbar} \varphi_{i,\text{g.s.}}(\mathbf{r}), \qquad E_{\text{boost}} = \frac{N_{\text{el}}}{2m_e} |\mathbf{p}_0|^2 \qquad (3.6)$$

where $\mathbf{p}_0$ is the boost momentum related to the boost energy E_{boost}. The time
extension of this excitation covers with equal weight all frequencies (Dirac δ
pulse). As a consequence, the response of the system involves all eigenfrequencies. The only remaining parameter is the amplitude of the boost. Actually,
very small boosts are used to probe the optical absorption spectrum; see section 3.4.2.

3.3   KEY PHYSICAL MECHANISMS TO BE ANALYZED

Sections 3.1 and 3.2 have provided a short panorama of typical irradiation scenarios and associated physical mechanisms. We have also seen how to access
theoretically dedicated excitation mechanisms via, in particular, lasers and

charged projectiles. Before discussing how to evaluate experimental observables for possible comparisons between modeling and experiments in section 3.4, we want to briefly summarize some generic features in terms of the basic physical mechanisms involved in typical irradiation scenarios.

3.3.1 THE VARIETY OF PHYSICAL MECHANISMS

We have identified several important mechanisms and ways to address them theoretically, with more or less accuracy. It is important, before stepping further, to summarize and possibly sort these various physical mechanisms. The aim is here to switch from a vertical/specific viewpoint, directly linked to a given example, to a more horizontal/generic viewpoint, as several mechanisms turn out to be present in numerous situations. This will allow us to sort the various mechanisms at work and analyze them in terms of relevant theoretical observables. The idea here is thus to identify a few key ideas/questions to be discussed later on in detail in Chapter 4. For the sake of clarity we structure this short section in a sequence of bullet points.

- **The key role of electrons**
  Key physical mechanisms involve electrons, ions, and possibly the environment. Electrons are the first component of the system to react to the excitation because of their small mass. Their response can be dominantly collective, for example via plasmonic effects (section 3.1.1) or individual, as when a hole is created locally (section 3.1.2). Typical observables associated to such electronic effects are extracted from VMI in terms of PES and PAD (section 1.2.3). Photoabsorption signals, when available, are also quantities bringing important information on underlying mechanisms. For example, resonance effects between system's eigenfrequencies and an external perturbation (typically the laser frequency) are an important aspect of irradiation scenarios, which we shall specifically address in section 4.1.
- **Interplay between electrons, ions and environment**
  Coupling of electron response to ions takes place on a longer time scale as seen in all cases. It can manifest itself in terms of vibrations (section 3.1.2), fragmentation (section 3.1.3) or full explosion (section 3.1.1). Typical observables then involve ionic degrees of freedom such as mass spectra (section 3.1.3), but electronic observables are also impacted by such a coupling and thus provide an indirect way to analyze it.
  The same holds true for all what concerns coupling to environment. The term may look quite generic and vague. We have considered it in a specific case in section 3.1.3 but it has to be understood in a more general way. Indeed most realistic situations involve environment and gas phase studies of irradiation in general correspond more to a principle study than to a realistic one. The account of the environment is

thus compulsory as soon as one wants to address realistic situations even if "model studies" in the gas phase may help understand what is actually going on. Furthermore, there exist specific situations in which an environment is present by construction such as in the case of deposited or embedded clusters (see for example section 4.2.3). Finally, as also pointed out in section 3.1.3 the notion of environment may refer to a description splitting a given system (itself in gas phase) in two subsystems because of the localized nature of some excitation processes.

- **The key role of eigenfrequencies**
  The response of a system to irradiation involves its eigenfrequencies such as plasmon or ionic vibrations. However, eigenfrequencies may also refer to single particle transitions rather than collective states. The presence of such eigenfrequencies leads to resonance effects between the excitation mechanism itself and these eigenfrequencies. This brings two kinds of information, a structural and a dynamic one.

  Eigenfrequencies are ground state properties and knowing them gives information on a system's ground state, and the vicinity thereof. But they are also a key to understand the dynamical response of a system even far off equilibrium, precisely because of the above-mentioned resonant couplings. The example of section 3.1.1 is very telling in this respect, in the case of a collective plasmon. In the longer term, such an eigenmode is attenuated because of couplings with other modes, or because of internal heating of the system via energy redistribution, or because of a change in energy balance following further ionization. When excited in the non-linear domain an eigenmode will tend to relax rather quickly. Relaxation thus provides key dynamical information, which provides a complement to the static information delivered by the eigenfrequencies themselves. In other words, the width of an eigenfrequency is also an essential characteristic of a system, which may furthermore depend on its degree of excitation. Such general considerations in fact concern both electronic and ionic degrees of freedom. They may point out missing components in the description, such as dynamic correlations beyond mean field, but not necessarily, such as for example in Landau damping [LP88].

- **System properties and generic features**
  Generally speaking, one wants to access system's properties, which means primarily static quantities. But dynamical quantities as well are quite important and need to be addressed via the response of the system. This may furthermore allow to access generic dynamic mechanisms which will deliver an improved understanding of the response of systems to irradiation.

For example, eigenfrequencies characterize processes close to equilibrium, in the "linear" domain of excitation. But they can have a major impact on dynamics, even in the non-linear domain. A laser, for example, will couple resonantly to them when its frequency is close enough to the plasmon frequency. This induces enhanced ionization (see the example of section 1.1.1) which allows to trace back the whole dynamic evolution. The relaxation of this coupling furthermore provides valuable information both on the dynamics and on the structure (single-particle levels, couplings...) of the system itself.

3.3.2   IONIZATION MECHANISMS WITH LASERS

The study of laser irradiation of atoms and molecules has a long history which has allowed to identify different regimes of ionization, which can be identified by compact quantities. Because of their huge versatility, lasers play a very special role in the studies dedicated to irradiation processes. This holds true both from the experimental and theoretical point of view. It is thus important to recall a few basic properties of ionization dynamics in the case of laser irradiations as we shall encounter these mechanisms at various places in Chapter 4.

The first quantity of interest is the ponderomotive potential defined as the cycle averaged kinetic energy of a freely oscillating electron (pure quiver motion):

$$U_p = \frac{e^2 |\mathbf{E}_0|^2}{4 m_e \omega_{\text{las}}^2} \qquad (3.7)$$

where $\mathbf{E}_0$ is the laser field strength, ω_{las} the laser frequency (section 3.2.1) and m_e the electron mass. Roughly speaking as long as U_p remains below the typical binding energy of an electron one speaks of a photon-dominated regime of ionization: what counts is the photon energy. On the contrary, when U_p becomes larger than the typical binding energy of an electron one speaks of field dominated ionization: what makes ionization is the intensity, not the frequency. This condition is related to the Keldysh parameter γ [Kel65] which compares the Ionization Potential (of energy E_{IP}) to the peak kinetic energy of a freely quivering electron ($2U_p$)

$$\gamma = \sqrt{\frac{E_{\text{IP}}}{2U_p}}. \qquad (3.8)$$

When $\gamma \gg 1$ ionization is dominated by vertical excitation of a bound electron by absorption of one single or many photons via the rapidly oscillating laser field, leading in the latter case to Multi-Photon Ionization (MPI). This mechanism lasts many laser cycles and is dominant for weak and moderate fields in the so called perturbative domain. MPI may be enhanced when intermediate resonant states can be attached which can push electrons high up in the continuum, an effect directly visible in the Photo Electron Spectra (sections 1.2.3 and 3.4.4.5). One then speaks of Above Threshold Ionization (ATI).

When $\gamma \leq 1$ the binding energy can be overcome within a single laser cycle and one then speaks of Optical Field Ionization (OFI). In this case, the laser field can be viewed as quasi stationary. Bound electrons can then tunnel through the barrier resulting from the ionic potential complemented by the laser field. One can introduce the tunnel time $\tau_{\text{tunnel}} = \sqrt{2E_{\text{IP}}m_e/(e|\mathbf{E}_0|)^2}$ which allows to re-express Keldysh parameter as $\gamma = \omega_{\text{las}}\tau_{\text{tunnel}}$. In the OFI dominated regime the tunneling time τ_{tunnel} becomes comparable to or smaller than the optical period. MPI is the leading process otherwise.

The above discussion is well established for atoms, where the laser parameters govern the dynamics. In molecules and clusters structural aspects progressively take their share. Ionization barriers, in particular, are affected by neighboring ions. An example of the impact of such effects is provided by the so called Charge Resonance Enhanced Ionization (CREI) in which a particular internuclear distance within a dimer leads to enhanced ionization. One can also cite here the well studied Interatomic Coulombic Decay (ICD) occurring in weakly bound matter, for a recent review see [JHW+20].

In larger systems the concept of inner and outer ionization also provides a convenient and physically meaningful description of the process. We already identified it in the example of section 1.1.1. In a given system, cluster or molecule, electrons can be separated into tightly bound electrons and quasifree electrons, the latter being mostly the ones ensuring bonding. Inner ionization then labels the process of promoting tightly bound electrons into the "conduction band" of quasifree electrons. In turn, the final excitation into the continuum, delivering the net ionization, is quoted outer ionization. For example, in metal systems at moderate laser intensities outer ionization strongly prevails. In rare gas clusters, inner ionization is a key mechanism to understand coupling to lasers. In such cases, the net charge acquired by the systems confines inner ionized electrons within the system, at least for some amount of time. This leads to the formation of a plasma of nanometric dimensions, quoted a nanoplasma [FMBT+10], as observed in the example of sections 4.1.3.4 and 1.1.1.

3.4 THEORETICAL EVALUATION OF MAJOR EXPERIMENTAL OBSERVABLES

This section is to be considered as sort of a mirror of section 1.2.3 discussing experimental observables relevant for irradiation. We briefly outline here the formal aspects allowing to access these quantities. The focus is of course on electronic quantities which constitute the core of this book but ions cannot be overlooked and we shall thus also briefly discuss them.

3.4.1   IONIC QUANTITIES

Even if electrons are the focus of this book ions play a major role in the response of a system. In the vast majority of cases, ions are treated classically in the widely used TDDFT approaches at the level of Ehrenfest dynamics (section 2.5). In practice one thus has access to their positions $\mathbf{R}_I$ and momenta $\mathbf{P}_I$ as a function of time. This gives access to the associated kinetic energy

$$E_{\text{kin,ions}} = \sum_I^{N_I} \frac{\mathbf{P}_I^2}{2M_I} = \frac{3}{2} N_I k_B T_I \tag{3.9}$$

where M_I's are ionic masses, N_I the number of ions and k_B Boltzmann's constant. This allows to evaluate the associated ionic kinetic temperature T_I.

The ionic temperature can be an initial temperature, such as for example when considering clusters formed at finite temperature. In this case, it is an initial temperature and the initial state of the system is created with an initial thermal distribution of ionic momenta. One then has to let the system relax to reach its own "thermal" ground state before launching an irradiation process. This furthermore implies that relevant results can only be attained by considering several samplings of this thermal distribution. An example of the impact of such an initial temperature on Photo Electron Spectra (an electronic quantity) is given is section 4.1.2.4.

Even when starting from zero temperature ground state an ionic temperature may result from the irradiation process. Two effects pile up there: a natural relaxation of the deposited energy toward thermalization at the side of ions themselves and an effect of the coupling to electrons. The electrons remaining in the system usually thermalize on the medium to long term. Energy exchange can thus occur between two objects of potentially different temperatures, electrons and ions, up to reach on the even longer term a global thermal stage at one unique temperature (see section 2.5 on ionic motion and electron-ion coupling).

3.4.2   PHOTO-ABSORPTION SPECTRA

Optical response is a key observable in cluster physics. It can be computed from the small-amplitude limit of TDDFT. Systems with high symmetry (ideally spherical) allow to deal practically with an explicit linearization of the TDDFT equations which leads to the much celebrated Random-Phase Approximation (RPA), section 2.2.4.1 [MRS10]. For general systems, one better uses full TDDFT with very small boost. One then propagates electrons and samples a protocol of the dipole moment

$$\mathbf{D}(t) = \int d\mathbf{r}\, \mathbf{r}\, \varrho\,(\mathbf{r}, t). \tag{3.10}$$

The dipole moment is usually "measured" with respect to the ionic center of mass.

In standard mean field or extended mean field computations based on TDLDA (section 2.2.3.4) or kinetic equations (section 2.3.7) the one body density $\varrho\,(\mathbf{r}, t)$ reads

$$\varrho\,(\mathbf{r}, t) = \sum_{i=1} w_i(t) \int d^3r\,|\varphi_i(\mathbf{r}, t)|^2 \qquad (3.11)$$

where the summation runs over all states included in the computations. At TDLDA level occupation numbers w_i are independent of time and stay at their original values (1 for occupied levels, 0 for unoccupied ones) while they become time dependent and fractional in the case of kinetic equations (section 2.3.7).

After a sufficient time, say $T_{\max}$, one multiplies the dipole signal $\mathbf{D}(t)$ with an appropriate window function $\mathcal{W}(t)$ [PTVF92], Fourier transforms it into $\tilde{\mathbf{D}}(\omega)$, and finally obtains the spectral strength $S_{D_i}(\omega)$ for $i \in \{x, y, z\}$ and corresponding spectral power $\mathcal{P}_{D_i}$ as:

$$S_{D_i}(\omega) = \frac{\mathrm{Im}\,\{\tilde{D}_i(\omega)\}}{p_{0,i}}, \quad \tilde{D}_i(\omega) = \int dt\,\mathcal{W}(t)\,e^{i\omega t}D_i(t), \qquad (3.12)$$

$$\mathcal{P}_{D_i}(\omega) = \frac{|\tilde{D}_i(\omega)|^2}{p_{0,i}^2}, \qquad (3.13)$$

where $\mathbf{p}_0$ is the initially applied boost momentum. The maximum possible spectral resolution is given by $\delta\omega = 2\pi/T_{\max}$. The window function $\mathcal{W}(t)$ serves to attenuate the dipole signal toward the end point $T_{\max}$ and to avoid artifacts from non-zero $\mathbf{D}(T_{\max})$ [PTVF92]. It is interesting to note that this treatment in connection with absorbing boundary conditions (see section 3.4.4.2) allows one to compute correctly the escape width of spectral states lying in the electron continuum. For details of spectral analysis and variants thereof, see [CRS97].

Spectral analysis in the regime of $\hbar\omega$ below the ionization threshold shows a series of discrete peaks with a strength function

$$S_D(\omega) = \sum_n |D_{n0}|^2\delta(\omega - \omega_n), \qquad (3.14)$$

for a spectrum of excitations n with frequency (energy) ω_n and transition strengths $|D_{n0}|^2$. It can obviously be interpreted as a spectrum of excited states with energy $E_n = \hbar\omega_n$. All of a sudden, we swept from a time-dependent picture of *excitations* to a stationary picture in terms of *excited states*. We are trained from traditional quantum mechanics courses to think in (stationary) energy space. Short time physics forces us to think in time representation. The mix of both pictures causes often confusion. We thus try a few clarifying words and do that for the case of small amplitude motion where both pictures are still strictly equivalent. Linearized TDDFT (section 2.2.4.1) describes excitation dynamics as oscillations in $1ph$ space using the Thouless representation (2.42). It can be reformulated as the matrix equations of RPA [MRS10], which are stationary equations producing the spectrum of excited eigenstates Ψ_n. The bridge between the two pictures is established by expressing the eigenstate through an excitation operator $\hat{C}_n^\dagger$ as $\Psi_n = \hat{C}_n^\dagger \Psi_0$, where Ψ_0 is the (correlated) ground state of the system and by expanding the generator $\hat{G}$ of the Thouless representation Eq. (2.42) in terms of the $\hat{C}_n^\dagger$. The comparison of the two pictures reads then formally

energy picture	time picture
$\Psi_n = \hat{C}_n^\dagger \Psi_0$	$\Psi(t) = \left(1 + \eta \sum_n \gamma_n \hat{C}_n^\dagger e^{-i\omega_n t} + \text{h.c.}\right)\Psi_0$
$E_n = \hbar\omega_n$	ω_n

where the γ_n regulate the relative contribution of the mode n according to the initial condition (e.g., boost). The η is an overall scale factor which plays no role for the analysis as long as the dynamics stays in the linear regime. This is almost a 1:1 mapping, except for one detail: the energy picture does not know η. An excitation energy appears always as one full quantum $\hbar\omega_n$ while the TD state exists for any (small) amplitude η. TDDFT produces in that sense a classical motion. The step from time to energy-picture with $\omega_n \longrightarrow \hbar\omega_n$ so to say re-quantizes TDDFT. Thus far for the exact mapping in the linear domain. Nonlinear TDDFT dynamics cannot be mapped exhaustively. It still covers some of the multiple excitations as, e.g., higher harmonics (section 4.1.3.3), but cannot resolve all many-particle-many-hole states in a fully correlated description which are present; e.g., in TDCI (section 2.3.4.1). Still, the lesson from the linear domain is also helpful to navigate through the non-linear domain.

3.4.3   ENERGETIC ASPECTS

Global energies are not directly accessible experimentally but they may leave printouts in various indirect quantities such as the energies of emitted electrons, themselves accessible through Photo Electron Spectra (PES) (section 1.2.3). It is thus important, from a theoretical viewpoint, to have access to them and to define the most important ones. The excitation process will deposit a certain amount of energy in the system. The system's energy $E(t)$ will thus evolve in time accommodating the absorbed energy to transform it into various components.

In the case of a laser irradiation, for example, the absorbed energy can be expressed from the local current $\mathbf{j}(\mathbf{r}, t)$ and laser amplitude $E_0(t)$ as

$$E_{\text{abs}}(t) = \int_0^t dt' \int d^3r \, E_0(t') \cdot \mathbf{j}(\mathbf{r}, t'). \tag{3.15}$$

The real space integration in principle extends over the whole space. In practice computations in a box imply for example absorbing boundary conditions which generate a small correction to this integral [VSR17]. The absorbed energy is in turn shared into two basic components associated to electron emission and internal energy of (still bound) electrons. Charge loss via electron emission means energy loss via the electrons leaving the system and energy rearrangement due to the resulting net charge. The remaining internal energy is shared between collective motion of electrons E_{coll} and "intrinsic" excitation energy of the electron cloud itself directly at the origin of the thermal part E^*_{therm} of the excitation energy, see section 2.3.7.2 and [VSR17] for more details.

The collective-flow energy E_{coll} reads

$$E_{coll}(t) = \frac{1}{2m} \int d^3r \, \frac{\mathbf{j}^2(\mathbf{r}, t)}{\varrho\,(\mathbf{r}, t)} \tag{3.16}$$

where m is electron mass. To properly extract E^*_{therm} at the quantal level is a bit involved, but one can use a semi-classical expression which is both simple and rather accurate. At Extended Thomas Fermi (ETF) level it reads, following [BB97]

$$E^*_{therm} = E^{TDLDA}_{kin} - \int d^3r \, ([\frac{2}{3}(3\pi^2)]^{2/3}\rho(\mathbf{r}, t)^{2/3} + \frac{(\nabla\rho)^2}{18\rho}) - E_{coll} \tag{3.17}$$

where E^{TDLDA}_{kin} is the TDLDA electron kinetic energy. An example of energy share within a laser irradiation will be given in section 4.1.1.3. For details, in particular when dissipation is included on top of TDLDA, see for example [VSR17].

3.4.4 OBSERVABLES FROM ELECTRON EMISSION

3.4.4.1 General Considerations

The theoretical description of electron dynamics relies on a finite basis of s.p. states or on a finite basis of grid points in a numerical box. Electron emission is related to probability flow across the bounds of the basis or box. In the following, we will address the description of emission in connection with grid representation. Absorbing boundary conditions are here the key to attain electron emission. This will be addressed in section 3.4.4.2. With that tool given, one can evaluate several observables associated with emission from (photo-)excited systems: total ionization, ionization probabilities, Photoelectron Angular Distribution (PAD), and Photo-Electron Spectra (PES). We already discussed them from an experimental perspective in section 1.2.3. We now want to address them from a practical theoretical viewpoint.

A most detailed access to ionization is attained via VMI (Velocity Map Imaging) techniques (section 1.2.3) which provide a double differential cross section $d^2\sigma/d\theta dE$ resolving both energy E and angle θ. The latter angle is measured with respect to geometry of the excitation process (laser polarization or projectile direction). The transverse directions are averaged out. By energetic or angular integration one obtains the PES and PAD as single differential cross sections ($d\sigma/d\theta$ for the PAD and $d\sigma/dE$ for the PES, respectively).

An integration over the remaining variable finally provides the total number of emitted electrons but the latter can experimentally be attained without access to a VMI. The same holds true from the theoretical point of view. Actually VMI, PES, PAD and total ionization can be evaluated independently. VMI results are in practice not always easy to interpret quantitatively. PES, PAD and total ionization being integrated quantities are often simpler to analyze and allow quantitive comparisons between various systems/scenarios. The case of ionization probabilities is a bit special: they allow to reconstruct a distribution around the average net ionization, which might provide a better link to experimental results. In the following we shall focus on these "integrated" quantities and briefly outline how they can be attained numerically, in particular within TDDFT-based approaches. Note finally that all these quantities, PES, PAD, total ionization, can be evaluated in a time-resolved manner which adds an additional interest to these observables.

3.4.4.2   Absorbing Boundary Conditions

Ionization is a central issue in the course of irradiation processes. Physically speaking ionization corresponds to the fact that an electron has left the system. From an energetic point of view this means that it has reached continuum. From a spatial point of view it means that the electron will asymptotically reach a region where it is not anymore (or negligibly) sensitive to the fields created by its parent species. In both pictures one is facing a major difficulty as both address an account of a practically infinite spatial or energy region. Such an extension is obviously raising a major computational issue, so that ionization and especially differential observables like PES (section 3.4.4.5) and PAD (section 3.4.4.4) remain touchy to evaluate and possibly biased by approximations.

The simplest, while very efficient way, is the mask function method which is illustrated here because of its simplicity. In that case the outer region is "empty" and no computations are performed there. It can be shown that the method is equivalent to the other well known approach consisting in adding an imaginary potential inside the computational box [RSA$^+$06]. When applying a mask function TDLDA one full time step proceeds as

$$\varphi_i(\mathbf{r},t) \longrightarrow \tilde{\varphi}_i(\mathbf{r}) = \hat{U}_{\mathrm{KS}}(t+\delta t, t)\, \varphi_i(\mathbf{r},t), \tag{3.18a}$$
$$\varphi_i(\mathbf{r},t+\delta t) = M(\mathbf{r})\, \tilde{\varphi}_i(\mathbf{r}), \tag{3.18b}$$

$$M(\mathbf{r}) = \begin{cases} 1 & \text{for} \quad |\mathbf{r}| < R_{\mathrm{in}}, \\ \cos^{\gamma_M}\left(\dfrac{\pi}{2}\dfrac{|\mathbf{r}| - R_{\mathrm{in}}}{R_{\mathrm{out}} - R_{\mathrm{in}}}\right) & \text{for} \quad R_{\mathrm{in}} < |\mathbf{r}| < R_{\mathrm{out}}, \\ 0 & \text{for} \quad |\mathbf{r}| > R_{\mathrm{out}}. \end{cases} \tag{3.18c}$$

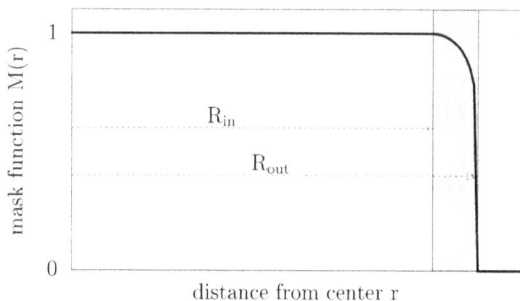

Figure 3.1 Mask $M(\mathbf{r})$ plotted as a function of r, the radial distance to the center of the numerical grid. The absorbing zone extending between the two radii R_{in} and R_{out} is indicated in gray.

First comes one standard KS step expressed here in terms of the unitary TDLDA propagator $\hat{U}_{\text{KS}}$ which yields the intermediate wave function $\tilde{\varphi}_i(\mathbf{r})$. This is followed by the action (3.18b) of the mask function M defined in Eq. (3.18c) and sketched in Figure 3.1. The action of the mask steadily removes electrons trying to flow toward the bounds of the box. The mask zone on a 3D grid can be chosen with arbitrary geometry, rectangular, ellipsoidal, or spherical. The latter choice is to be preferred. The spherical profile is helpful to minimize grid artifacts, e.g., when computing angular distributions.

The limiting radii need yet to be specified. The outer radius is the smallest distance from the origin of the box to the bounds, formally $R_{\text{out}} = \min_{\mathbf{r}_\nu \in \text{bounds}}(|\mathbf{r}_\nu|)$ where the index ν indicates that $\mathbf{r}_\nu$ is a grid point. The inner radius is then

$$R_{\text{in}} = R_{\text{out}} - N_M \delta x, \qquad (3.19)$$

where δx is the grid spacing. The N_M characterizes the number of absorbing points in one direction. It is a crucial numerical parameters.

The mask technique looks simple and straightforward. However, it is not perfect. One will always encounter a small amount of reflected flow, particularly for electrons with low kinetic energy. One can minimize the back-flow by proper choice of the exponent γ_M entering the mask profile Eq. (3.18c). This depends, however, on the actual numerical parameters (number of absorbing points, size of time step). Typical values of γ_M are of order $1/8$ or lower. A detailed description and discussion of this approach and its proper choice of numerical parameters is found in [RSA$^+$06].

Absorbing boundary conditions introduce a subtle difficulty in the time propagation: ortho-normality of s.p. wave functions is gradually lost if absorption is active. This can be ignored when computing one-body observables in TDLDA. Extra measure are needed if one augments the propagation by dynamical correlations, for details see [LvdHB09].

3.4.4.3   Total Ionization

The most obvious observable is the total ionization, in other words the number of escaped electrons denoted by N_{esc}. Absorbing boundaries attenuate the s.p. wave functions φ_i if electron flow hits the bounds. This leads to a time-dependent single particle occupation $n_i(t)$, total charge due to electron emission $\sum_i w_i n_i$ and its complement, the net ionization N_{esc}

$$n_i = \int d^3r \, |\varphi_i(\mathbf{r}, t)|^2 \tag{3.20}$$

$$N_{esc}(t) = \sum_{i=1}^{N} (1 - w_i n_i) = N - \int d^3r \, \varrho(\mathbf{r}, t) . \tag{3.21}$$

where N is the initial number of active (valence) electrons in the system and the w_i are occupation weights arising in theories beyond TDLDA which include dynamical correlations, see section 2.3. The actual total ionization stage of the system is composed of the initial ionization and the number of emitted electrons as $Q(t) = Q(0) + N_{esc}(t)$.

Having the detailed n_i at hand, one may also compute the distribution of ionization probabilities (probability to find a certain, integer, ionization stage Q) [UG97, Ull00, RSU98, CRSU00]. This allows to establish a better link to experiments by explicitly displaying the possible charge states k to which the system can ionize. Let us briefly outline the method in the TDLDA case, namely for fixed occupation numbers $w_i = 0$ or 1. The starting point is the single particle bound-state occupation probabilities $n_i(t)$ defined above. The simplest example of a Helium atom with a single, doubly occupied level $i = 1$ is quite illustrative here. The probabilities for possible charge states are then simply $P^{(0)}(t) = n_1(t)^2$, $P^{(1+)}(t) = 2n_1(t)(1 - n_1(t))$, $P^{(2+)}(t) = (1 - n_1(t))^2$.

To generalize the above considerations, we start with the relation

$$1 = \sum_{k=0}^{\Omega} P^{+k}(t) = \prod_i (n_i + (1 - n_i))^{\nu_i} = \prod_i (n_i + \bar{n}_i)^{\nu_i}, \tag{3.22}$$

where ν_i is the (time independent) i-orbital degeneracy and Ω is the maximum possible charge state (equal to the initial number of active (valence) electrons N). One can then explicitly perform the multiplication on the right-hand side of Eq. (3.22) and sort the resulting terms according to the number of occupied and unoccupied orbitals (factors n_i and $\bar{n}_i = 1 - n_i$, respectively) which directly allows an identification with the $P^{+k}(t)$. This estimate works completely within TDLDA exploiting its s.p. occupations n_i. The probability distribution $P^{+k}(t)$ may also be affected by correlations. Coherent many-body correlations require an extension of the above algorithm. In the most cases more relevant dynamical correlations, see sections 2.3.7.1 and 2.3.7.2, can be characterized by a change in the s.p. weights w_i which then allows to evaluate the $P^{+k}(t)$ similar to in TDLDA after extending the above scheme to occupation weights $0 \le w_i \le 1$.

3.4.4.4 Photo-Angular Distribution (PAD)

The Photo-emission Angular Distribution (PAD) $d\sigma/d\Omega(\vartheta, \varphi)$ can be evaluated by resolving electron emission in angular segments defined usually with respect to the system's center. One then collects all electron loss induced by the absorbing bounds in each segment separately yielding finally the total PAD, for details see e.g., [PRS04, WDRS15, DVC+22].

The procedure outlined is immediately applicable if one knows the system's orientation with respect to the laboratory frame. For then one may for example study the shape of the PAD with respect to laser polarization and properly define longitudinal and transverse components of the emission pattern. Such a situation is possible if the target system is deposited on a surface or embedded into a matrix. In the gas phase, one may also use a first laser pulse to align molecules before the actual irradiation process [SS03], but the latter recipe is not simple. In general, in gas phase, one does not know the orientation of the target system with respect for example to laser polarization axis. Experimental results then pile up electron emission from any orientation of the target with respect to laser polarization axis. In order to make a relevant comparison with experimental results, computed PAD's thus need to be averaged out over any orientation of the system. In practice one can only consider a finite number of such orientations. One usually takes a regular set thereof, each orientation being attributed an elementary solid angle [WFD+10a]. The total, Orientation Averaged PAD (OAPAD) is then attained as the sum of all individual PAD's, each weighed by the elementary solid angle.

The case of single photon ionization allows to make one step more in the characterization of an OAPAD. In that case, indeed, one can make a perturbative estimate of the OAPAD which can then be written as

$$\frac{d\sigma}{d\Omega} \propto (1 + \beta_2 P_2(\cos\theta)) \tag{3.23}$$

where Ω is the solid angle along direction θ measured with respect to laser polarization axis and P_2 is Legendre polynomial of order 2. The parameter β_2 is known as the anisotropy parameter and is bound to vary between -1 and +2 [CZ68]. Positive values correspond to prolate shapes of the PAD along the laser polarization axis, negative ones to oblate shapes. The anisotropy parameter fully characterizes the OAPAD with one single number. It is thus a quite useful quantity to study the impact of laser parameters, such as for example laser frequency, on a given system [WFD+10b]. Outside the truly perturbative regime corrective terms have to be added and the direct numerical orientation averaging outlined above may provide a simpler practical solution.

3.4.4.5 Photo-Electron Spectrum (PES)

The PES provides the energetic content of the electrons which have left the system. This means that one has to recur either to an energy representation or to a momentum one. Most dynamical approaches such as most of the ones based on TDDFT rely on a spatial grid representation of the wavefunctions (section 3.5.1.3) and thus practically limited to a finite computational volume. This eliminates simple access to continuum states, which, at best, would correspond to box states. Switching to momentum representation a priori requires a space to momentum Fourier transform. To attain sufficient accuracy, in turn, requires very large grids which rapidly become computationally prohibitive. Difficulties become even more demanding in the case of high intensity lasers in which emitted electrons may have developed large amplitude trajectories before being actually released from the system which acts as an accelerator on top of the laser potential, following the ideas of the so-called three-step model [Cor93]. There has thus been numerous strategies developed to overcome this difficulty.

Traditional approaches to compute PES rely on (multi-)photon perturbation theory [Fai87] but these approaches are naturally limited to low laser intensities and thus marginally applicable to the irradiation processes of interest here. For one-electron systems, PES and PAD can be calculated exactly by directly solving the TDSE (Time Dependent Schrödinger Equation, section 2.3.3), for example projecting the wavefunction obtained from the TDSE at the end of the pulse onto continuum states [23]. But for more than two electrons TDSE cannot be used in full 3D (section 2.3.3), and severe approximations such as the SAE hypothesis (section 2.2.4.4) are often used.

The surface flux method of [TS12] was developed to explore the case of high laser intensities. It relies on an evaluation of the flux of particles through the box boundaries. It can be seen as a rather direct extension of the simple mask technique presented just below, with the addendum that on top of the number of electrons emitted one also evaluates the velocity thereof. It provides in principle a more accurate access to PES and PAD but its complexity makes practical applications comparable to the mask method. Other approaches rely on a geometrical splitting of real space into an inner box and an outer space. Emitted electrons are then treated in terms of Volkov states which are the eigenstates of a free electron in a laser [CFB98, GVM$^+$12]. This makes sense far away from the irradiated species and a proper treatment thus requires very large external boxes which finally make the transformation of wavefunctions from internal box to outer ones potentially difficult [WGR17] Most of these methods thus remain computationally demanding and one has to accept some approximations to reach simpler methods.

A simple approach, which also delivers good quality results, but at a low computational price, consists of deducing the PES from the temporal phase oscillations of the s.p. wave functions $\varphi_i(\mathbf{r}_\mathcal{M}, t)$ at measuring points $\mathcal{M}$ close to the absorbing bounds [PRS00]. This method directly relies on the mask used to absorb emitted electrons and we thus quote it the mask method. Fourier transformation of the collected $\varphi_i(\mathbf{r}_\mathcal{M}, t)$ from time to frequency domain yields the spectral strength $\mathcal{Y}_i(E_\text{kin}, \Omega_\mathcal{M})$ of kinetic energies E_kin of electrons emitted from state i in direction $\Omega_\mathcal{M}$ (the latter being the space angle of $\mathbf{r}_\mathcal{M}$ relative to the center of the system). The resulting PES becomes then

$$\mathcal{Y}(E_{\text{kin}}, \Omega_{\mathcal{M}}) = \sum_{i=1}^{N} \mathcal{Y}_i(E_{\text{kin}}, \Omega_{\mathcal{M}}) \,. \qquad (3.24)$$

In case of strong fields, boundary corrections are required which modify the Fourier transformation by field factors [DRRS13]. A more detailed description of the evaluation of PES is found in [PRS00, DRRS13, WDRS15].

Even more instructive is the simultaneous measurement of kinetic energy and angular distributions yielding combined PESPAD information which can be represented as a 2D plot in the plane of kinetic energy and emission angle. An alternative representation for the same thing is a Velocity Map Imaging (VMI) where the information is plotted in polar coordinates as function of angle and electron velocity. We will show results in both forms later on.

As a final word it should be reminded that the energy content of emitted electrons can also be evaluated in the case of collisions with a charged particles (section 1.2.3). While the term PES then becomes a bit inappropriate its physical content, namely the kinetic energies of emitted electrons remains unchanged. Conversely its evaluation remains unchanged.

3.4.5 MORE ELABORATE OBSERVABLES

So far, the observables discussed in this section are one-body observables, except for the detailed ionization probabilities evaluated with Eq. (3.22). One-body observables comply nicely with mean-field theories (TDHF, TDLDA) because these are designed to describe one-body observables, in particular the local one-body density. Not only that, mean-field theories let the system evolve along a common mean-field which averages over all reaction channels. This is perfectly fine for low excitations and/or early times of dynamical evolution where different reaction channels are still rare. Examples for that regime are optical response, dynamical polarizability at small amplitudes, or observables of electron emission (average ionization, PES, PAD) typically below one charge unit. Beyond that, predictions for high energy impact and/or long time evolution are risky, not necessarily wrong, but not for sure right.

We have said that the detailed ionization probabilities $P^{+k}(t)$ for ionization stage k lie outside the realm of one-body observables. This looks, at first glance, surprising because Eq. (3.22) is composed from s.p. information. The point is that the outcome produces states with different charge and thus different mean fields. Consider as example a dimer molecule. The neutral charge state is usually stable. The cation $k = 1$ may already dissociate and a doubly ionized molecule ($k = 2$) dissociates even faster. Even though the configurations may still stay close to each other initially, they deviate dramatically in the further course of time evolution. Coupling of electronic and ionic motion serves here as magnifying glass for the fluctuations of the mean-field. The situation is similar for electron transfer in atomic collisions and multi-fragmentation of highly ionized large clusters. Even for an atom, we could have visible differences for the further evolution because different charge states have different ionization potentials and thus different electron emission in the further evolution. All these examples call for theories which can deal with many different mean fields which are MCTDHF/MCTDDFT, see section 2.3.4.3, and stochastic approaches, see section 2.3.8. A small system as a dimer molecule has only a few open reaction channels and may be tractable by MCTDDFT. Larger molecules and clusters require stochastic methods or, what is usually done, classical approaches.

There exist many detailed studies of two-electron systems, the He atom or H_2 dimer. Correlation effects can already be assessed in the double-ionization signal, see e.g., [PDMT03, FNP+08] which is computed from time-dependent treatments as described in Eq. (3.22). Coincidence experiments can deliver explicit information on two-electron correlations, e.g., the correlation between forward and backward-emitted electrons from an excited atom [WGW+00, UMD+03]. This requires at the side of theory the computation of correlated observables of emission. In a perturbative theory close to stationary states, one can evaluate the two-body density in the space of outgoing states $\langle\Psi|\hat{\rho}(\mathbf{k}_1)\hat{\rho}(\mathbf{k}_2)|\Psi\rangle$ where $\hat{\rho}$ is the density operator in momentum space and $\mathbf{k}_1$, $\mathbf{k}_2$ characterize the outgoing states. In a fully-fledged time-dependent treatment, one has to extend the strategy for PES from section 3.4.4.5 to record the time-dependent wavefunctions at pairs of measuring points. Appropriate schemes have yet to be developed.

3.5 NUMERICAL CONSIDERATIONS

So far, we have presented many of the most used theoretical approaches. In this section, we briefly address possible numerical realizations of theories. For simplicity, we concentrate on mean-field models because they are the most widely used approaches and because the principles of numerical representations remain basically the same also for many more elaborate theories. Note, nevertheless, that even within mean-field theories, we shall give here only a brief overview, to give the reader an idea about typical numerical strategies.

3.5.1 REPRESENTATION OF WAVEFUNCTIONS AND FIELDS

3.5.1.1 Basis Sets

Mean-field theories describe the state of a system by a set of single-particle wavefunctions $|\psi_i\rangle$. DFT is formulated in a couple of local fields, preferably the local density distribution and the local KS potential. Starting point for any numerical realization is a set of basis functions $\{|B_\nu\rangle, \nu = 1...N_B\}$, whose properties, namely overlaps $\langle B_\nu|B_\mu\rangle$ as well as matrix elements $\langle B_\nu|B_\kappa|B_\mu\rangle$ and $\langle B_\nu|\nabla^2|B_\mu\rangle$ are all known analytically. The s.p. wavefunctions are then expressed in terms of the basis set as

$$|\psi_i\rangle = \sum_\nu |B_\nu\rangle \phi_{\nu i} \qquad (3.25)$$

where $\phi_{\nu i}$ is a matrix of complex numbers constituting the numerical representation of $|\psi_i\rangle$. A similar expansion holds for all one-body operators and local fields. Everything can then be mapped to numerical operations between the expansion coefficients of wavefunctions and fields.

One of the most used basis sets are Gaussians times polynomials (in other words harmonic oscillator wavefunctions)

$$\langle \mathbf{r}|B_\nu\rangle \propto \exp\left(-\frac{\lambda_\nu}{2}(\mathbf{r}-\mathbf{r}_\nu)^2\right)\sum_n c_n^{(\lambda)}(\mathbf{r}-\mathbf{r}_\nu)^n \qquad (3.26)$$

where λ_ν is the width, $\mathbf{r}_\nu$ the center, and the $c_n^{(\lambda)}$ describe the polynomial. The wavefunction centers are usually taken at the ionic positions, while width and polynomial order are chosen close to the expected extension and node structure of the wavefunction field. The static Kohn-Sham (KS) equation (2.33) then becomes a matrix equation

$$\sum_\nu h_{\mu\nu}\phi_{\nu i} = \varepsilon_i \sum_\nu I_{\mu\nu}\phi_{\nu i} \qquad (3.27a)$$

$$h_{\mu\nu} = \frac{\hbar^2}{2m}\langle B_\mu|\nabla^2|B_\nu\rangle + \sum_\kappa \langle B_\mu|B_\kappa|B_\nu\rangle U_{\mathrm{KS},\kappa}, \qquad (3.27b)$$

$$U_{\mathrm{KS},\kappa} = \int d^3r\, B_\kappa U_{\mathrm{KS}}, I_{\mu\nu} = \langle B_\mu|B_\nu\rangle. \qquad (3.27c)$$

At first glance is the appearance of the norm matrix $I_{\mu\nu}$ may look surprising. It becomes the unity matrix $I_{\mu\nu} = \delta_{\mu\nu}$ if the basis constitutes an ortho-normal set. But this is not always guaranteed. For example, the Gaussian basis Eq. (3.26) is not orthogonal and has a non-unit norm matrix $I_{\mu\nu}$. The time dependent KS equation looks similar with $\varepsilon_i I_{\mu\nu}\phi_{\nu i}$ replaced by $\langle B_\mu|i\hbar\partial_t|B_\nu\rangle\phi_{\nu i}$. The complication is that the basis states $\phi_{\nu i}$ do also depend on time in connection with ionic motion. This is by no means trivial, but efficient solutions have been developed; see e.g., [SS96].

The basis sets in terms of localized wavefunctions, as e.g., Gaussians, are not particularly well suited to represent general distributions. The evaluation of the local density and subsequent local KS potential in DFT is thus rather involved. One has to find a suitable real-space grid $\{\mathbf{r}_n\}$ with sufficiently precise integration rules (e.g., Gaussian integration points in the simple case of one reference position $\mathbf{r}_\nu$). The local density on that grid is then accumulated by triple summation as $\varrho\,(\mathbf{r}) = \sum_{i\mu\nu}\phi_{\mu i}^*\phi_{\nu i}\langle B_\mu|\mathbf{r}\rangle\langle\mathbf{r}|B_\nu\rangle$. Having that, the local KS potential $U_{\mathrm{KS}}(\mathbf{r})$ is evaluated at each point $\mathbf{r}_n$ in real space, and finally, the matrix elements $h_{\mu\nu}$ are computed from the matrix elements of the kinetic energy and local integration of the KS potential.

The direct Coulomb contribution can be handled in two ways, either through coordinate space similar as the part from LDA or directly through two-body matrix elements of $(\mathbf{r}_n - \mathbf{r}_{n'})^{-1}$. It is not the place here to present and discuss that in detail. But it is interesting to note that the use of Gaussian sets is partly motivated by the fact that Coulomb integrals in such basis take a close to analytical form in terms of error functions, which represents a significant simplification.

3.5.1.2 Grid Methods

The alternative is an immediately local representation. To give an impression, we discuss it here for the most straightforward cases of a Cartesian grid in three spatial dimensions. Formally, this can be realized by a basis of box functions $\langle \mathbf{r}|B_\nu \rangle = \chi(x - x_\nu)\chi(y - y_\nu)\chi(z - z_\nu)$ with $\chi(x) = \Delta^{-1}\Theta(x - \Delta/2)\Theta(\Delta/2 - x)$. The Δ is the grid spacing and θ the Heaviside function. In practice, the coordinate-space representation can be explained in intuitive and obvious manner. Starting point is a grid $\{\mathbf{r}_n\}$ in coordinate space. This reads for the simplest case of an equidistant grid

$$\mathbf{r}_n = (x_{n_x}, y_{n_y}, z_{n_z}), \ x_{n_x} = n_x\Delta, \ n_x \in \{1, 2, ...N_x\}, \tag{3.28}$$

and similarly for y_{n_y} and z_{n_z}. The s.p. wavefunctions are represented by their values on the grid

$$\varphi_i(\mathbf{r}) \equiv \varphi_i(\mathbf{r}_n) \tag{3.29}$$

and similarly for the other local fields as density and KS potential. The static KS equation on the grid then reads

$$\left(-\frac{\hbar^2}{2m}\nabla^2 + U_{KS}(\mathbf{r}_n)\right)\varphi_i(\mathbf{r}_n) = \varepsilon_i\varphi_i(\mathbf{r}_n) \tag{3.30}$$

where $\hat{\nabla}^2$ is the Laplacian operator, composed from second derivatives in x-, y-, and z-direction. The collection of local density is now trivially $\varrho\,(\mathbf{r}_n) = \sum_i |\varphi_i(\mathbf{r}_n)|^2$ and the KS potential is computed simply point by point. There remains to specify the strategies for integration, differentiation (to evaluate $\hat{\nabla}^2$, and solving the Coulomb problem. There are several integration algorithms available [PTVF92]. In practice, the simple trapezoidal integration formula is sufficiently precise, fast, and robust. What differentiation is concerned, equidistant grids cooperate very well with finite differences or definition in momentum space via Fourier transformation. The long range of the Coulomb field renders the Coulomb problem the most expensive part in DFT calculations whatever solution strategy one chooses.

So far for a brief tour through 3D Cartesian grid representation. There are similar strategies for cylindrical grids in cases of axial symmetry and radial grids in case of spherical symmetry. An example for a detailed comparison of a couple of grid techniques is found in [BLMR92] and a detailed presentation of the handling of 3D grids in [MRSU14, DVC+22].

3.5.1.3 Which Representation to Choose?

The great variety of numerical strategies and basis representations makes it impossible to cover all that in depth within this book. The above examples illustrate the two basically different option for basis sets. We close this section with a comparison of strengths and weaknesses of either choice.

Localized basis sets, as the Gaussians Eq. (3.26), are extremely flexible in accommodating different length scales (useful, e.g., for all-electron calculations), node structure, and level of precision. Gaussian sets furthermore provide an especially efficient evaluation of Coulomb field. However, localized basis sets, in general, become cumbersome for time-dependent problems, particularly if ionic motion is involved, and they are very inefficient if the electron continuum starts to play a role (as typically in case of electron emission).

Grid techniques, on the other hand, do an extremely good job for electron dynamics at large excitation energies, particularly for electron emission. For the latter, it is important to note that it is technically simple to design outgoing boundary conditions by adding an electron-absorbing zone around the bounds of the box associated with the ability to compute detailed observables of electron emission as Photo-Electron Spectra (PES, section 3.4.4.5), and Photo-electron Angular Distributions (PAD, section 3.4.4.4), for details see also [Ull00, RSA+06, WDRS15, WGR17]. As a consequence, wavefunction expansions dominate in structure calculations while grid techniques are the method of choice for simulating electron dynamics the more so the stronger the excitation.

3.5.2 SOME TYPICAL SOLUTION SCHEMES

We concentrate in this section on the numerical solution of the static and dynamic KS equations because this is still the most widely used approach and variety of solution schemes is limited. We will sketch briefly also a few samples from solution to many-body theories out of the enormously rich choice of models and solutions.

3.5.2.1 Statics

We start with the static KS equation (2.33) as matrix equation (3.27) in a given basis. This looks at first glance like a simple diagonalization problem. One calls a standard diagonalization routine from the library and obtains the expansion coefficients $\phi_{\nu i}$ together with the s.p. energies ε_i. But the KS Hamiltonian $\hat{h}_{\mathrm{KS}}$ and with it the Hamiltonian matrix $h_{\mu\nu}$ depends on the local density which, in turn, depends on the $\phi_{\nu i}$. Having changed the $\phi_{\nu i}$, we obtain a changed matrix $h_{\mu\nu}$ and have to diagonalize again. We are thus ending up with an iterative procedure of determining the set $\{\phi_{\nu i}, i = 1-...N\}$ and corresponding KS Hamiltonian. Repeating that often enough, converges eventually to a final solution where the $\phi_{\nu i}$ and ε_i do not change much anymore between two iterative steps. The number of iterations needed depends very much on the actual system and precision demands. A typical number is about 50 iterations for a quality solution.

Although conceptually simple, the just described iteration scheme is inefficient. We solve the matrix equation exactly and yet have to discard the solution for the next iteration as soon as we start diagonalization from scratch. It is more efficient to employ an iterative solution scheme for the diagonalization problem and to terminate that prematurely in order to feed the preliminary solution into the computation of local density and subsequent evaluation of the KS Hamiltonian. These interlaced iterations converge as well and save considerably computing time at the side of diagonalizations.

There is a variety of iteration schemes for the static Schrödinger or KS equation. Most of them are refinements of the basic gradient (or downhill) step. Recall that the KS equation can be derived by variation of the total energy with respect to the s.p. wavefunctions φ_i, see section 2.2.3. The gradient in the energy landscape, $-\delta E/\delta \varphi_i^* = -\hat{h}_S \varphi_i$, represents the steepest descent to the minimum (to be precise, the nearest local minimum). We just have to follow this path. This amounts to the simple gradient step

$$\varphi_i^{(n+1)} = \hat{O}\left\{\varphi_i^{(n)} - \delta_{\mathrm{grad}}\left(\hat{h}_{\mathrm{KS}} - \langle\varphi_i^{(n)}|\hat{h}_{\mathrm{KS}}^{(n)}|\varphi_i^{(n)}\rangle\right)\varphi_i^{(n)}\right\} \qquad (3.31)$$

where n is the number of iteration, $\hat{O}$ stands for the ortho-normalization of the whole set of s.p. wavefunctions, $\hat{h}_{\mathrm{KS}}^{(n)}$ is the KS Hamiltonian evaluated with the $\varphi_i^{(n)}$, and δ_{grad} is a numerical step-size parameter chosen just small enough to avoid divergence by overshooting. The latter is achieved safely by $\delta_{\mathrm{grad}} < \varepsilon_{\mathrm{max}}^{-1}$ where $\varepsilon_{\mathrm{max}}$ is the maximal energy in the actual numerical representation of the wavefunctions. This creates a conflict. The numerical representation becomes the better the large $\varepsilon_{\mathrm{max}}$ which, in turn, worsens the speed of the gradient step.

There exist several strategies to overcome that hindrance. We present here one example for an efficient and transparent method. A very versatile choice is to augment the gradient step by a pre-conditioner, i.e. to replace the number δ_{grad} by an appropriately chosen operator $\hat{D}$ which suppresses the high-energy components in $\hat{h}_{\mathrm{KS}}$. In coordinate-space representation (see section 3.5.1), one can identify the kinetic-energy operator $\hat{T}$ as being responsible for the high-energy components. This suggests to damp the kinetic energies yielding an accelerated gradient step as

$$\varphi_i^{(n+1)} = \hat{O}\left\{\varphi_i^{(n)} - \hat{D}\left(\hat{h}_{\mathrm{KS}} - \langle\varphi_i^{(n)}|\hat{h}_{\mathrm{KS}}^{(n)}|\varphi_i^{(n)}\rangle\right)\varphi_i^{(n)}\right\}, \quad \hat{D} = \frac{\delta_{\mathrm{acc}}}{E_{\mathrm{acc}} + \hat{T}},$$
$$(3.32)$$

where δ_{acc} and E_{acc} are numerical parameters for the accelerated step. The E_{acc} is typically of order of the lowest s.p. energy $|\varepsilon_1|$ and $\delta_{\mathrm{acc}} \approx 1/2$. This may have to be revised if one of the pseudo-potentials has extraordinarily high spikes and large non-local contributions. Anyway, the detailed choice is a matter of experience. For more detailed explanations and examples, see [RC82, BLMR92, MRSU14, DVC+22].

3.5.2.2   Dynamics

Now we turn to the solution of the time-dependent KS equation (2.34). A formal solution for the step from t to $t + \delta t$ reads at first glance

$$|\varphi_i(t + \delta t)\rangle = \exp\left(-i\hbar \delta t \hat{h}_{\mathrm{KS}}\right) |\varphi_i(t)\rangle.$$

However, recall that $\hat{h}_{\mathrm{KS}}$ depends on the set of $|\varphi_i\rangle$ on which it acts such that it is not the same all along the way from t to $t + \delta t$. The formally correct solution is a time-evolution operator $\hat{\mathcal{U}}_{\mathrm{KS}}$ involving an integral over $\hat{h}_{\mathrm{KS}}(t)$,

$$|\varphi_i(t + \delta t)\rangle = \hat{\mathcal{U}}_{\mathrm{KS}}|\varphi_i(t)\rangle = \hat{\mathcal{T}}\left\{\exp\left(-i\hbar \int_t^{t+\delta t} dt' \hat{h}_{\mathrm{KS}}(t')\right)\right\}|\varphi_i(t)\rangle, \quad (3.33)$$

where $\hat{\mathcal{T}}$ is the time-ordering operator [FW71]. The practical solution is to take a sufficiently short time step such that $\hat{h}_{\mathrm{KS}}$ remains effectively constant and to sew together these short-time steps to the finally wanted time span.

One has still to decide at which time between t and $t + \delta t$ one should take the KS Hamiltonian for the step. It is intuitively clear that the most fair choice is at mid-time $t + \delta t/2$ and one can prove by numerical error analysis that this choice is optimal concerning precision and stability (e.g., of energy conservation). Of course, the s.p. wavefunctions at mid-time have first to be estimated by what is called a predictor step. This amounts finally to a double-step procedure. First, we estimate the mid-time wavefunctions

$$|\varphi_i(t + \delta t/2)\rangle = \exp\left(-i\hbar \frac{\delta t}{2} \hat{h}_{\mathrm{KS}}(t)\right) |\varphi_i(t)\rangle \qquad (3.34a)$$

and evaluate the mid-time KS Hamiltonian from that. Then we can go for the final full step

$$|\varphi_i(t + \delta t)\rangle = \exp\left(-i\hbar \delta t \hat{h}_{\mathrm{KS}}(t + \delta t/2)\right) |\varphi_i(t)\rangle. \qquad (3.34b)$$

The evaluation of an operator exponential is another numerical task. In practice, one approximates it by a Taylor expansion

$$\exp\left(-i\hbar t \hat{h}_{\mathrm{KS}}\right) \approx \sum_{m=1}^{M} \frac{(-i\hbar t)^m}{m!} \hat{h}_{\mathrm{KS}}^m \qquad (3.34c)$$

with M typically between 4 and 8. Smaller values for the second step (3.34b) endanger norm conservation and larger values do not gain much as compared to the enhanced expense.

The stepping scheme Eq. (3.34) belongs to the family of the much celebrated predictor-corrector algorithms for the solution of one-parameter differential equations, see e.g., [PTVF92]. The advantage of this scheme is that
it can be applied in straightforward manner to any Hamiltonian in whatever
representation (coordinate-space, matrix, ...). It is the method of choice in
nuclear DFT with its rather involved KS Hamiltonian [MRSU14]. The disad-
vantage is that one must employ the rather expensive computation of the KS
Hamiltonian twice per step.

3.5.2.3 Time Splitting Propagator

The electronic KS Hamiltonian has very often the simple structure as a sum
of kinetic energy and a local KS potential, i.e. $\hat{h}_{KS}(t) = \hat{T} + V_{KS}(\mathbf{r}, t)$. This
allows to deduce from the exact propagator Eq. (3.33) a more efficient step
by splitting the exponentials of kinetic and potential energy

$$\hat{U}_{KS} = \hat{T}\left\{\exp\left(-i\hbar\int_t^{t+\delta t} dt' \hat{h}_{KS}(t')\right)\right\}$$

$$\approx \exp\left(-i\hbar\delta t\, V_{KS}(t+\delta t/2)/2\right)\exp\left(-i\hbar\delta t\hat{h}_{KS}\right)\exp\left(-i\hbar\delta t\, V_{KS}(t)/2\right).$$

Note that you can easily recognize the time ordering from the ordering of
potentials therein which renders this step nicely symmetric in t and $t + \delta t$.
This approach is coined a time-splitting scheme [FFS82] and it has the advantage that one needs to evaluate the KS potential only once per step. That
is not immediately obvious. But it becomes so when unfolding the step into
its successive sub-steps:

$$|\varphi_i^{(1)}\rangle = \exp\left(-i\hbar\frac{\delta t}{2}V_{KS}(t)\right)|\varphi_i(t)\rangle, \tag{3.35a}$$

$$|\varphi_i^{(2)}\rangle = \exp\left(-i\hbar\delta t\hat{T}\right)|\varphi_i^{(1)}\rangle, \tag{3.35b}$$

$$V_{KS}(t+\delta t/2) = V_{KS}[\{\varphi_i^{(2)}\}], \tag{3.35c}$$

$$|\varphi_i(t+\delta t/2)\rangle = \exp\left(-i\hbar\frac{\delta t}{2}V_{KS}(t+\delta t/2)\right)|\varphi_i^{(2)}\rangle. \tag{3.35d}$$

The key point is that the exponential with a local potential changes only the
local phases of the s.p. states which, however, does not have any effect on the
local density $\varrho(\mathbf{r}) = \sum_i \varphi_i^* \varphi_i$. Thus one can evaluate the new KS potential
already after step (3.35b) and recycle this right for the first operation (3.35a)
in the next step. This is why we need to evaluate the KS potential only once.

The most expensive step is the exponential of the kinetic energy in sub-step
(3.35b). An elegant and exact procedure can be used in connection with the
definition of the kinetic energy in Fourier space in which case the exponential
becomes a simple c-number operation in momentum space. The initial wave

function $|\varphi_i^{(1)}\rangle$ is thus first Fourier transformed into momentum space, the kinetic propagator is applied as simple product with the exponential of the kinetic energy, and finally the result is Fourier transformed back into coordinate space. This achieves an exact kinetic propagator.

In case one uses finite differences for the kinetic energy, then one can employ a rational approximation to the exponential, coined the Crank-Nicholson propagator [CN92]. Both ways provide a unitary step. The same holds for the exponentials of the potential. The time-spitting step thus has the additional advantage that it is strictly unitary thus preserving the ortho-normality of the set of s.p. wavefunctions all time. The time-splitting step (3.35) cannot be immediately applied if non-local pseudo-potentials are involved. One can extend the scheme by one further splitting delivering interlaced exponentials for kinetic energy, local potential, and now also non-local pseudo-potential part. Details for the practical implementation of the time-splitting step with and without non-local pseudo-potentials can be found in [DVC+22].

3.5.2.4 A Coulomb Solver in Real Space

The most time consuming part in building the KS potential is the computation of the Coulomb potential

$$V_C(\mathbf{r}) = \int d^3r' \, \frac{e^2}{|\mathbf{r} - \mathbf{r}'|} \, \varrho(\mathbf{r}'). \tag{3.36}$$

Direct evaluation of the integral over $\mathbf{r}'$ is overly expensive because it has to be done for every space point $\mathbf{r}$ on the grid. It is standard to evaluate V_C as solution of the Poisson equation

$$\nabla^2 V_C(\mathbf{r}) = -4\pi e^2 \, \varrho(\mathbf{r}) \tag{3.37}$$

where ∇^2 is the Laplacian operator. As this is a standard problem appearing in many areas of physics and technology, there exists a waste amount of Poisson solvers in mathematical libraries. In connection with finite differences for the Laplacian on regular grids, one finds often iterative solutions like successive over-relaxed gradient step or conjugate-gradient step, see e.g., [PTVF92], applied once the Coulomb field at the boundaries of the box has been fixed. The solution looks formally extremely simple in connection with the Fourier definition of the Laplacian. The Poisson equation in momentum space reads

$$\tilde{V}_C(\mathbf{k}) = -\frac{4\pi e^2}{k^2} \tilde{\varrho}(\mathbf{k}) \tag{3.38}$$

where the tilde indicates the Fourier transformed quantity. This looks deceivingly simple, but is still plagued by a strong singularity at $k \to 0$ which is related to the very long range of the Coulomb field in coordinate space.

One strategy to overcome this is to single out the long-range components in $\tilde{\varrho}(\mathbf{k})$ such that the remaining short-range density starts with $k^n, n > 2$, to treat the short-range and long-range parts separately, and to recombine the resulting V_C [LR94]. The other strategy is to map the density to a doubled grid in coordinate space, to compute the Green's function of the Poisson problem on the doubled grid once and forever, to apply the Green's function in Fourier space (simple multiplication), and to remap the resulting Coulomb potential onto the original grid [EB79, MRSU14]. The second strategy is exact but somewhat more expensive. The first strategy is about twice as fast, but imprecise if too much electronic density comes close to the boundaries.

3.5.2.5   The Case of Many-Body Theories

Finally, a few words about the numerical realization of many-body theories. There are a lot of different strategies to solve static problems. Most of them (CI, coupled cluster, $\exp(S)$, Green's functions, density-matrix hierarchies, ...) rely on basis expansions which can be handled by standard diagonalization, possibly interlaced diagonalization steps, or iterative techniques like the (re-fined) gradient step. This can drive easily to very large matrix problems for which particular algorithms have been developed as, e.g., the Lanczos method [Lan50]. Even more complex problems are often attacked with stochastic approaches as, e.g., the path-integral Monte-Carlo method[Cep95].

The field is still less well developed for dynamical problems. Standard propagation techniques (as discussed above for TDDFT) apply similarly for theories formulated in basis expansion (TDCI, section 2.3.4.1, and density-matrix hierarchy, section 2.3.5) where, however, the more expensive time-stepping algorithms limit the size of the expansion compared to what can be afforded in static calculations.

Similarly straightforward is the treatment of MCTDHF, see section 2.3.4.3. The different mean-field trajectories are propagated with the mean-field time steps as explained above. What remains is a matrix equation for the time-dependent coupling coefficients between the trajectories which, again, can be handled with standard predictor-corrector techniques.

Stochastic methods for time propagation in general can raise problems with long-time stability. However, they can be applied to evaluate the collision term in the semi-classical VUU equation, see section 2.4 for theory and [BD88] for numerical realization, or the quantum Boltzmann equation within STDLDA (section 2.3.8, [SR14, RS15]). This works because in the collision term there is a small correction to a leading mean field which guarantees that the overall propagation cannot run out of control.

4 Analysis of Irradiation Induced Dynamics

The present chapter aims at confronting the formal theoretical tools described in Chapter 2, and their practical implementation (Chapter 3), to actual examples such as the ones described in Chapter 1 and many other ones. The idea is here to consider well identified examples and discuss how to analyze them theoretically. By this we mean a choice and an implementation of a relevant theory and an understanding of the limitations of the chosen theoretical setup. These limitations may be formal or practical. In other words, the goal here is to give the reader the capability of a critical assessment of available theoretical tools, ultimately to be able to realize that some quantities are still beyond the reach of available theories.

This chapter thus provides a bunch of examples, involving most of the mechanisms and situations identified in section 3.3. In order to analyze these various physical mechanisms in a systematic way and see how they can be addressed from a theoretical point of view we shall sort them according to time, which is the natural input of any real time description of irradiation. This also makes sense, because most theories are able to address only a limited time span.

We shall thus first analyze electron-dominated processes in section 4.1. This will cover most phenomena such as plasmon resonances and resonant coupling to it, resonant ionization effects, hole migration... We shall also analyze there the damping of such motions, for example via electron heating.

Electrons are coupled to ions and possibly to an environment. As a second step, we shall thus focus on the role of ionic degrees of freedom and environment (section 4.2). The dynamics of one species does affect the others. This will include electron-ion coupling of course but also other aspects such as ionic temperature and finally long term evolution with possible explosion or fragmentation.

In section 4.1, and, to a large extent in section 4.2 we discuss electron dynamics at a quantum mechanical level. This means, in the regime of high excitation energies, almost exclusively a description in terms of TDDFT (section 2.2). Nevertheless, even this very efficient approach becomes too expensive at very high energies and/or large system sizes. This is then the regime for semi-classical approaches as, e.g., collective flow in a fluid dynamical picture or phase-space dynamics. We shall thus dedicate a section to large systems at large excitation energies in which detailed quantum mechanical aspects are expected to play a secondary role. These systems, analyzed in section

DOI: 10.1201/9781003127949-4

4.3, constitute the workhorses of the flourishing studies gathered under the acronym plasmonics.

4.1 ELECTRON DOMINATED PROCESSES

4.1.1 BASIC EFFECTS

4.1.1.1 Optical Response: A Key Guiding Tool

All irradiation processes amount to a time-dependent Coulomb field impinging on a system. The reaction of the system is dominated by the optical response (or photo absorption cross-section, section 3.4.2) which is the entry door to any photo-excitation mechanism. It gives access to electromagnetically driven transitions, both individual and collective ones. It is thus always the first step in any investigation to look at the optical response spectra. From the theoretical side, it is computed either by RPA (formally linearized TDLDA, see section 2.2.4.1) or by spectral analysis following a very small initial boost, see section 3.2.3. Two typical optical spectra are shown in Figure 4.1, in the cluster Na_{41}^+ as example for a metallic system and in H_2O for a covalently bound molecule.

The experimental spectrum of Na_{41}^+ is dominated by one strong peak which exhausts nearly all photo-absorption strength. This is the Mie surface plasmon resonance [Mie08, KV95] which is already well described by collective models

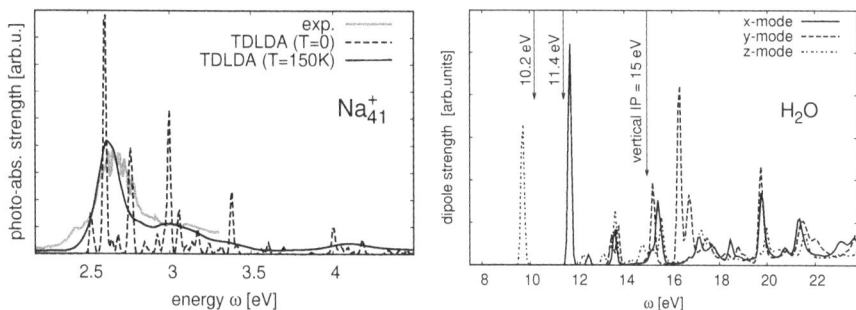

Figure 4.1 The photo-absorption strength for the metal cluster Na_{41}^+ (left) and for the covalent molecule H_2O. The spectra for Na_{41}^+ (left) are compared with experimental data from [HS99] which were measured at finite temperature $T = 105K$. The theoretical results were computed with TDLDA+ADSIC using the QDD code with standard parameters [DVC$^+$22]. They are shown for the ionic ground state configuration ($T = 0$) as well as for ensembles at finite temperature as indicated. The spectra for H_2O are shown for the ionic ground state configuration. But here, we distinguish the modes along the three principal axes of the molecule. Vertical arrows indicate the two frequencies at which the detailed time evolution is studied later on (see Figures 4.3 and 4.6) and the ionization potential (IP) which indicates the onset of the electron continuum.

and which plays the leading role in plasmonics, see section 4.3. It is located well below IP in this cluster. The peak is rather broad and the theoretical spectra reveal why. In the Na_{41}^+ ground state (temperature 0) the spectrum is almost exclusively concentrated at the plasmon peak, but much fragmented over a dense series of peaks. This is called Landau fragmentation in analogy to Landau damping in bulk plasma [LP88]. It arises because the plasmon resonance in heavy metal clusters lies in a region of high density of $1ph$ states to which the resonance couples. The dipole strength is now much larger than the original single-particle strength of the $1ph$ states. It is essentially brought in and distributed by the plasmon resonance. However, metal clusters have a high ionic mobility (low melting point) associated with considerable fluctuations of ionic configurations already at rather low temperatures. Furthermore, most experiments are conducted at finite temperature, as metal clusters are usually formed at finite temperature, before entering into the measurement devices of the experimental setup. Averaging the spectra over a thermal ensemble of configurations at the appropriate temperature then yields a smooth and broad Mie plasmon peak rather well fitting the experimental distribution, which naturally contains thermal effects. Besides smoothing from ionic ensemble averaging, there is also a dynamical effect from coherent coupling of electronic to ionic motion which spreads electronic peaks by phonon satellites. This leads to a smooth spreading for large systems with dense phonon spectrum (i.e. ionic vibrations). It can be resolved as separate side-peaks in small systems. For an example see Figure 4.20 and discussion thereof.

The right panel of Figure 4.1 shows the spectrum of H_2O. It is much more fragmented and spreads over a much wider energy range (mind the energy scale), than in the metal cluster case. Unlike the metallic case, the dipole strength of the various spectral states is basically the strength of the underlying $1ph$ states. Moreover, the density of states below the continuum threshold (see vertical dashed line with IP) is very low. The many peaks above threshold are further smoothed by the escape width related to the lifetime of a continuum state within the molecules vicinity. Another difference to metal clusters is that molecules, particularly organic ones, have pronounced spatial structures with the consequence that the dipole spectra in different spatial directions come out very different as seen here by comparing the spectra from the three principal axes of the molecule. Finally, we want to point out that we do not compare with experimental data in this case of H_2O because there the spectra are much affected by ionic motion leading to a detailed fragmentation of each electric peak by coupling to the oscillations of the very mobile H atoms.

TDLDA (section 2.2.3.4) employs a time-dependent Kohn-Sham (KS) Hamiltonian which takes up the actual electron density at every instant of time. This update of the KS Hamiltonian, particularly its Coulomb field, is the expensive part of TDLDA. But only this way it accounts for the dynamical polarizability. A simplification is the independent-particle model where KS Hamiltonian of the ground-state solution is kept frozen. For example, the SEA, see last paragraph of section 2.2.3, belongs to this class. In linearized

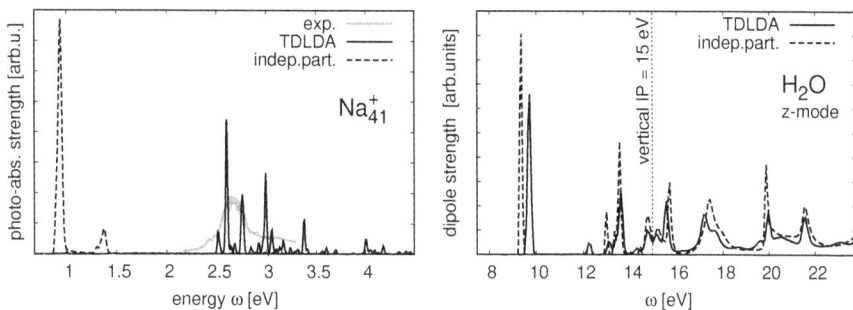

Figure 4.2 Comparison of the photo-absorption strength from TDLDA (full lines) and from independent particle motion (dashed lines) for the metal cluster Na_{41}^+ (left) and for the z mode of the covalent molecule H_2O. The spectra for Na_{41}^+ (left) are also compared with experimental data from [HS99]. All theoretical results are shown for the ionic ground state configuration ($T = 0$). They were computed with TDLDA+ADSIC using the QDD code with standard parameters [DVC+22].

TDLDA, alias RPA, it corresponds to take the difference of s.p. energies as model for the excitation energy. Figure 4.2 compares TDLDA with the independent particle model for the same test cases as in Figure 4.1. The covalent H_2O molecule shows very little difference which means that the independent-particle model is acceptable, at least in the linear response regime. However, the metal cluster Na_{41}^+ shows a dramatic difference. The dynamical polarizability in TDLDA shifts the peak of photo-absorption strength far up in frequency. The different behavior stems from the localization of the electrons. Electrons states in metals are delocalized and fill the whole volume. This causes long-range oscillations of the excited electron cloud (surface plasmons) which generates a strong Coulomb response associated with large energy shifts. Covalent molecules, on the other hand, have well-localized electron states whose excitation produces only small oscillations and subsequently small energy shifts. The example shows that one cannot easily formulate a rule-of-thumb for the applicability of the independent-particle model. Each new case requires new checking.

Experimentally, photo-absorption spectra are obtained by scanning photon frequencies in fine steps. We have simulated this procedure in TDLDA. The result will be discussed in connection with measuring ionization in Figure 4.6.

4.1.1.2 Real-Time Dynamics After Excitation by A Photon Pulse

Metal clusters and covalent molecules thus exhibit rather different optical responses. They will consequently couple to light in different ways. For example, H_2O is transparent in the visible domain because of lack of optical response in this spectral range (around a few eV). In turn Na metal clusters have a color, which is globally characteristic of their size (see Eq. (1.1) in section 1.1.1), as their optical spectrum is dense in this spectral range.

In addition to that, the structure of the response also implies different couplings to a laser pulse. Let us consider a laser with a well defined spectral content. This means that we assume there is no chirp (frequency drift tuned during the pulse) and the pulse is long enough to deliver a well-defined photon frequency. Provided the frequency matches a dense spectral region, it will nevertheless more easily couple to a Na cluster via its spread plasmon rather than to water where transitions may be much sharper and where on-off resonance effects are thus easier to spot (Figure 4.1).

Spectral analysis looks at a system from the viewpoint of frequency space. This is appropriate for weak excitations and was for long time the only access to dynamical properties of quantum systems. Strong and short photon pulses as they are well available today allow to explore dynamics directly in the time domain. For a comparatively early example see the pump and probe analysis of ionic motion in small molecules [Zew94]. With time going on, ever shorter pulses become available opening the path to time-resolved analysis of electron dynamics [KI09, BVLN22], see also the example of section 1.1.2.

With Figure 4.3 we start to look at dynamics in time domain. It shows the time evolution of three key observables (dipole moment, energy absorbed from

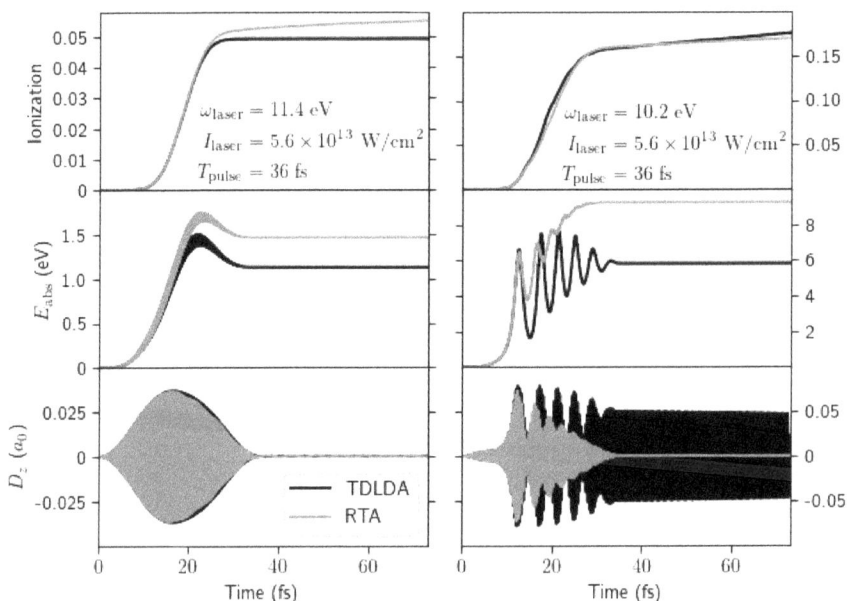

Figure 4.3 Time evolution of three key observables (dipole moment, absorbed energy, ionization) for laser excitation of H_2O for an off-resonant (left) and resonant case (right). Compared are results from mere TDLDA(+ADSIC) with a TDLDA(+ADSIC) augmented by dissipative dynamics (in relaxation-time approximation = RTA). Figure adapted from [DVC$^+$22].

the photon field, ionization) for H_2O excited by short photon pulses at two frequencies indicated in the right panel of Figure 4.1. The lower frequency (left panels) is off resonance while the higher frequency (right panels) is resonant, i.e. close to an eigenfrequency of the system.

Resonant or off-resonant excitation leads to large differences in the time evolution of every observable. In the off-resonant case (left panels), the envelope of the dipole signal follows exactly the envelope of the external pulse and dies out as soon as the pulse is over. There is only little energy absorption and consequently only little electron emission. Resonant excitation (right panels) is associated with much more energy absorption and subsequently more electron emission. The dipole amplitude is much larger, it starts soon to deviate from the photon pulse, and it continues to oscillate considerably long after the pulse is over. Part of the absorbed energy is invested in electron emission, the other part is stored in the system and visible in the surviving dipole oscillations. This lasts extremely long in the case of pure TDLDA evolution and accordingly, ionization continues as well. Emission exports energy out of the system and this way cools down the dipole motion. In fact, electron emission is the by far dominant channel for damping of dipole oscillations in pure TDLDA.

However, the emission is not the only means of dissipation at electronic level. This calls for electron correlations beyond TDLDA, in which only part of the (static) correlations are included via the exchange-correlation functional (section 2.2.3). It is known from homogeneous systems, that dynamical correlations from electron-electron collisions deliver the right amount of dissipation [KB62, PN66]. We have learned in section 2.3 that a fully detailed description of dynamical correlations in finite Fermion systems is still at the edge (or beyond) present days computational capabilities. A manageable scheme for a fully quantum mechanical description of electronic dissipation exists in the RTA, see section 2.3.7.2. Figure 4.3 shows also RTA results for direct comparison with TDLDA. There are only small differences in the off-resonant case (left panels), very little for the dipole signal and ionization, a bit more energy absorption for RTA. The reason for the latter effect will become more transparent in the resonant case (right panels). Here we see large differences.

RTA opens a new channel for dissipation, namely conversion of collective flow energy (mostly dipole) into internal (thermal) energy. This induces a considerable attenuation of the dipole signal and correspondingly suppresses ongoing ionization. As before, dissipation leads to more energy absorption. Particularly interesting is the pattern for E_{abs} as a function of time. The result from TDLDA shows strong oscillations which amounts to a repeated change between photon absorption and emission. TDLDA limits the energy intake because all energy is kept in the dipole mode which turns to stimulated photon emission if sufficient energy is available for that. RTA damps these oscillations efficiently because it converts a large part of the dipole energy to internal energy thus suppressing stimulated emission and giving room for more absorption.

The simple distinction resonant versus off-resonant is a qualitative description of expected signals and dissipation. However, it is also interesting to have quantitative criteria to estimate the ranges of validity of TDLDA. To that end, one has to look at the typical relaxation time for a given system and process in relation to pulse length and key time scales of the system. It is impossible to provide general rules for all these times. There is one rule-of-thumb which helps to estimate the trends for the relaxation time associated with dissipation. It is inversely proportional to in-medium electron-electron scattering cross section σ_{ee} and squared excitation energy E_{exc}^2 [SRS15]. This at least helps with some trends. Longer pulses and/or high excitation energy render dissipation more important. The trends are similar in all systems. The time scales vary substantially depending on the electronic spatial densities (which, in turn, determine scattering cross sections) and spectral densities. These have to be explored for each family of molecules anew (see also the discussion around Figure 1.4).

4.1.1.3 Energetic Considerations

In order to dig deeper into the energy balance of an irradiated system, we disentangle the energy into different contributions. To keep the presentation simple, we ignore the small correction from the bounds of the box (for details see [VSR17]). The basic, initially given, quantity is the total energy of the propagated state $E(t)$ either from TDLDA or from RTA. From this, we obtain the excitation energy $E^*(t) = E(t) - E_{\mathrm{g.s.,initial}}$ relative to the initial static ground state (g.s.) energy $E_{\mathrm{g.s.,initial}}$. The aim is now to separate E^* into its various contributions.

The energy absorbed from the laser is shared into two major contributions, energy associated with ionization and internal energy of the remaining electrons (see section 3.4.3). To better analyze this energy sharing we define two auxiliary energies, the charge equivalent ground state energy $E_{\mathrm{g.s.}}(Q(t))$ and the density-constrained (DC) ground state energy $E_{\mathrm{DC}}(\varrho, \mathbf{j})$. Recall that the charge state $Q(t)$ changes in time due to ongoing ionization. The $E_{\mathrm{g.s.}}(Q)$ is the energy of a ground-state to given, mostly fractional, charge Q. The state of charge Q is a mixed state where the HOMO is only partially filled to match Q. The DC state is the state with lowest energy $E_{\mathrm{DC}}(\varrho, \mathbf{j})$ which has exactly the same local density $\varrho(\mathbf{r}, t)$ and current $\mathbf{j}(\mathbf{r}, t)$ as the actual dynamical state.

We can now define the major contributions to the total excitation energy E^*.

1. The charging energy is shared between a kinetic and a potential part: $E_{\mathrm{charge,loss}}$ is the energy exported immediately by ionization and represents the kinetic energy carried away by the emitted electrons; $E_{\mathrm{charge,pot}}(Q) = E_{\mathrm{g.s.}}(Q) - E_{\mathrm{g.s.,initial}}$ accounts for the excitation energy invested for charging the system.

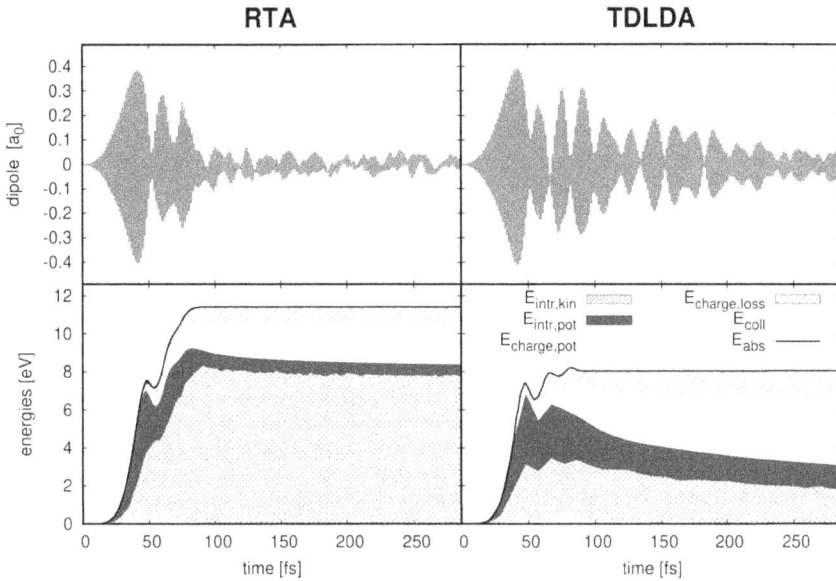

Figure 4.4 Time evolution of the dipole (upper panel) and the energy balance (lower panel) in a strong laser excitation of Na_{40}. Compared are results from pure TDLDA(+ADSIC) (right panels) with RTA (left panels), both obtained with the QDD code using standard parameters [DVC+22]. Totally absorbed electronic energy E_{abs} is sorted charging and intrinsic energy contributions, see text. Figure adapted from [VSR17].

2. The internal energy contains the collective energy E_{coll} (Eq.(3.16)) which is the kinetic energy contained in collective flow $\mathbf{j}(\mathbf{r},t)$ (section 3.4.3). The rest of the internal energy is the intrinsic energy, itself shared between a kinetic and a potential component: $E_{intr,kin}$ is the intrinsic kinetic energy (i.e. the kinetic energy minus E_{coll}), which becomes the thermal energy near equilibrium; $E_{intr,pot} = E_{DC}(\varrho, \mathbf{j} = 0) - E_{g.s.}(Q)$ is the intrinsic potential energy which is the potential energy difference between lowest energy to given density distribution $\varrho(\mathbf{r},t)$ ground state at charge $Q(t)$.

Figure 4.4 (lower panels). compares the time evolution of the five major energies to the excitation energy for TDLDA (right panels) and RTA (left panels). The test case is the metal cluster Na_{40} excited by a resonant photon pulse. Upper panels furthermore display the dipole response for completeness, showing again that RTA dynamics strongly damps dipole oscillations for such a laser frequency close to the plasmon resonance of the system. Most interesting here are the lower panels with energy sharing. At first glance, we see a substantial difference in total E^*, the same effect as we had seen already in the previous figure. But the partial contributions differ even more, quantitatively as well as qualitatively. Particularly pronounced is the difference between charging energy (kinetic plus potential contribution) and intrinsic energy. TDLDA generates little intrinsic energy and predominantly charging energy, a feature which we had already expected from the previous figure. RTA has a much greater bias toward intrinsic energy, which complies with the strong attenuation of the dipole signal seen above. It is important to note that the strong conversion to intrinsic energy starts early, already during the (rather short) pulse. This explains why RTA can suppress so effectively the induced emission seen in pure TDLDA (see discussion above). Note that the kinetic energy from collective flow is very small in that case and restricted to the early phase in both cases.

4.1.1.4 Field Enhancement Effects

It is known already from classical electrodynamics that external electrical fields in contact with matter induce polarization effects which yield finally a different effective field [Jac62]. The high polarizability of metallic systems can lead to large field amplification factor, particularly when driven near the surface plasmon resonance frequency. This effect is well known from metal clusters [RS98] and has found meanwhile several practical applications [SFY05, GZS+06, DPH21, PNDH21]. We illustrate the effect here with TDLDA simulations for the Na_{93}^+ cluster. To measure the effective field, we build the dynamical part of the Kohn-Sham (KS) field by subtracting the static KS field as irrelevant background. The derivative of the dynamical KS field is the searched effective field strength and the ratio to the external photon field amplitude E_0 then constitutes the actual field amplification factor. This quantity varies in time. We are interested on the peak effect and thus take the maximum value appearing some time. The ampification factor thus reads for the field along laser polarization axis z

$$\eta_{ampl}(z) = \frac{1}{E_0} \max_t \left[\frac{\partial (U_{KS}(\mathbf{r},t) - U_{KS}(\mathbf{r},t=0))}{\partial z} \right] \tag{4.1}$$

Figure 4.5 shows the result for the local field amplification factor η_{ampl} (Eq. 4.1) for various laser conditions. The test case is the metal cluster Na_{93}^+ because metals have the largest dynamical polarizabilities, best suited for exhibiting field amplification. We use the jellium model (section 2.1.1) for the ionic background to avoid field fluctuations from detailed ion structures, thus better focusing on field effects. The left panel shows the electronic ground state density distribution in one particular direction. We note that it looks the same in all directions because the system is spherically symmetric. The faint vertical line indicates approximately the surface region of the cluster. The right panel shows the amplification factor η_{ampl} (Eq. 4.1) along laser

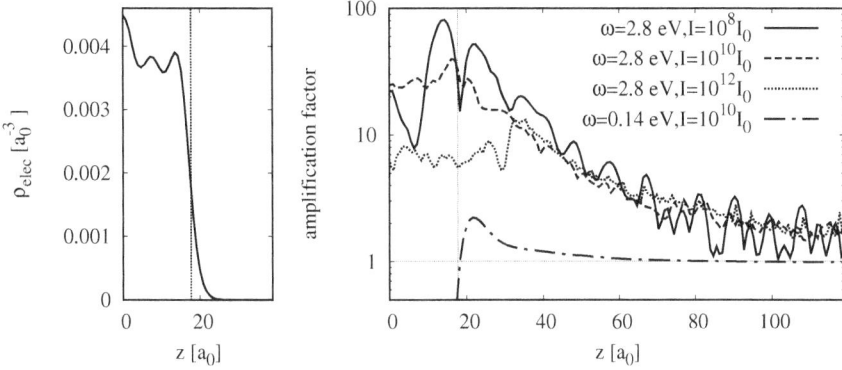

Figure 4.5 Left panel: Electron density distribution $\rho_{\mathrm{elec}}(z)$ along the z-axis for the ground state of Na_{93}^+ computed with soft spherical jellium background (Wigner-Seitz radius $r_s = 4$ a_0, surface thickness $\sigma = 1$ a_0). The approximate half-density radius is indicated by a vertical dashed line. Right panel: Local field amplification factor η_{ampl} according to Eq. (4.1) along laser field axis (z-axis) for laser pulses of total length $T_{\mathrm{pulse}} = 100fs$ with frequency and intensity as indicated. The reference intensity is $I_0 = 1$ W/cm^2. All results were computed with TDLDA+ADSIC using the QDD code with standard parameters [DVC$^+$22].

polarization axis (z-axis). The very low frequency $\hbar\omega = 0.14$ eV is close to static polarizability. There is a moderate factor two enhancement close to the surface region.

The situation is much different for a laser with frequency near the plasmon resonance $\hbar\omega = 2.8$ eV. The high collectivity of the Mie surface plasmon produces a large dynamical polarizability which, in turn, leads to huge field amplification factors, up to factors of about 100. Amplification is, of course, largest close to the cluster, but decays rather slowly such that it reaches out systems radii. The effect is largest for moderate intensities and becomes somewhat degraded for high intensities because the plasmon resonance is broadened.

Recall that polarizability grows with system size such that even larger amplification factors can be found for larger systems. This effect of field amplification has found many practical applications to generate extremely well localized, strong pulses in the near-field region of a metal cluster or tip [SFY05, GZS$^+$06, DPH21, PNDH21].

4.1.2 ANALYSIS OF IONIZATION

4.1.2.1 Global Analysis of Electron Emission

We start with the simplest, global observable of electron emission, the number of emitted electrons N_{esc} (section 3.4.4.3). Figure 4.6 collects results for the dimer Na_2 as an example. The left panels show the energy E_{abs} absorbed from

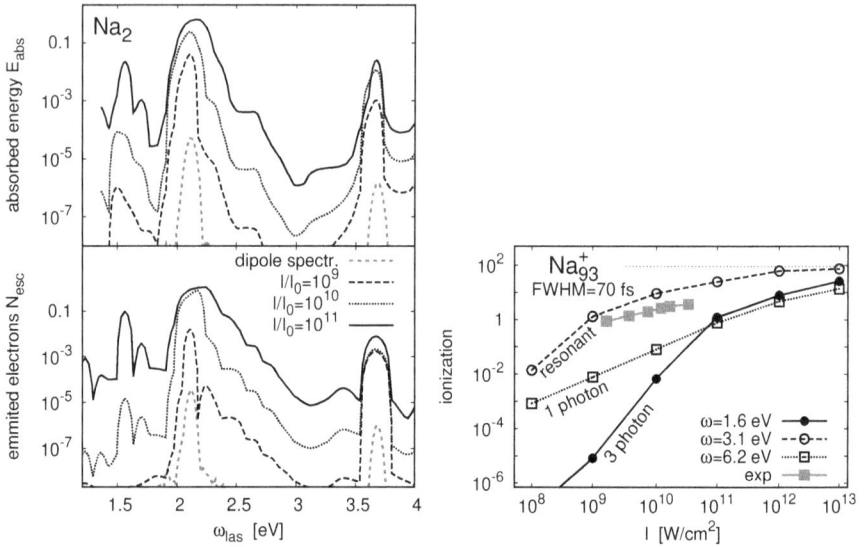

Figure 4.6 Lower left panel: Total ionization (average number of emitted electrons) for Na$_2$ after irradiation by laser pulses polarized along the cluster axis, with duration of 290 fs, as a function of laser frequency $\omega_{\rm las}$, for four different intensities as indicated and drawn in logarithmic scale. The dashed curve shows the optical absorption strength. Upper left panel: Energy absorbed from the laser field for the same cases as in the lower panel. Right panel: Average number of emitted electrons after ionization of the Na$_{93}^+$ cluster by a laser pulse with cos^2 profile and FWHM (Full Width at Half Maximum) of 70 fs drawn as function of laser intensity for three laser frequencies as indicated. The frequency 3.1 eV is slightly above and close to the plasmon resonance while the two other frequencies are off resonance. All results were computed with TDLDA+ADSIC using the QDD code with standard parameters [DVC$^+$22]. Experimental data from [SKvIH01].

the photon field (upper) and $N_{\rm esc}$ (lower) as a function of photon frequency. They are obtained running TDLDA for photon pulses with systematically varied frequency $\omega_{\rm las}$, but the same pulse length and intensity, and recording final values of $E_{\rm abs}$ and $N_{\rm esc}$. This is exactly the same strategy as in an experiment where photon frequencies are scanned and the reaction of the system is recorded. The $E_{\rm abs}$ in the regime of weak pulses is a measure for the photo-absorption spectrum. The $N_{\rm esc}$ measures the average photo-electron yield. For higher intensities, the profile of the curves is changed by non-linear effects. This is particularly well visible at the dominant peak around 2.1 eV (out of which the Mie plasmon resonance develops with increasing system size). The peak becomes successively broader with increasing intensity and the average moves slightly toward higher frequencies. This is due to two effects which exist already at the level of TDLDA: first, charging of the system by electron loss up-shifts the resonance frequency and so stretches the spectrum toward higher frequencies, and second, strong fluctuations of the Kohn-Sham mean field broaden the resonance.

Both quantities, E_{abs} and N_{esc}, show very similar patterns, but not exactly the same. For example, the ratios between two curves at two different intensities differ for each peak. They are generally larger for N_{esc} and the more the lower the frequency. This is due to the photo-emission threshold (also coined Ionization Potential, IP), which lies at around 5.1 eV. For lasers with frequency in the 1.5 eV range, it then requires 4 photons to emit one electron. This has consequences for the trend of N_{esc} with pulse intensity.

The right panel of Figure 4.6 exemplifies this trend of ionization with laser intensity on the test case Na_{93}^+ for three frequencies as indicated. The middle frequency 3.1 eV lies at the Mie plasmon resonance and the upper frequency is above emission threshold (IP=4.6 eV for Na_{93}^+). The lower frequency at 1.6 eV thus requires 3 photons to emit one electron. We can read this off from the trend $N_{esc}(I) \propto I^3$ in the figure. Above threshold, one photon suffices to emit one electron and accordingly, we find the trend $N_{esc}(I) \propto I^1$ for $\hbar\omega = 6.2$ eV. The resonance energy corresponds to emission as two-photon processes which we see here only between the two lowest intensities in the plot. Above $N_{esc} \approx 1$ the trend levels off until it reaches slowly the limit of full ionization. That is the turning point between frequency-dominated regime and field dominated regime [RS98] which is also responsible for the reduction of field amplification with increasing intensity as discussed in connection with Figure 4.5. The same bent to a slower increase of $N_{esc}(I)$ with increasing intensity I is also seen for the two other frequencies and it appears at the same critical value $N_{esc} \approx 1$. Note, however this critical emission applies to the present test case Na_{93}^+. It may change with system size and, more importantly, system constituents.

The experimental data in the right panel of Figure 4.6 stay already above the critical point and show the slower increase as typical there. It is satisfying that theory and experiment shows the same slope and only a slight mismatch in absolute yield. In fact, one cannot expect a perfect match because the time profile of the laser pulse differs from the theoretical modeling and the intensity may not be too well known due to inhomogeneities in the space profile. Moreover, dynamical correlations tend to turn part of the excitation energy to intrinsic heat, see section 4.1.1.3, which reduces direct electron emission somewhat.

We emphasize that N_{esc} is the average number of emitted electrons as mean-field calculations generally are describing average properties of one-body observables. Actually, the Kohn-Sham states allow also to compute the ionization stage of each s.p. state in terms of the actual electron content $n_i(t)$ of electron state i, see section 3.4.4.3. The complement is the depletion $1 - n_i(t)$ of the state. An example for Na_{93}^+ is shown in the upper panel of Figure 4.7. The laser frequency of 6.2 eV is rather high. Thus the distribution of depletions reaches far toward lower levels. Higher frequencies would deplete even deeper down and low frequencies remove only electrons near the HOMO.

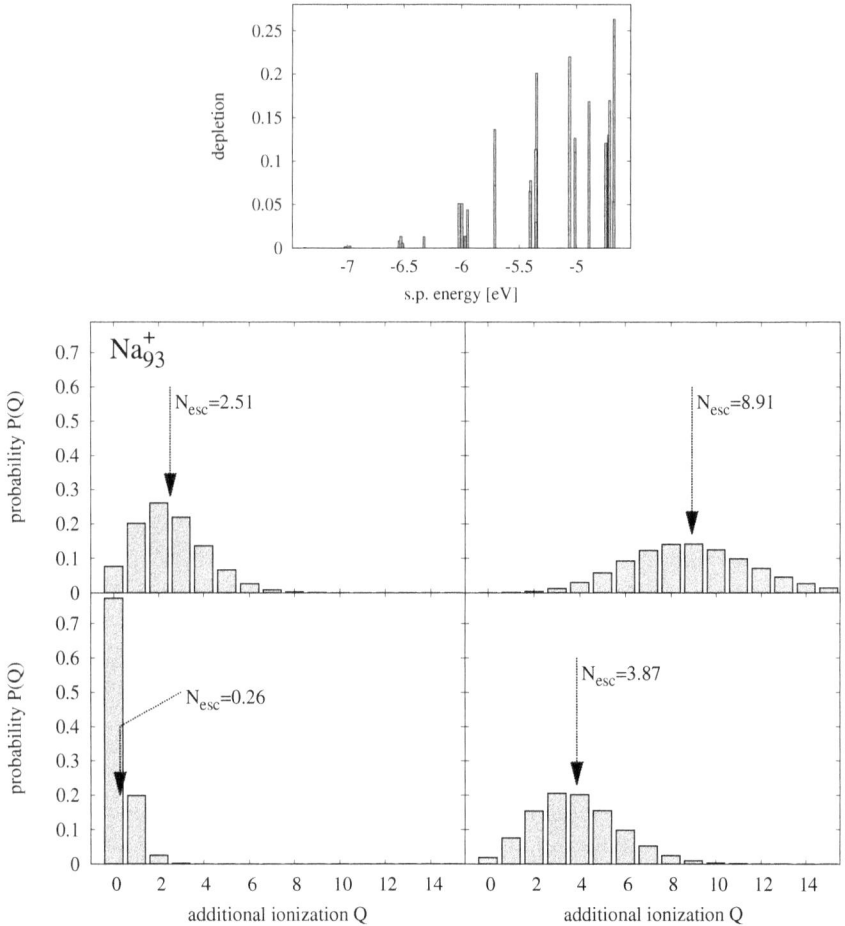

Figure 4.7 Detailed ionization observables for Na_{93}^+ after irradiation with laser pulses of frequency $\hbar\omega = 6.2$ eV and overall pulse-length of 200 fs. Upper panel: Final depletion $1 - n_i$, as defined in Eq. (3.21), for laser intensity $I = 10^{11}$ W/cm². Middle and lower panels: Probability $P(Q)$ to find the final ionization stage Q for intensities $I = 5\,10^{10}, 5\,10^{11}, 8\,10^{11}, 2\,10^{12}$ W/cm² respectively leading to average electron emission $N_{\mathrm{esc}} = 0.26, 2.51, 3.87$ and 8.91 as indicated. All results were computed with TDLDA+ADSIC using the QDD code with standard parameters [DVC⁺22].

The information on the detailed n_i allows to deduce by some combinatorial analysis the probability $P(Q)$ to find a certain total ionization stage Q as outlined in section 3.4.4.3. The result of the test case for Na_{93}^+ irradiated by a photon pulse with frequency 6.2 eV with various intensities (thus various values of N_{esc}) is shown in Figure 4.7. It illustrates nicely the multitude of ionization stages for one given electron emission N_{esc}. Not surprisingly, the higher the net (average) total ionization the larger the attainable charge states. Furthermore, the larger N_{esc}, the broader the distribution.

4.1.2.2 PES and PAD in The One Photon Regime

We now turn to the more involved, but also much more informative observables of Photo-Electron Spectra (PES) (section 3.4.4.5), Photo-electron Angular Distributions (PAD) (section 3.4.4.4), and both in combination as PES-PAD. Figure 4.8 shows experimental and theoretical PES-PAD for the cluster anion Na_{58}^{-}. The PES corresponds to angular averaged PES-PAD and is shown

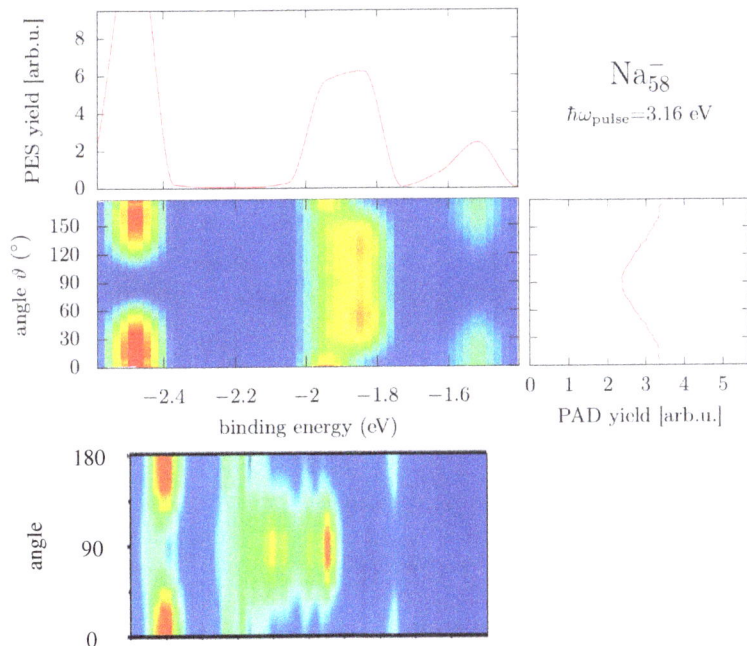

Figure 4.8 Lower panel: Experimental results for combined PES-PAD after laser excitation of Na_{58}^{-} with a strong photon pulse of frequency 2.5 eV (adapted from Bartels et al [BHH$^+$09]). The energies are down-shifted by the photon energy to indicate the binding energy of the s.p. states from which the structures originate. The energy scale is the same as for the theoretical PES-PAD right above the experimental one. Middle left panel: Theoretical results for combined PES-PAD of Na_{58}^{-} after a pulse with frequency 3.16 eV, again with the energies down-shifted by the photon energy. The ionic background was approximated by spherical jellium. The results were computed with TDLDA+ADSIC using the QDD code with standard parameters [DVC$^+$22]. Middle right panel: The computed PAD as such. Upper panel: The total computed PES obtained by angular integration of the PAD part.

above the theoretical PES-PAD. The PAD, on the other hand, is obtained from integrating PES-PAD over kinetic energies and is shown to right of the theoretical PES-PAD.

The theoretical calculation was done with a slightly larger photon energy to lift the deepest bound states safely above emission threshold to avoid the region of low kinetic energies where the absorbing boundary conditions are still reflecting some electron flow [RSA$^+$06]. But the results can be compared in spite of the different photon energies because we look at the PES-PAD in terms of of s.p. binding energies by using relation (4.2). It is satisfying to see that theory produces the same structures as data, though not perfectly at the same s.p. energies. Theory allows to tell the nature of the s.p. states. The uppermost peak stems from a 2d state, below that comes the group of 1g states, and the lowest is a 2p state. The 1g states are much more bundled in the calculations due to some extent to the spherical jellium approximation. The experiment, on the other hand, was done at finite temperature and embraces a thermal mix of configurations.

The angular distributions for the separate s.p. states in PES-PAD provide detailed and useful information which is lost in PES and PAD alone, although more state-specific information can also be retrieved from PAD when having it as a function of photon frequency [PRS04]. It is to be noted that the angular pattern in PES-PAD does not allow to conclude directly on the underlying angular structure of the emitting state. This requires support from theoretical calculations.

The PES alone allows immediately to conclude on the s.p. energies of the emitting states. Given a peak at some energy E_{kin}, we associate it to a s.p. state at energy

$$\varepsilon_i = E_{\text{kin}} - \hbar\omega_{\text{las}} \tag{4.2}$$

as long as we deal with one-photon emission processes, i.e. for photon energies $\hbar\omega_{\text{las}}$ larger than all relevant s.p. energies. Multi-photon processes produce pattern of their own in PES-PAD and PES. This will be discussed later in section 4.1.2.3.

Looking deeper into PES from one-photon processes, one finds sometimes deviations from simple association of the peaks in PES with the s.p. energies in the ground state of the emitting system, particularly for deep lying states and/or loosely bound systems as, e.g, cluster anions. This is demonstrated for the extremely sensitive case of Na_7^- in Figure 4.9. The positions of the experimental peaks (lower panel) for the lowest bound states agree still acceptably well with TDLDA (upper panel), although the HOMO-1 is fragmented into two peaks in data, but only one peak in TDLDA. A marked difference shows up for the deep-lying state. Data place it less deep and exhibit a broad structure (lower panel) while TDLDA shows only one strong peak at slightly deeper s.p. energy and no further fragmentation whatsoever.

The discrepancy may be resolved when seeing PES from the finally remaining cluster with charge state enhanced by one unit as compared to the initial

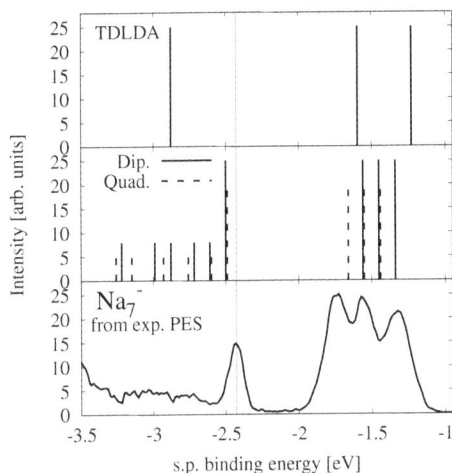

Figure 4.9 Lower panel: Measured PES for Na_7^- [MHH$^+$03]. Upper panel: Computed PES from Kohn-Sham calculations with ADSIC to deal with correct s.p. separation energies. Middle panel: Theoretical result deduced from the excitation spectrum of neutral Na_7 [MK07].

ground state [MK07]. The middle panel shows the dipole and quadrupole excitation spectrum of neutral Na_7 computed with TDLDA about the ground state of Na_7 (not about Na_7^- as in the upper panel) and subsequent spectral analysis (identifying "binding energy" with excitation energy). That reproduces the fragmentation structure of the lowest bound group of states, it places the deep-lying state correctly, and it shows appropriately a dense series of highly excited states which complies nicely with the broad structure in data.

What is then the reason that TDLDA plus explicit photon excitation from the ground state of Na_7^- missed some structures? The answer for the deep lying states is found by counting the particle-hole structure (assuming weak excitations). TDLDA from Na_7^- produces a $1ph$ state about Na_7^- from which the particle state escapes. This leaves a spectrum of one-hole state as seen in the upper panel and with one-hole energies $\varepsilon_i = E_{kin} - \hbar\omega_{las}$. TDLDA about neutral Na_7 produces a $1ph$ state about Na_7 which is a $1p2h$ (one-particle two-hole) state about Na_7^-. This is already a structure beyond a mean-field excitation when viewed from the starting point Na_7^-. Such correlated structures appear typically at higher energies, i.e., for holes in deeper bound states. This indicates that a TDLDA description of PES in the one-photon regime is likely to underestimate fragmentation broadening of deep-lying states.

As discussed in section 3.4.4.4 evaluation of PAD in gas phase requires an orientation averaging which has to be taken into account for a proper comparison between experiments and theory. This requires piling up computations

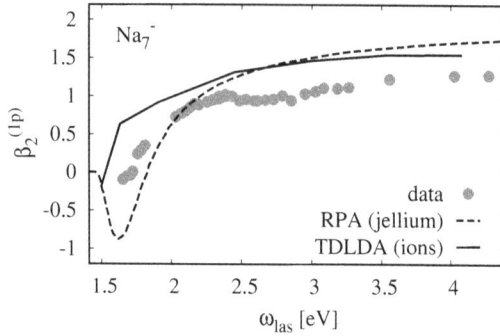

Figure 4.10 Anisotropy β_2 versus laser frequency for Na_7^-. A comparison is done between experiments [BHH$^+$09], results of RPA (section 2.2.4.1) calculations with spherical jellium background [SPI10] and TDLDA calculations with detailed ionic structure. See [WDSR12] for details.

performed varying the orientation of the target system with respect to laser polarization.

In the one-photon regime one can furthermore compactify the PAD content into one single quantity called the anisotropy parameter β_2 (see Eq. (3.23)). This allows to study in a simple and compact way the evolution of PAD with laser parameters. An example of such a study is presented in Figure 4.10 where β_2 is plotted as a function of laser frequency in the small anionic cluster Na_7^-. Figure 4.10 provides a comparison between experimental β_2 and the ones obtained by a full real-time TDLDA calculation (with detailed ionic background) and an estimate at RPA level (with spherical jellium background). The latter approach is justified in that case of one-photon processes in the perturbative regime. The TDLDA, in turn, would allow to address situations at much higher excitation energies. Note that the anion Na_7^- is a demanding test case, as it has a vanishingly small ionization potential because of its extra electron. This requires huge computational boxes to properly resolve the long range tail of the HOMO wavefunction. The comparison between both theories and experiment is fair with a good reproduction, to a large extent quantitative, of the trend of β_2 with laser frequency. The calculation with ionic background reproduces the experimental trend, but slightly over-estimates anisotropy which could be due to missing correlations. The calculation with jellium background is comparable in order of magnitude, but differs in the trend at low energies. Here we can see the impact of ionic structure.

4.1.2.3   PES and PAD in The Multi-Photon Regime

The case of Multi-Photon Ionization (MPI) enters a domain where dynamics induces more and more non-linear effects. Analysis of ionization then exhibits even more complex pattern. Figure 4.11 exemplifies PES-PAD for MPI in the

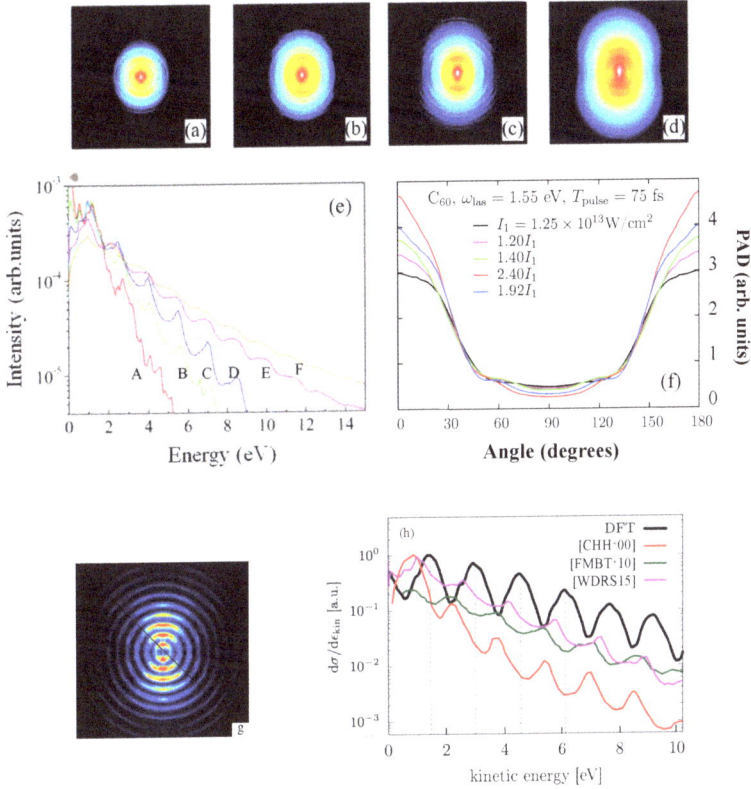

Figure 4.11 Panels (a) to (e): Experimental results taken from [HCC⁺13]. Upper panels (ad): Experimental photo-electron momentum distributions (VMI) from the ionization of C_{60} with a 30-fs IR laser pulse of various intensities : (a) 1.1×10^{12} W/cm^2, (b) 9×10^{12} W/cm^2, (c) 2.8×10^{13} W/cm^2, and (d) $\times 10^{14}$ W/cm^2. Panel (e): The angle-integrated photo-electron spectra for intensities of (A) 1.1×10^{12} W/cm^2, (B) 9×10^{12} W/cm^2, (C) 2×10^{13} W/cm^2, (D) 2.8×10^{13} W/cm^2, (E) 6.7×10^{13} W/cm^2 and (F) $\times 10^{14}$ W/cm^2. Panels (f) to (h): Results from TDLDA+ADSIC calculations using the QDD code [DVC⁺22]. The basic pulse parameters are $I = 1.25 \times 10^{13}$ w/cm^2, $\hbar\omega_{las} = 1.55$ eV, and FWHM=20 fs. Panel (f) shows PAD from C_{60} after irradiation with a photon pulse of various intensities in the $\times 10^{13}$ W/cm^2 range as indicated. Panel (g) shows VMI distributions in similar fashion as the experimental plots in panels (a)-(d). Panel (h) shows the PES in comparison with three experimental results. Red line: for pulse parameters $I = (8 \pm 2) \times 10^{13}$ w/cm^2, $\hbar\omega_{las} = 1.57$ eV, FWHM= 25 fs [CHH⁺00]. Green line: for pulse parameters $I = 10^{13}$ w/cm^2, $\hbar\omega_{las} = 1.55$ eV, FWHM= 60 fs [FMBT⁺10]. Purple line: for pulse parameters $I \simeq 1.25 \times 10^{13}$ W/cm^2, $\hbar\omega_{las} = 1.55$ eV, $T_{pulse} = 60$ fs [WDRS15].

case of the C_{60} cluster, comparing experimental and theoretical results. The upper panels ((a) to (d)) show experimental VMI (Velocity Map Imaging, section 1.2.3), thus in polar coordinates (energy in radial direction and angle in angular direction), for a 30 fs IR laser at 4 intensities between 10^{12} and 10^{14}W.cm^{-2}. The angular distributions start out radially symmetric at low electron energies and become increasingly forward peaked with increasing energy. Furthermore, the VMI images become more and more elongated along laser polarization with increasing laser intensity which reflects an increasing alignment of electronic emission along laser polarization. This trend is confirmed in panel (f) by the PAD from TDLDA calculations at step-wise increased laser intensity. The VMI in panels (a)-(d) exhibit fringes of maxima along the radial velocity direction. This trend is better visible in the angular integrated results yielding the PES. The latter is shown in panel (e) for different laser intensities. A typical ATI (Above Threshold Ionization) pattern is observed. When the light intensity increases, the peak-to-valley contrast of the peaks is reduced and the slope becomes smaller indicating that higher intensities produce more high-energy particles. Particular good contrast is seen for medium intensities. The maxima follow an equidistant series with an energy difference 1.5 eV corresponding to the laser frequency.

Panel (h) shows the PES from the calculations (black line labeled TDLDA) in comparison with a couple of experimental results using more or less similar laser conditions. They all have the same basic structure of a regular sequence of broad, single peaks. Slope and background differ a bit because detailed experimental conditions (cluster temperature, spatial and temporal pulse width, frequency, ...) slightly differ, though not affecting the basic pattern. The TDLDA calculations allow to explain these [WGB+15]. The sequence of peaks from multi-photon emission follows the rule

$$E_{\text{peak},i\nu} = \varepsilon_i + \nu\hbar\omega_{\text{las}} \tag{4.3}$$

where ε_i are the s.p. energies of the initially occupied states and ν is the number of photons. The above formula holds true for small net ionization. In the case of large ionization one has to account for the fact that net ionization downshifts ε_i which leads to a red shift of PES peaks.

In the case of C_{60} shown in Figure 4.11 there is one surprise in the pronounced pattern of the PES. Thinking of a spectrum of the 240 valence electrons in C_{60} and many different ν with small energy spacing of $\hbar\omega_{\text{las}} = 1.55$ eV, one would expect a real mess in the PES while Figure 4.11 (panels (e) and (h)) shows a nice and orderly distribution. One part of the answer is the observation that low-frequency photons remove electrons only from weakly bound states [VWD+10], in the present case from HOMO, HOMO-1, and HOMO-2. Still, three levels in combination with different photon numbers ν could look rather irregular. The second, and crucial, part of the answer is that the level distance of HOMO, HOMO-1, and HOMO-2 complies very well with the photon energy 1.55 eV such that $E_{\text{peak,HOMO}\nu} \approx E_{\text{peak,HOMO}-1\nu+1} \approx E_{\text{peak,HOMO}-2\nu+2}$

form one peak giving rise to the impressive regular sequence seen in panel (h) of Figure 4.11.

Finally, panel (g) shows VMI from the TDLDA calculations. The intensity $I \simeq 1.25 \times 10^{13}$ W/cm^2 used in TDLDA comes closest to the case in panel (c) which, indeed, looks similar. Again, we see that theory produces better contrast between the fringes (see also panel (h)). Ionic motion, thermal effects, and ensemble average tend to smooth the PES while the present TDLDA calculations were done with fixed ionic configuration at zero temperature.

4.1.2.4   Electronic Temperature Effects on PES and PAD

Electronic emission requires energy to be pumped from the deposited excitation energy. There is thus a natural competition between direct emission and energy rearrangement inside the system. In the medium to long term, the electron cloud will furthermore thermalize, as already discussed in section 4.1.1.3. Indeed, as we have seen above (e.g., Figure 4.4), any excitation leaves some internal thermal energy in the system which also leads to electron emission, however, with large delays of nanoseconds and more.

Emission from a thermalized source has the specific feature that it is isotropic. Direct emission, following, for example, laser irradiation, bears a trace of the direction of the perturbing field, for example, the laser polarization axis, and is thus partly directional. It is thus expected that emission signals contain both components, a directional one and an isotropic one, the relative weights thereof depending on details of the excitation. The mixture of isotropic and anisotropic components is clearly most simply visible on the PAD but the shape of the PES (exponential slope) may also provide an indicator of thermalization. It should finally be noted that directional and isotropic emissions do not occur on the same temporal range while experiments collect all emitted electrons without distinguishing time scales.

Figure 4.12 shows the PES of a C$_{60}$ irradiated by lasers of various frequencies but tuned such that only one photon is absorbed by each C$_{60}$ (see [HRA$^+$17] and the followup investigation in [ASR$^+$23]). The case is thus illustrating one-photon processes (section 4.1.2.2). The interesting feature of this systematic investigation is the appearance, for all laser frequencies, of a strong low-energy peak in the PES (gray shaded area). A more detailed analysis (including also the PAD) points out toward a thermal origin of this structure. In addition, energetic considerations support the idea that only part of the electron cloud has been thermalized. One thus sees sort of a hot spot. The case is nevertheless mixed, exhibiting sort of an intermediate stage of full thermalization. In any case it points out a thermal component in the emission.

Electronic temperature effects can also be seen in the case of multi photon processes (section 4.1.2.3) as illustrated in Figure 4.13 in terms of PAD. Indeed thermalization means that the information on the initial laser polarization direction is, at least partially, lost and thermal emission is then isotropic.

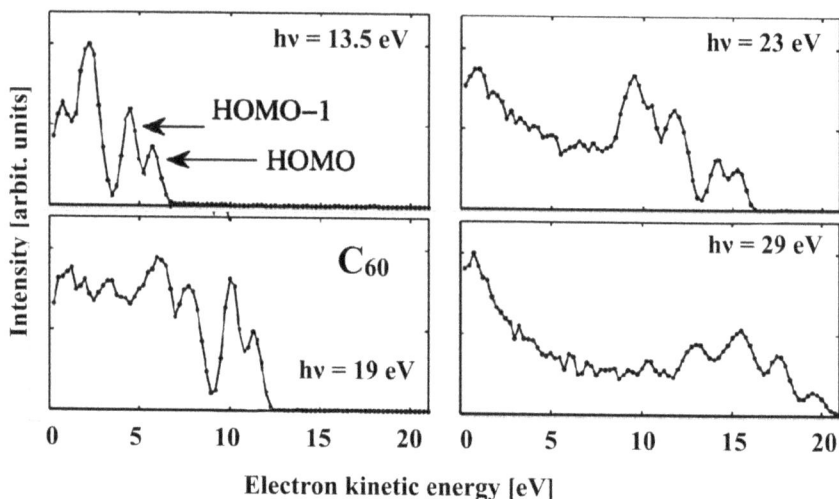

Figure 4.12 PES in C_{60} irradiated by lasers of various frequencies (as indicated). Other laser characteristics have been tuned such that only one photon is absorbed by the target, so that one is in the one-photon regime. Adapted from [HRA$^+$17].

The emerging PAD thus contains contributions from direct (dominantly directional) and from thermal emission, the latter delivering an isotropic background to the PAD.

This mixing of contributions is demonstrated in Figure 4.13. The TDLDA result is much more forward peaked than the data (factor about 2 for the

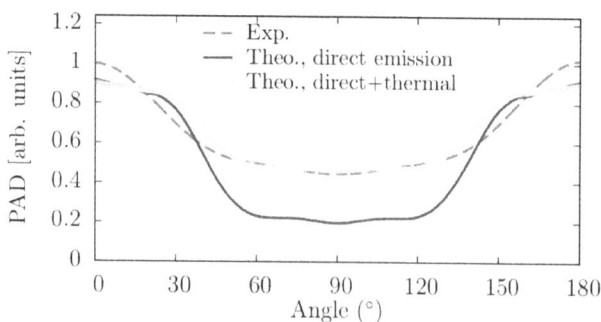

Figure 4.13 Experimental PAD from HOMO and HOMO-1 states (dashed) of C_{60} in the same multi-photon regime as in Figure 4.11, compared to total theoretical PAD from TDLDA+ADSIC (black) and from TDLDA plus contributions from thermal electron emission (grey). From [WGB$^+$15].

ratio of forward to sideward emission). To estimate the thermal background, we have recorded the remaining intrinsic energy left over after excitation and cooling by direct emission, assumed full thermalization of the remaining intrinsic energy, and converted all thermal energy into isotropic electron emission (ignoring the competing channel of monomer emission which is very weak for the tightly bound fullerene structure). The result shows a very satisfying agreement with data. This example demonstrates once again that TDLDA delivers a pertinent description of the early phases of irradiation dynamics, but misses dissipation at longer simulation times.

4.1.2.5 The Complexity and Richness of PES and PAD

In the one-photon regime (section 4.1.2.2) we have seen that the PES delivers a printout of the sequence of single particle (s.p.) energies from the occupied states, following the rule Eq. (4.2). This is a clean case in which interpretation of the PES is direct and gives access to structure properties. But PES do also allow to access dynamical features for example in the multi-photon regime (section 4.1.2.3), where the PES show successive copies of the s.p. spectrum (Eq. (4.3)), dressed with dynamical features such as shifts due to ionization (section 4.1.2.3).

Simple laser setups are characterized by one well-defined photon frequency ω_{las}. This makes the rules Eqs. (4.2) and (4.3) simple enough to use and thus the PES rather easy to interpret, even if, at high laser intensity, one should also take care of the ponderomotive potential U_p representing the average kinetic energy of a free electron in the laser field. For the sake of simplicity, we shall discuss here only cases for which U_p remains negligible (section 3.3.2).

The advent of a large variety of coherent light sources allows to access more complex dynamical scenarios, using several frequencies and/or pulse combinations. Pump and probe setups (section 3.2.1) are a typical example thereof (see for instance section 1.1). These setups provide a time-resolved access to dynamics, at ionic pace with femtosecond pulses (section 1.1.1) down to electronic pace with attosecond (as) pulses (section 1.1.2). In such cases, though, while the PES remains a highly valuable tool of investigation of the dynamics via ionization characteristics, its analysis in simple terms (Eqs. (4.2) and (4.3)) has to be modified. For instance, in a pump-and-probe experiment with two different laser colors, one has to consider linear combinations of the two laser frequencies to match the PES peaks, which may become very complicated in the multi-photon regime. In fact one has to take a wider perspective to understand the PES in the course of such irradiation processes. One must integrate into the analysis other mechanisms of emission such as for example plasmon enhanced ionization (Figure 4.6). Clearly such a strategy may quickly become impossible to control. But it points out an important physical mechanism: the fact that the PES, reflecting the full dynamical response, is to be interpreted on the basis of the full response of the system, including dipole and other multipole responses.

One is thus bound to generalize the simple rule of analysis given by Eqs. (4.2) and (4.3) to include contributions from the electronic response in its dominant channels, typically accessed as the first moments of the electronic cloud (monopole, dipole, and quadrupole). Altogether this delivers an expression

$$\mathcal{Y}(E_{\text{kin}}) \leftrightarrow \sum_i [\eta_M P_M(E_{\text{kin}} - \varepsilon_i) + \eta_D P_D(E_{\text{kin}} - \varepsilon_i)$$
$$+ \eta_Q P_Q(E_{\text{kin}} - \varepsilon_i) + ...] \tag{4.4}$$

where weights η's will depend on the actual situation, and where P_X's are the responses of the first moments of the electronic cloud. Using at least the quadrupole is for example compulsory to understand some pump probe scenarios on the basis of an IR pump followed by an XUV train [GDRS15]. Of course the generalized expression Eq. (4.4) contains the rules Eqs. (4.2) and (4.3) as special cases for which the response is dominated by a simple laser dipole $P_D(\omega) \propto \delta(\omega)$. All in all, one can thus conclude that the PES reflects the typical frequencies present in the system, either those delivered from outside by a laser field and/or intrinsic ones and/or possibly any combination thereof which makes it both rich and complex.

Equation (4.4) is interesting as it gives access to a wide range of scenarios, including situations without lasers. Indeed collisions with highly charged swift ions also lead to energy-resolved ionization spectra, which have the same form as standard PES. The effect of a charged projectile is to deliver a short electromagnetic pulse covering a broad band of frequencies, which thus may be seen as a photon of mixed frequencies. It turns out that the analysis of the PES obtained in such fast collisions can also reveal structures which can be attributed to eigenfrequencies of the system. An example is shown in the left panel of Figure 4.14 in the case of C_{60} irradiated by a Si^{12+} projectile of energy 3.25 MeV per nucleon. Experimental data are compared to TDLDA dipole and quadrupole responses. The experiment clearly reflects the multipole response. This is a typical example of application of the mechanisms contained in Eq. (4.4) but now visible in a collisional scenario. The right panel of Figure 4.14 shows corresponding PAD recorded at various electron energy. The PAD is fitted to a multipole expansion in the spirit of Eq. (3.23) but now including the next term in the expansion, namely the quadrupole. Agreement between experimental data and fit is very good but only once one includes the quadrupole component, while the pure dipole fit (not shown on the figure) fails badly to reproduce the dips around 45–60° and 120–135°. Both PAD and "PES" thus clearly demonstrate the excitation of collective multipole modes in this irradiation scenario by a charged projectile.

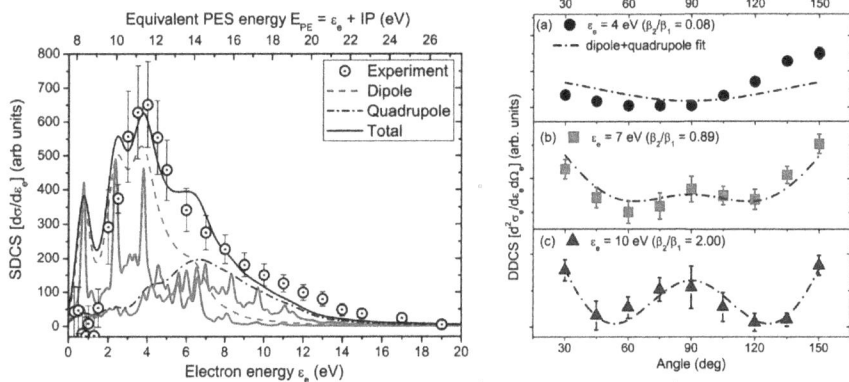

Figure 4.14 Impact of multipole modes on Single Differential Cross Section (SDCS, equivalent to PES, see upper scale, left panel) and PAD (right panel). A C_{60} has been irradiated by a Si^{12+} projectile of energy 3.25 MeV per nucleon. Left panel. TDDFT results after pure dipole or quadrupole excitation are shown in gray line. They have been convoluted (dashed for dipole, dotted-dashed for quadrupole) with detector resolution for better comparison with experiments. Right panel. PAD for a few selected electron energies. Experimental results have been fitted with a Legendre polynomial expansion including dipole and quadrupole components. Adapted from [KGD$^+$22].

4.1.3 EXTREME LASER PULSES

The steady development of laser technology has paved the way to physics at femtosecond down to attosecond scale, high-pulse intensities, and high-photon energies in the XUV regime [KI09, CPHL$^+$17, NDC$^+$17]. Many new experiments come in reach with that, e.g., time resolved measurements at electronic scale, access to deep lying, or high energy deposit in atoms and small molecules. We address here a few examples out of this rich developing field, in complement to the example of section 1.1.2.

4.1.3.1 Short and Intense Pulses: Down to The Few Femtosecond Domain

The extremely short pulses create sensitivity to one pulse parameter which was negligible for long pulses, namely the phase of the fast oscillations relative to the pulse envelope, coined the Carrier Envelope Phase (CEP). This is illustrated in the left panels of Figure 4.15 which shows the pulse profiles for four pulses with two different pulse length and two different CEP. It is obvious for the shorter pulse (left upper panel) that the sequence of forward-backward pushes looks much different, depending on the CEP. Furthermore this occurs at the time scale of the electronic response for this system. The CEP thus

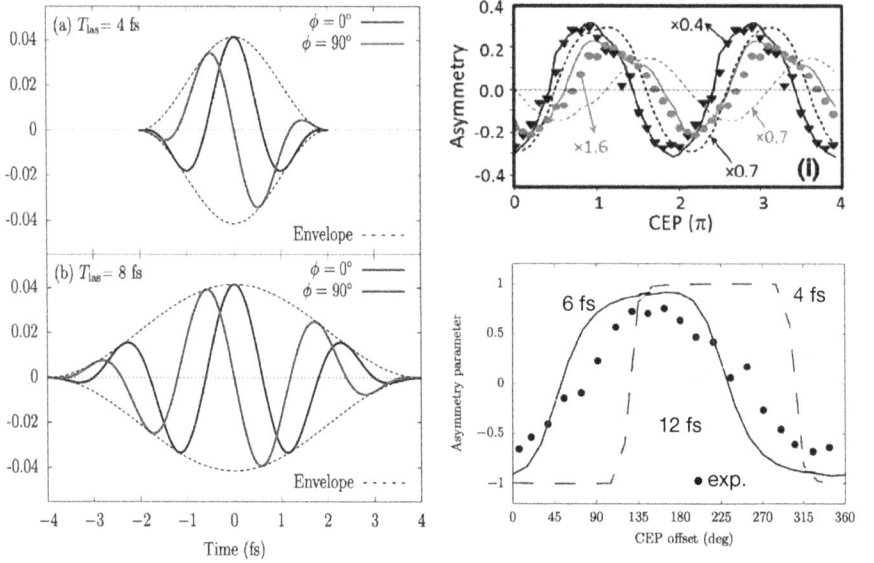

Figure 4.15 Illustration of the impact of the Carrier Envelope Phase (CEP) of the photon pulse. Test case is C_{60}. Theoretical results were computed with TDLDA+ADSIC using the QDD code with standard parameters [DVC+22]. Left panels: The pulse profiles for CEP $\phi = 0°$ and $90°$ for a 4 fs pulse (upper) and an 8 fs pulse (lower). The pulse envelope is indicated by dashed lines. Right lower panel: Asymmetry parameter η Eq.(4.5), computed at TDLDA level [GDR+17], as a function of the CEP for different pulse lengths compared with the experimental results [LMW+15]. Right upper panel (from [LMW+15]): Comparison between angle- and momentum-integrated asymmetries from experiment, for radial momenta p_r (perpendicular to spectrometer axis) for high-momentum electrons ($1.3 \leq p_r \leq 1.5$ $\hbar/a_0$, black triangles) and for low-momentum electrons ($0.4 \leq p_r \leq 0.6$ $\hbar/a_0$, grey circles), and two theoretical approaches, a TDSE (section 2.3.3) denoted QD (solid black and grey curves), and MC simulations (dotted black and grey curves). For better comparison in the oscillatory behavior, the experimental and MC curves were multiplied by the indicated factors.

lets us expect a dramatic effect on the direction of emission seen in the PAD. The difference is still visible for the somewhat longer pulse. But the envelop covers more oscillations which tends to average out the impact of CEP.

In order to investigate the effect as a function of the CEP, we need to compress the PAD to one relevant number. We expect an effect on the direction of emission. An appropriate observable is then the asymmetry [RSM18], which can be expressed from the yield $\mathcal{Y}(\theta)$ along a direction θ measured with respect to the laser polarization axis $\theta = 0$

$$\eta = \frac{\int_0^{\theta_0} d\theta \, (\mathcal{Y}(\theta) - \mathcal{Y}(\pi - \theta))}{\int_0^{\theta_0} d\theta \, (\mathcal{Y}(\theta) + \mathcal{Y}(\pi - \theta))}. \tag{4.5}$$

The angle θ_0 defines a cone along laser polarization axis in which the emission signal is gathered. In the right lower panel of Figure 4.15 $\theta_0 = \pi/12$ and η is shown as a function of CEP for a couple of different pulse lengths. The shortest pulse length (4 fs) delivers a sharp transition between regimes of forward and backward preferences (asymmetries). The transition regimes becomes broader with increasing pulse length and almost disappears already for 12 fs pulses. The experimental pulse length is not well known. But the data still show a pronounced forward-backward change with CEP. By comparison with TDLDA results, we can argue that the experimental pulse length comes close to 8 fs.

Alternatively, one can consider any solid angle aligned along laser polarization axis to gather emission and define η, for example, filtering transverse momenta. This is the way η has been measured (and computed) in the right upper panel of Figure 4.15. This comparison with theories is also instructive as both the approaches, TDSE (section 2.2.4.4), labeled QD in the figure, and the statistical treatment, labeled MC (section 2.3.8) provide a reasonable agreement with experiment. This indicates that the asymmetry is a robust signal, basically a protocol of collective flow driven by the external fields which obviously is reproduced properly in different approaches.

4.1.3.2   Short and Intense Pulses: Down to The Attosecond Domain

Laser technology has managed to deliver ever shorter photon pulses now reaching regularly the attosecond regime [KI09]. This opens a world of new possibilities for time resolved tracking of molecular reactions, for reviews, see e.g., [NDC+17, YUG+18]. The ultimate aim is, of course, to resolve electronic processes at their short time scale, around and below 1 fs. The motivation is the overwhelming success of pump-and-probe analysis of ionic dynamics [Zew94]. However, the time resolved analysis of electronic processes is much more demanding because electronic time scales are not that well separated as ionic ones from electronic times. Nonetheless, there exist already a couple of investigations of electronic dynamics which, however, are much more involved than conventional pump-and-probe analysis. The interpretation of the signals has usually to be supported by theoretical models which renders the measurements being somewhat indirect. But the short pulses bring also great progress in analyzing ionic dynamics in the realm of fast ionic processes.

We exemplify that here with an example from Ref. [TKGV+15], measuring the dissociation of the N_2 dimer after photo-ionization by a sub-300 attosecond pump pulse (covering photon frequencies between 16 and 50 eV) combined

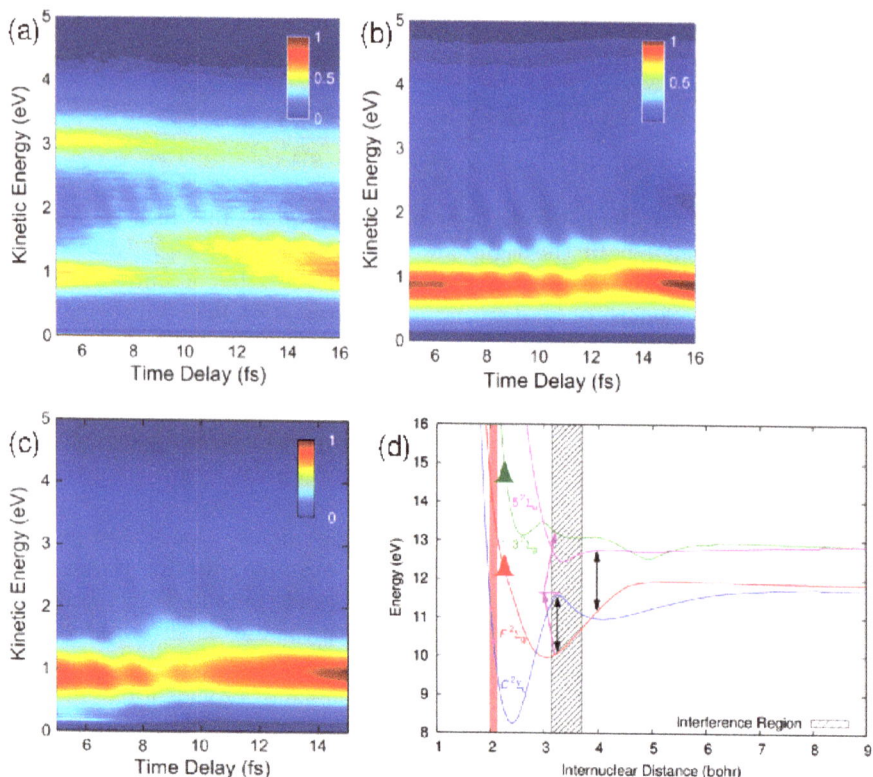

Figure 4.16 (a): Experimental kinetic energy spectra of the N^+ ion emerging from dissociation of N_2 triggered by a pump-probe excitation; spectra are drawn as a function of time delay. (b): Same as (a), but from a theoretical simulation with the TDSE. (c): Same as (a), but from simulating the dynamics in a simple four-states model as sketched in panel (d). (d): Energy curves as a function of inter-nuclear distance for the four electronic states of N_2^+ considered in the simple dynamic model. Adapted from [TKGV+15].

with a waveform-controlled sub-4 fs IR probe pulse. The three-dimensional angular and energy distribution of the emerging N^+ ions was measured with VMI for a dense series of delay times between XUV and IR pulse. Panel (a) of Figure 4.16 shows the kinetic energy spectrum (accumulated in a small cone along laser polarization axis) as function of time delay. The signal around energy of 1 eV shows interesting pattern: it disappears around 8 fs delay and around the same time one observes a fast modulation with periodicity of about 1.22 fs together with a tilt of the signals due to energy dependent phase shift. Panel (b) shows for comparison the result of a theoretical simulation with a TDSE coupled to ionic MD. The pattern and time scales are qualitatively reproduced. A deeper analysis of the underlying mechanisms could pin down four major players which are responsible for the effect. There are four electronic states whose energy curves are shown in panel (d). A model calculation describing the electronic dynamics within a finite Hamiltonian which uses as basis only these four electronic states (coupled to ionic motion) delivers the result in panel (c) which reproduces pattern and time scales as well as a full TDSE does. The mechanism can be read off from panel (d). The initial XUV pulse lifts molecule from the $C^2\Sigma_u$ ground state to the three relevant states in the model. At a time where the subsequent ionic expansion has reached the gray area in the plot, the 1 eV IR pulse can trigger transitions which kick the molecule out of its Born-Oppenheimer path. Transition back to the ground state configuration is seen as loss in the signal. Transition between excited state produces the oscillating pattern. The example demonstrates nicely the high information content of time resolved measurement, now possible at rather short scales. We also see that even this transparent case requires model calculations for a proper interpretation.

4.1.3.3 High Laser Frequencies and High Harmonic Generation (HHG)

Nonlinear response in photon-matter interaction can produce higher harmonics of the impinging photon frequency. This is known even for weak photon pulse as second or third harmonic generation depending on the symmetry of the material [New11]. A new feature comes up in strong photon fields: High Harmonic Generation (HHG). An electron is catched by the strong and fast oscillating photon field and carried around thus colliding several times forth and back with its mother ion. The many encounters add up coherently to high-frequency photon emission and/or high energy electrons. Figure 4.17 demonstrates HHG and generation of high-momentum electrons in the PES for an Ar atom side by side. Both signals show sort of a plateau reaching up

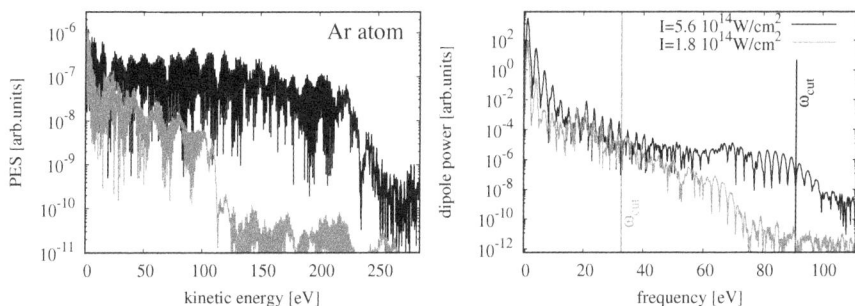

Figure 4.17 PES (left) and dipole power spectrum (right) from the Ar atom irradiated by a strong IR pulse with frequency $\hbar\omega_{\text{pulse}} = 1.4$ eV, $\sin^2$ envelope with overall pulse length $T_{\text{pulse}} = 24$ fs, and two different intensities as indicated. Both spectra were computed with TDLDA+ADSIC using the QDD code with standard parameters [DVC$^+$22].

to a limiting value at which the yield breaks down very quickly. This is more pronounced for the PES and somewhat optimistic for HHG. Therefore, the cutoff frequency for HHG, eq. (4.6), is indicated by a vertical dashed line in the plot of HHG (right panel).

The mechanisms beyond both phenomena is the same. The high frequencies, or momenta respectively, are produced by emitting electrons which are forced to recollide with the source by the strong electromagnetic field. The peak momentum and peak frequency are related to the maximal ponderomotive potential U_p from the driving IR field. The maximum HHG frequency, i.e. the frequency where the steep decrease starts, is driven by the ponderomotive potential Eq. (3.7) as

$$\hbar\omega_{\text{cut}} = E_{\text{IP}} + 3U_p \ , \ U_p = \frac{e^2 E_0^2}{4m_e\omega_{\text{phot}}^2} \ , \tag{4.6}$$

see [KSK92]. This estimate predicts $\hbar\omega_{\text{cut}} \approx 46$ eV for the higher intensity. But we find about 70 eV. Moreover, the lower intensity should show the cutoff at three times smaller energy which would be around 20 eV. The problem is here that the cutoff is not very pronounced. The cutoff becomes steeper for larger systems.

The cutoff energy in PES is given by $E_{\text{kin,max}} = 3.2U_p$ after first return and increases with cycles. We find the cutoff around 200 eV for the stronger intensity which corresponds to 4.4 cycles, well in accordance with the pulse structure. Moreover, the cutoff for the lower intensity stays about factor 3 below the higher intensity which complies with the proportionality to U_p in case of the same pulse profile.

It is known that the typical plateau in the HHG spectra becomes more pronounced for larger molecules. The transition down from the plateau is generally less steep for a single atom [FMBT$^+$10].

4.1.3.4 Toward Nanoplasmas: Strong Fields and Classical Modeling

Although TDLDA offers a very efficient, quantum mechanical approach to the electronic dynamics of molecules, it reaches its limit for very large system and/or high excitations. Large systems are demanding simply by the number of single-electron wavefunctions to be handled. High excitations go beyond a mean field approach because dynamical correlations become important. Fortunately, these are exactly the regimes which validate (semi-)classical approaches because large electron numbers and high excitation energies render shell effects unimportant [Bra93]. Closer to DFT is VUU (section 2.4.1) being valid already in intermediate regimes. Dynamical correlations are brought in here through electron-electron collisions carefully blocked to obey the Pauli principle (for details see section 2.4). This approach still becomes cumbersome for still larger systems and energies. Here we come into the domain of classical MD for electrons (see section 2.4) where particle-particle correlations are naturally included as in any MD simulation.

The classical regime is characterized as a situation where the (effective) temperature is comparable or larger than the relevant energy differences. Typically, the classical limit is well reached for the valence electrons, but by no means for the core electrons (which imply much larger energies). This is harmless if we can assume that the core electrons do not participate in the process. However, it happens often that strong fields also shake the core electrons. One of the outcomes of this is the mechanism of inner ionization (section 3.3.2) as already discussed in connection with Figure 1.1. The strong external field triggers ionization of a core electron which can end up either directly in the particle continuum (outer ionization) or in the range of the valence electrons becoming mobile throughout the molecule, but still bound to the system as a whole (inner ionization). A full quantum-mechanical treatment of core ionization is even more out of reach. But the process is much faster than time scales of valence electrons. This allows to treat it simply by transition probability in terms of creation rates for valence electrons from core ionization.

We continue the discussion around Figure 1.1 with another typical example for treating strong field dynamics. Figure 4.18 shows the theoretical results for the dynamics of a large Ar cluster under the impact of combined strong XUV and IR fields. The modeling is done in a combination of rate equations and classical MD for the electrons and ions as very often used in strong field dynamics [RPSWB97, SR03, Bau04, SSR06, AF10, APF14]. The release of electrons from core into valence space and the reverse process of electron re-capture back into bound core states is described by rate equations which are fed by the respective transition probabilities computed for the actual time-dependent external fields. The electrons thus lifted into valence space (and resisting re-capture) are then treated together with the ions by classical MD as described in section 2.4.2). Large clusters as the Ar_{3871} in Figure 4.18 are associated with several thousands of valence electrons which provides good enough statistics for dealing with rate equations. Smaller systems have to resort to a phase-space description of the electron distribution in terms of pseudo-particles, in practice then to the Vlasov equation. The immediate description of electron-electron correlations is lost in that approach and has to be added explicitly leading eventually to a VUU treatment (section 2.4.1).

The dynamics thus simulated are demonstrated in Figure 4.18. We start with panel (b) which shows the time evolution of ionization and cluster radius. Long after the XUV pulse is over, we see a steep increase of ionization and with considerable delay a subsequent increase of cluster radius. The insert shows a crucial detail, the initial stages of ionization in log scale to amplify the faint, but decisive, ionizing effect of the XUV pulse. There is not much charge release into valence space. But it delivers at least some handle for the IR pulse. Now we should look simultaneously at panel (c). The solid lines show the Mie frequency as estimated by Eq. (1.1) which is to be compared with the IR frequency indicated by the horizontal dashed line. The valence cloud is off resonance in the early phases. Still, the IR pulse shakes the valence

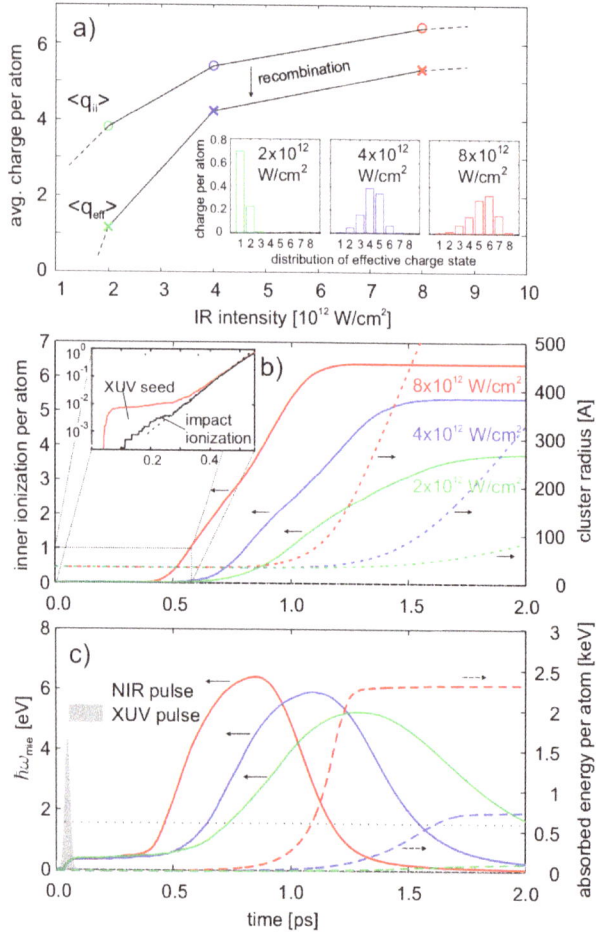

Figure 4.18 Results of a MD simulation of electronic and ionic dynamics of Ar_{3871} after irradiation by an initial XUV pulse having FWHM 30 fs, frequency $\hbar\omega_{XUV} = 20$ eV, and intensity $I_{XUV} = 2.5\,10^{10}$ W/cm^2 and a subsequent IR pulse with FWHM 1 ps, $\hbar\omega_{IR} = 1.55$ eV, and various intensities as indicated. The maximum of the IR pulse is reached about 1ps after the XUV pulse (see the shaded areas in panel c). Taken from [SAMB$^+$16]. Panel a: Final charge state from inner ionization $\langle q_{ii} \rangle$ and effective charge state $\langle q_{\mathrm{eff}} \rangle$ after electron-ion recombination. The insets show the distribution of effective charge states for the three IR intensities. Panel b: Time evolution of inner ionization (solid lines) and cluster r.m.s. radius (dashed lines) for three intensities as indicated. The inset shows the evolution for early times in log scale. Panel c: Time evolution of the estimated Mie frequency $\hbar\omega_{\mathrm{Mie}}$ of the nanoplasma (solid curves) and of total energy absorbed from the XUV+IR pulses. The dash-dotted horizontal line indicates the IR frequency. The shaded areas indicate the pulse envelopes.

cloud. This serves to release steadily more and more core electrons which, in turn, increases the Mie plasmon frequency. At a certain time, depending in the IR intensity, the Mie frequency comes into resonance with the IR pulse which gives rise to the steep increase in ionization seen in panel (b). Inner ionization is always accompanied by outer ionization such that charge state of the cluster as a whole grows in time. The thus increasing Coulomb pressure leads to cluster expansion as seen from the increasing cluster radius (dashed lines in panel (b)). This, in turn, reduces the Mie frequency and we see in panel (c) a second transition through resonance which delivers the dominant portion of energy absorption, dashed line in panel (c), because at that stage the electron cloud has grown largest. A summary of the final charges and distribution as function of IR intensity is shown in panel (a). Recombination is a small effect for large intensities, but develops to a severe hindrance the lower the intensity becomes. The inset shows the detailed charge distributions which resemble those from Figure 4.7 although the latter emerged from a much different process.

4.2   COUPLINGS TO IONS AND ENVIRONMENT

4.2.1   IONS AND ELECTRONS I: EHRENFEST DYNAMICS (TDLDA-MD)

In this section, we discuss the well-settled, simple approaches to the interplay of electrons and ions: first, the impact of ionic configurations on electronic properties, and second, the regime of electron-ion dynamics which can be treated within Ehrenfest dynamics, see section 2.5.3. The best examples have already been brought forward in earlier parts of this book. We do not want to repeat ourselves and thus we will use this section to illustrate basic principles on relatively simple examples.

4.2.1.1   Thermal Effects on Photo-Electron Spectra and Angular Distribution

Measurements on free molecules are usually done at some finite temperature. This is mostly negligible for rigid molecules, but can have important consequences for softer ones. We had seen that already for the optical absorption spectra of Na_{41}^+ cluster in Figure 4.1 where temperature effects had been incorporated to simulate the experimental situation. This large cluster has already a highly fragmented photo-absorption spectrum such that thermal ensemble mix just serves to smooth the fuzzy zero-temperature spectra.

We continue temperature studies here with a smaller cluster, Na_{11}^+, and for different observables, namely Photo-Electron Spectra (PES) and Photoelectron Angular Distributions (PAD). Technically, a thermal ensemble is generated from stochastic initialization of ionic velocities according to the wanted temperature and long TDLDA propagation of the thus excited system. We allow for an initial phase of thermalization and record after that ionic configurations in large time intervals to grab a representative ensemble of configurations. The wanted observables are computed by TDLDA for each sample

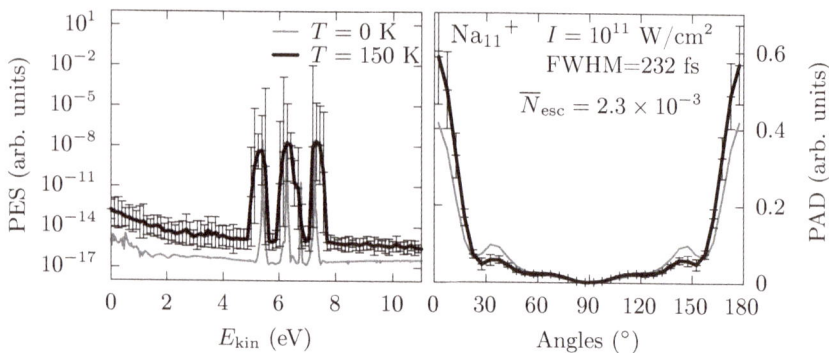

Figure 4.19 Photo-electron spectra (left) and photo-electron angular distribution (right) in Na_{11}^+ after irradiation by an XUV pulse with frequency $\hbar\omega_{las}=$ 13.6 eV, peak intensity $I = 10^{11}$ W/cm^2, and sin^2 envelope with FWHM of 232 fs. Compared calculations for 0 K (ground state) and for a thermally excited ensemble at 150 K. The excitation is comparatively weak and produces a total electron emission of 5.3×10^{-4} for 0 K. Data from [GDRS16], computed with TDLDA+ADSIC.

of this ensemble with frozen ionic positions. The results are finally averaged incoherently. This also yields the statistical fluctuations as the variance over the ensemble.

That strategy goes well beyond the optical spectra in Figure 4.1 and it was used to produce the PES and PAD in Figure 4.19 here. The variance of the results from statistical sampling is indicated by error bars. Comparing results at 0 and 150 K shows four major changes due to thermalization: 1.) a broadening of the PES peaks while leaving the peaks still distinct, 2.) global enhancement of the PES signal, 3.) a smoothing of the fluctuations in the PAD, and 4.) a stronger forward/backward orientation of the PAD.

Broadening of PES and smoothing of PAD is an expected and typical result of thermalization. Enhanced PES signal and enhanced forward signal in PAD is not obvious at first glance and requires more explanations. The mechanism is that some samples in the ensemble are elongated and have lower IP. This renders them particularly responsive to the given laser frequency and they thus emit considerably more electrons than the ground state configurations. This suffices to enhance the emission in the average and with it the global PES signal. Emission from an elongated cluster is also related to more pronounced forward/backward emission which, in turn, delivers the enhanced forward/backward peaks in the PAD. The example shows that temperature effects can be rather intricate. On the other hand, we see that the gross structure of PES and PAD remains valid even for a molecule as soft as a metal cluster.

4.2.1.2 Effect of Ionic Configuration on Photo-Absorption Spectra

Pump and probe setups are since decades powerful and versatile tools for time-resolved analysis of ionic dynamics in molecules [Zew94]. We have already discussed in this book typical examples for such measurements, see Figures 1.1 and 1.2. Many of the pump and probe scenarios are well tractable by TDLDA-MD, see e.g [ARS02, ARS04, PIT+12], as long as the corresponding ionic dynamics does not involve crossing APES (Adiabatic Potential Energy Surface) and diabolic points. The whole setup is involved, experimentally as well as in its theoretical simulations. We illustrate here a key ingredient for the pump and probe technique, namely the fact that the optical response of a molecule is highly sensitive to its structure. To that end, we consider Na_2 as the simplest example of a dimer molecule with a clean optical spectrum.

Figure 4.20 summarizes the necessary properties. The lower left panel shows the Born-Oppenheimer energy surface, the total energy as function of the distance of the two ions. It shows a nice, nearly quadratic shape around minimum. The lower right panel shows the photo-absorption strength at the ground state configuration, the molecular distance where the minimum in energy surface is found. There is one all dominant resonance peak which means that the spectrum can be characterized by one number, the peak frequency. Finally, the left upper panel shows the important result, the resonance peak

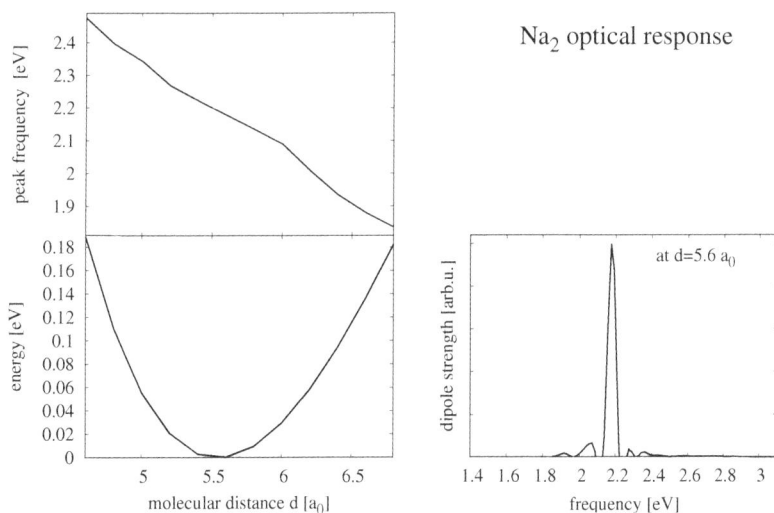

Figure 4.20 Observables in the Na_2 molecule computed with TDLDA+ADSIC using the QDD code with standard parameters [DVC+22]. Lower left: the Born-Oppenheimer energy surface, i.e. total energy relative to minimum as function of ion distance. Upper left: Frequency of the dominant dipole excitation as function of the ion distance. Lower right: The photo-absorption spectrum at the ionic ground-state configuration. The sattelite peaks stem from coupling to ionic motion.

Guidebook to Real Time Electron Dynamics

frequency as a function of molecular distance. This frequency depends sensitively and monotonously on the molecular distance. It is then clear that one can measure the molecular bond length through the optical response.

That is exactly the idea beyond pump and probe analysis. The pump pulse is designed to excite the molecule in a desired way. The probe pulse tracks the subsequent dynamics of the ionic configuration by measuring a key quantity sensitive to the optical response as, e.g., electron emission. The situation is, of course, more complex in complex molecules where the optical spectra are often less unique. This requires a careful choice of the appropriate working point. But the abundant literature on time-resolved analysis by pump and probe measurement proves that such hurdles can be tackled with.

Bio-molecules are especially prone to dramatic changes in ionic configuration, often at very low cost in energy. Accordingly, different configurations of the same molecule can exhibit very different photo-absorption spectra. We take here a crisp example from [CMV$^+$09]. Figure 4.21 shows the

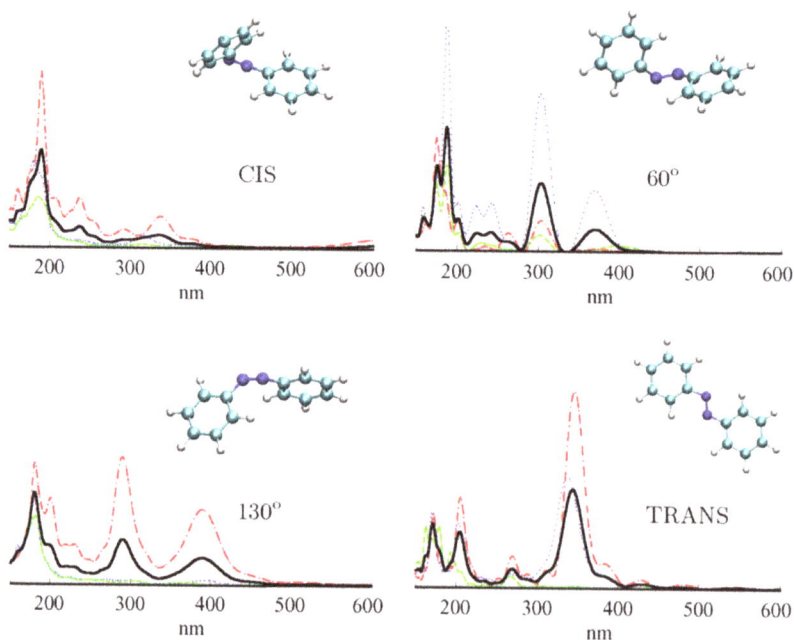

Figure 4.21 TDLDA results for the photo-absorption spectra of the isolated azobenzene chromophore, taken at four representative points along the isomerization path connecting the cis- with the trans-conformations. The corresponding ionic configurations are indicated in the inserts. The black curve shows the average absorption while the colored lines show the spectra along the three different spatial directions. From [CMV$^+$09].

optical absorption spectra of the azobenzene ($C_{12}H_{10}N_2$) chromophore for four ionic configurations along the transition from the cis configuration (upper left panel) to the trans configuration (lower right panel). The spectra are computed with TDLDA in the linear regime of small excitations. The chromophore is part of a huge bio-molecule and it is assumed that the electronic dynamics of the optical response is confined to the chromophore, the surrounding areas of the molecule remaining optically inert and contributing only to the structure of the chromophore, see section 1.1.3. The changes of the spectrum along the path are dramatic. Of course, the overall region where the peaks reside remains the same because the overall size remains the same. But fragmentation structure, peak positions, and share of strength differ significantly. The spectra demonstrate also the typical fragmentation structure of covalent systems which is more involved than the plasmon dominated spectra of metals. Imagine now designing a pump and probe experiment for such systems. It is a lot harder to figure out a sensitive and unambiguous working point than it is for metal clusters.

4.2.1.3   Effect of Ionic Dynamics on Photo-Absorption Spectra

Electronic and ionic motion run at very different time scales. Thus one employs Born-Oppenheimer dynamics, see section 2.5.2.1, for ionic motion on long time scales. This works as long as electrons make no transitions between APES (Adiabatic Potential Energy Surface) and stay close to the ground state. Off-equilibrium situations where the electron cloud is highly excited require a simultaneous dynamical propagation of electrons and ions. The technically most straightforward strategy for that is Ehrenfest dynamics (ED) for the ions. We have to be aware of its limitations, as argued in section 2.5, but it is the method of choice whenever it is applicable. ED in its TDLDA-MD version is numerically not much more expensive than purely electronic TDLDA simulations. The only extra expense comes from the fact that we often have to propagate longer to account properly for effects of ionic motion.

Here we continue the example of the optical response of the H_2O molecule from the right panel of Figure 4.1 to check the impact of ionic motion on the spectra. Figure 4.22 compares the experimental spectra with theoretical TDLDA results, once with frozen ionic background and once with treating ionic motion simultaneously at the level of TDLDA-MD. H_2O is a welcome test case because it is a polar molecule where ions couple willingly to a dipole boost which enhances the chance to see ionic effects in the spectra. Using the LDA functional of [PW92] and Goedeker type pseudo-potentials of [GTH96] with an ADSIC (section 2.2.3.5.3), the bond lengths and angle are about 5% off the experiment, which causes a slight global shift in the spectra. For a qualitative discussion as we aim at here, we compensate that with down-shifting the theoretical spectra by about 2 eV. Doing that we see a comfortable agreement between theory and data. Experimental peaks are generally somewhat broader mainly due to temperature effects (ignored in theory) and for higher

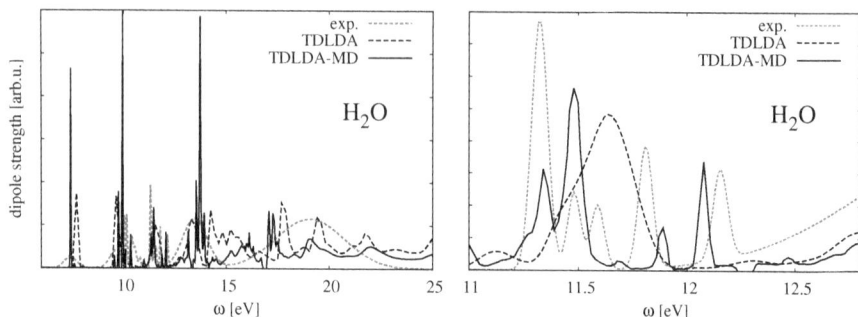

Figure 4.22 Comparison of experimental and theoretical photo-absorption spectra for the H_2O molecule. The theoretical spectra are down-shifted by 2 eV to match the experimental ones. Theoretical results are shown for TDLDA with frozen ionic background and for TDLDA-MD (both obtained, in fact, from TDLDA+ADSIC and the QDD code [DVC$^+$22]). The left panels show the spectra in a large energy window covering all relevant strength. The right panel zooms on a narrow energy window around the second peak in the spectrum. Experimental data taken from [KRMSS13].

frequencies also due to missing dynamical correlations. Particularly gratifying is the qualitative agreement in the vicinity of the two peaks (around 10 eV and around 12 eV) showing significant fine structure in experiment as well as in TDLDA-MD.

The right panel of Figure 4.22 zooms into the region around 12 eV. The difference between mere TDLDA and TDLDA-MD shows the impact of ionic motion. It is large. We see in TDLDA-MD obviously spectral satellites from coupling of an electronic resonance to ionic vibrations. The vibrations must be rather fast as the spectral distance of the satellites is of order of 0.1 – 0.2 eV. This points to oscillations of the light hydrogen ion relative to the heavy oxygen partner. The thus induced spectral fragmentation agrees nicely with the data. This example demonstrates that TDLDA-MD is capable of describing the coupling of electronic and ionic motion in molecules. The example was graceful to the extent that the fast hydrogen oscillations render the effect well visible at a 0.1 eV scale. Oscillations of significantly heavier ions are much slower and require much longer simulation times to be resolved. This is, in principle, possible with today's numerical capacities. But mind that the oscillations also have an intrinsic electronic damping due to electronic correlations. For too long time spans they thus tend to be wiped off and effects of coupling to ions may become more difficult to spot. Another problem comes from the experimental side. One needs very low temperatures and high energy resolution to see the faint energy differences associated with vibrations of heavy constituents of a molecule.

4.2.1.4   Effect of Ionic Motion on Fragment Energies

The case of photo-absorption spectra and many other published examples (e.g., pump and probe simulations) show that TDLDA-MD has a broad range of applications. And yet, it is limited whenever ionic motion undergoes bifurcation of trajectories as it can happen, e.g., in atomic collisions or fragmentation reactions. However, in the case of more violent processes TDLDA-MD can regain predictive power for average yields.

As an example, Figure 4.23 shows the total kinetic energies of the emitted heavy fragments in a hefty Coulomb explosion of (di-)iodimethane molecules, as a function of charge state [TNK$^+$17]. The molecules are excited by ultra-short and strong XUV pulses which push the systems immediately to high charge states. The thus generated large Coulomb pressure accelerates the highly ionized atoms apart from each other. Kinetic energies and direction of the ejected fragments collected in the simulations deliver a wealth of detailed observables. As one out of many, we show in Figure 4.23 the total kinetic energies of ejected fragments. They increase monotonically with the final charge reached in the reaction and with the number of heavy fragments (C and I ions). This is an obvious trend.

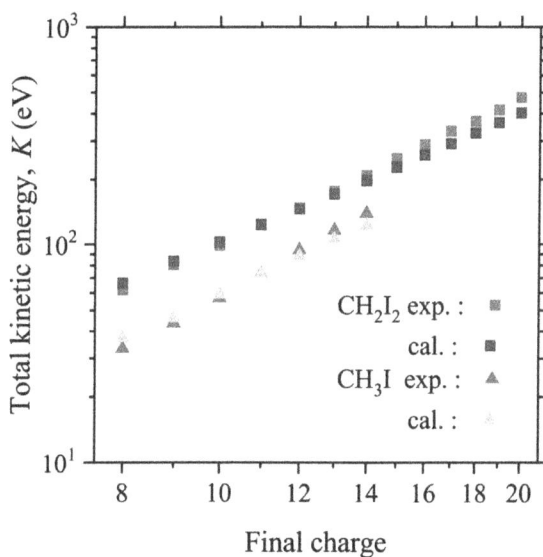

Figure 4.23 Comparison of experimental and theoretical (from TDDFT-MD) results for the total kinetic energy K of the carbon and iodine ions from Coulomb explosion of CH_3I and CH_2I_2 as function of final charge state Q_{fin}. The value of K shown is the average value obtained from the distribution function of the total kinetic energies. From [TNK$^+$17].

The remarkable aspect in the figure is the comparison with theoretical simulations carried out with ED with the electronic component described at the Self-Consistent-Charge Density-Functional Tight-Binding level (SCC-DFTB), see section 2.2.4.3. The agreement is satisfying showing that SCC-DFTB-MD is suited to describe average observables in very energetic reactions. It does even an acceptable job for more detailed observables such as energy distributions [TNK+17]. However, it ought to be mentioned that (semi-)classical approaches like VUU or electronic MD become valid at such high excitation energies. Most simulations in this regime thus prefer the numerically less demanding (semi-)classical treatment. For an example see, e.g., Figure 4.18 or Figure 1.1.

4.2.1.5 Electron-Ion Coupling in Photo-Induced Charge Transfer Dynamics

Ionic motion in stable molecules usually covers only small distances, and yet, this can have a huge effect on the electronic dynamics. An impressive illustration of the importance of ionic motion was presented in [RFS+13] for the model light-harvesting supramolecular triad consisting out of three molecular sub-units: porphyrin, carotene, and fullerene. The three moieties are covalently bonded, with the porphyrin acting as the chromophore which generates an electron-hole pair upon irradiation (a 10 fs pulse centered at 550 nm (2.25 eV) in the reference experiment). The fullerene behaves as an electron acceptor, while the carotene stabilises the hole left over by the separation of the primary electron-hole pair.

The triad is an artificial reaction center which has been devised to mimic the photo-induced charge transfer which occurs in natural photosynthesis. The porphyrin chromophore in the triad plays a role similar to that of chlorophyll in natural photosynthesis, and has the advantage of being more tractable by means of TDDFT-MD simulations. In fact, large scale real time (rt)-TDDFT modeling of chlorophyll in a light-harvesting complex (17,280 atoms) is possible [JSARM+15], but only for a few tens of fs and clamped atoms.

Figure 4.24 shows the result of a TDLDA-MD simulation of the dynamics following an instantaneous electron-hole excitation of the porphyrin chromophore. The atoms are modeled as classical particles and their velocities and position updated according to the Ehrenfest dynamics (ED) propagation scheme, see section 2.5.3. The electronic states are propagated according to the rt-TDLDA scheme within ALDA, see section 2.2.3.4.

The observable of interest is the charge transfer from the excited porphyrin to the fullerene. It can be followed by integrating excess electronic density, $\Delta \varrho (\mathbf{r}, t) = \varrho (\mathbf{r}, t) - \varrho (\mathbf{r}, 0)$, over the volume of the fullerene. Figure 4.24(a) shows the time evolution of the integrated charge transferred to the C_{60} moiety. The case will full ionic motion (green) shows a considerable transfer, up to one full charge unit, after approximately 70 fs. The charge transfer dynamics is however non-monotonic, and displays regular oscillations with period of about 30 fs, also observed in the reference experiment and linked to the

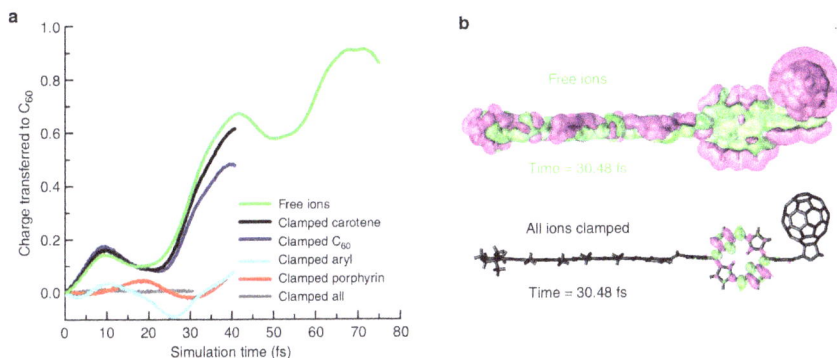

Figure 4.24 Results from Ehrenfest dynamical simulations of electronic and ionic motion after instantaneous excitation of a triad molecule composed of porphyrin, a carotene, and a fullerene moiety. Panel (a): Charge transferred to the fullerene under different conditions of ionic motion as indicated. Panel (b): Snapshots of the photo-induced change in charge density (charge difference relative to ground state density) at a time delay of 30.48 fs for the triad with freely moving ions (upper sub-panel) and with all atoms clamped (lower sub-panel). Purple indicates an increase in electronic charge density and green shows decrease. From [RFS$^+$13].

C-C symmetric stretch mode of the carotene [RFS$^+$13]. This correlation in the oscillations strongly suggests a transfer mechanism mediated by the ionic motion.

Clamping all ions (light gray curve) suppresses charge transfer completely. The major player in the blockade is the porphyrin: It suffices to clamp the aryl group connecting the porphyrin and C_{60} to suppress the charge transfer (light blue curve). The effect is dramatic. Even though the amplitude of the ionic motion is small, it suffices to change electron transport properties significantly.

Figure 4.24(b) illustrates the excess charge $\Delta \varrho (\mathbf{r}, t)$ density (purple, positive; green negative) 30.48 fs after the initial particle-hole excitation. The upper sub-panel shows the results of a simulation with ionic propagation (free ions), and the lower sub-panel shows the results of a simulation with clamped ions. In the first case, the excess charge density is spread over the whole triad, while in the second case, the excess charge is localized on the porphyrin.

Alert readers may wonder that we present here an example where TDDFT can describe electron transfer very well while we argued in section 2.3.2 that electron transfer goes beyond TDDFT. The point is that the case here considers transfer within one large molecule which remains as such intact. Where TDDFT fails is electron transfer in collisions or molecular fragmentation because this creates very different mean-field trajectories. The electronic localization into the fullerene is due to the electron-ion interaction. An electron is allowed to transfer from the porphyrin to the fullerene because of a favorable level alignment occurring periodically, following the ionic oscillations. The same electron is also prevented from "transferring back" due to the energy dissipated into localized molecular vibrations of the C_{60} moiety. Irreversible relaxation and thermalization of electronic degrees of freedom are however not guaranteed in a propagation scheme based on ED, see section 2.5.3.

A cooperative localization mechanism which involves both electron-electron *and* electron-ion interaction cannot be excluded. rt-TDDFT propagation based on a semi-local adiabatic functional like LDA is indeed not expected to model charge transfer reactions and electronic localization with clamped atoms, see section 2.3.2. Here the challenge comes from requiring a density functional approximation which is both more accurate than LDA or GGA and at the same time not much more computationally expensive, so that an adequate (i.e., several tens of fs) rt-TDDFT propagation can be still afforded.

The challenge posed by long-time irreversibility has different aspects: first, there is a fundamental aspect that ED is "too much coherent" so that a single-trajectory ionic propagation coupled to an initially pure electronic state will always return a pure state. However, thermalized electronic states are mixed, not pure. Averaging over an ensemble of trajectories starting from different, random initial conditions mitigates the issue. Owning to its efficiency, Trajectory Surface Hopping (TSH) (section 2.5.4.2) is becoming a standard approach to introduce electron-ion decoherence in finite systems. This approach has been successfully demonstrated in [YH21], where the TD-DFTB (section 2.2.4.3) has been used to make the electronic propagation more efficient. In this way, an overall propagation of several hundreds of fs becomes affordable, and confirms the charge transfer dynamics observed in [RFS+13]. Note that TSH still requires an average over an ensemble of ionic trajectories (e.g., 10^4 Surface Hopping trajectories per initial condition in [YH21]).

As more efficient approaches to electronic propagation, e.g., TD-DFTB, become available, another aspect of irreversibility should enter into the picture. The electronic energy initially lost to the localized molecular vibration will be eventually dissipated into the larger molecular environment in which a system is usually embedded. The challenge here is to describe the system-environment coupling in an efficient, yet accurate way (section 2.5.5.3).

4.2.2   IONS AND ELECTRONS II: BEYOND EHRENFEST DYNAMICS

Examples in the previous section highlighted the wide scope of a mean-field approximation for the electron-ion interaction, like Ehrenfest Dynamics (ED). ED is most commonly implemented along with a TDDFT description of the electron dynamics (section 2.5.3), and becoming more widely available in general-purpose electronic structure codes [NWB+20, COB18]. As discussed in section 2.5.4, ED suffers from some specific limitations, which manifest, e.g., in the relaxation of molecules from an excited state or in inelastic collisions.

In this section, a few prototypical examples of non-adiabatic molecular dynamics requiring a description beyond ED (section 2.5.4) are reported.

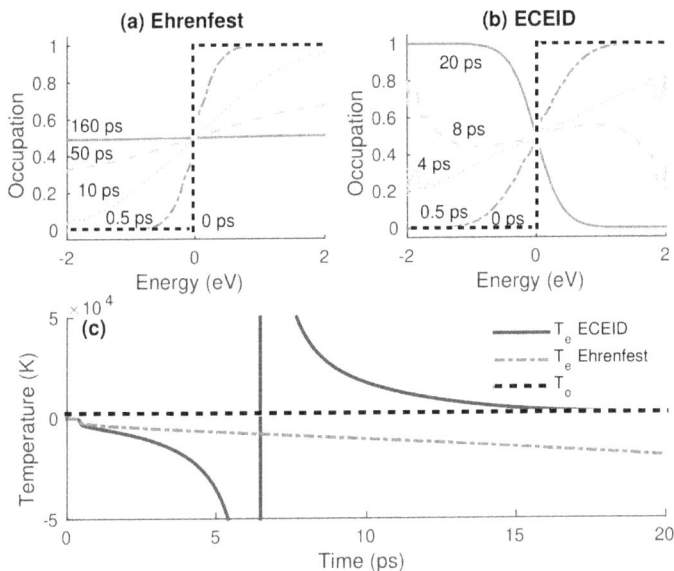

Figure 4.25 Population inversion ECEID simulation compared to ED. The figure shows snapshots of the population of the electronic states in (a) ED at 0 ps, 0.5 ps, 10 ps, 50 ps, 160 ps and (b) ECEID at 0 ps, 0.5 ps, 4 ps, 8 ps, 20 ps. (c) shows the time evolution of the electronic temperature for ED (dotted-dashed line) and ECEID (blue line) compared with the fixed oscillator temperature (red line). Initially the electronic temperature is negative because of the population inversion. In ECEID this trend reverts as time evolves, crosses over toward a non-inverted population. Temperature suddenly jumps to positive values, and then decreases eventually reaching the oscillator temperature, i.e. equilibrating electrons and phonons. Conversely, ED evolves toward a completely uniform distribution (grey line in panel (a)), corresponding to infinite temperature. From [RTKC16].

We shall first discuss in section 4.2.2.1 a principle problem with ED, solved with Effective Correlated Electron Ion Dynamics (ECEID, section 2.5.4.4). We shall then present examples based on Trajectory Surface Hopping (TSH, section 2.5.4.2), at the TSH level in section 4.2.2.2 and in the more elaborate Ab Initio Multiple Spawning (AIMS, section 2.5.4.3) in section 4.2.2.3.

4.2.2.1 A CEID Example

It was pointed out in sections 2.5.4 and 4.2.2 that Ehrenfest Dynamics (ED) has some difficulties to properly accommodate energy transfers between electrons and ions (or phonons in a solid). The point is illustrated in Figure 4.25 where we show the time evolution of the population of electronic states starting from an inverted population, in a simple tight-binding N-atom chain interacting with a phonon bath kept at constant temperature. The atoms are monovalent, so that there are N electrons in the chain. Inverted population means that the initial state is constructed by setting the occupation numbers of the lowest $N/2$ energy levels to zero and to 2 (due to spin degeneracy) the remaining (highest) $N/2$. While ED evolves, incorrectly, toward a flat distribution corresponding to an infinite electronic temperature, ECEID evolves toward a Fermi-Dirac distribution at a finite temperature that coincides with that of the phonon bath [RTKC16].

In [RTKC16] it was also shown that the evolution of the electronic occupations could be reproduced to an excellent extent using a kinetic model. This was very recently followed by a formulation of the problem purely in terms of rate equations for the coupled evolution of the electronic and phonon occupations [BTS+20], which in addition includes coherence effects that were absent in [RTKC16]. This further simplification, correct to first order in perturbation theory, is extremely efficient. There are two main limitations to the ECEID approach, though. Firstly, it applies to weak electron-phonon coupling, and secondly, it is spatially homogeneous and hence applicable to a laser, but not to an ion irradiation. This formulation is closely related to the Boltzmann transport equation and to non-equilibrium Green's functions approaches [KvLP+21], see also sections 2.3.6 and 2.4.

4.2.2.2 Molecular Scattering and Vibrational Relaxation on Metal Surface

In its paper introducing the Fewest Switches version of TSH [Tul90], Tully describes an interesting *gedanken* experiment of an atom impinging on a metal surface. Two qualitatively different outcomes are possible, namely scattering (possibly inelastically) or sticking of the atom on the surface. The case raises a few challenges to non-adiabatic molecular dynamics. The two diverging outcomes imply two qualitatively different potential energy surfaces which will quickly diverge after the initial atom-surface collision. Due to the underlying mean-field approximation, ED will instead generate a non-adiabatic trajectory somehow intermediate between the two extreme cases, with the atom either scattered or stuck on the surface. This is clearly unphysical, and the drawback of the mean-field approximation cannot be fixed by averaging over different initial conditions. Surface Hopping is better suited to model this *gedanken* experiment, as it propagates the non-adiabatic trajectories using the adiabatic forces from a single potential energy surface. As a consequence, averaging over different initial conditions *and* an *ensemble* of stochastically generated trajectories will provide a more realistic model.

A variant of the TSH algorithm, deemed Independent-Electron Surface Hopping (IESH) was introduced in [SRT09], and applied to model the vibrational relaxation and scattering of a nitrogen monoxide (NO) molecule on the Au(111) surface. The non-adiabatic modeling is simplified by using a model Hamiltonian fitted to reproduce the band structure from *ab initio*

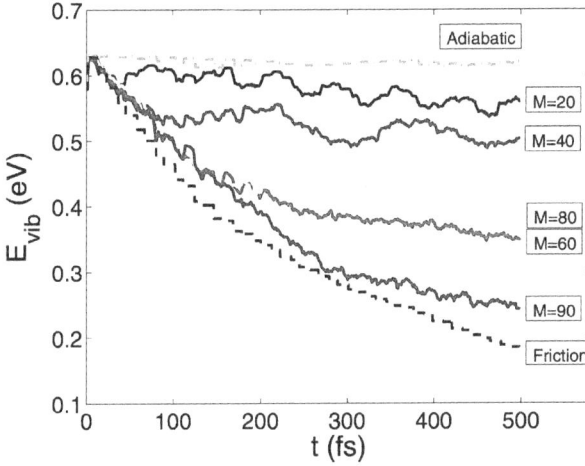

Figure 4.26 Non-radiative energy relaxation from an excited vibrational state of nitrogen monoxide, NO. Adiabatic molecular dynamics gives no relaxation. A mean-field approximation based on the "electronic friction" [HT95] (labelled "Friction" in the figure, see text for details) reproduces the expected relaxation. IESH with a variable number M of discretized states also gives relaxation and appears to converge as M increases. From [SRT09].

electronic structure calculations. To make the description of the Au surface more tractable, the dense continuum of the metallic orbitals is discretized. The key approximation of the IESH is to treat the electrons as non-interacting. The approximation is critical to include energy dissipation through the excitation of multiple electron-hole pairs of the metal surface.

As shown in Figure 4.26, the vibrational relaxation of NO predicted by the IESH converges with the number of states, M, discretizing the metallic continuum. The converged relaxation is compared against a model of "electronic friction" which includes a many-body damping term, K_{IJ}, into the Newton equations of motion [HT95]

$$m_I\ddot{\mathbf{R}}_I = -\nabla_{\mathbf{R}_I} V(\mathbf{R}) - \sum_J K_{IJ}\dot{\mathbf{R}}_J, \qquad (4.7)$$

where the index I labels the atoms, of mass m_I and positions $\mathbf{R}_I$, and $V(\mathbf{R})$ is the many-body potential energy. The damping term depends on the non-adiabatic couplings (section 2.5.1) as $K_{IJ} = (\pi/\hbar)A_{j_F,j_F+1}(\mathbf{R}_I)A_{j_F+1,j_F}(\mathbf{R}_J)$, where j_F is the highest occupied adiabatic orbital.

Figure 4.26 shows a good quantitative agreement between the 'electronic friction' and IESH findings, although the former gives a slightly faster relaxation. As in the case of ED, the modeling of 'electronic friction' is obtained from a self-consistent mean-field approach. It can then yield inaccurate—and even unphysical—non-adiabatic dynamics of a molecule impinging on a surface. For instance, it is known that the IESH provides a more realistic vibrational density of states in the NO molecule [SRT09] and good agreement between the experimental findings and the IESH predictions has been found.

The IESH approach has been generalized to include modeling at finite temperature with a Langevin thermostat [ST12]. The domain of application of IESH and potential alternatives have been assessed in [MOS18, DS20]: IESH agrees with the Marcus theory of electron transfer [SJL+16] in the low-coupling regime and works best for metals modeled in the wide-band limit. As the computational complexity of the IESH scales as M^2, fully converged results may be hard to achieve.

4.2.2.3 Prototypical Models of Non-Adiabatic Molecular Dynamics

In the same 1990's landmark article [Tul90], Tully devised three one-dimensional model problems to benchmark the Fewest Switches TSH algorithm (FSSH), see section 2.5.4.2. Those Tully's models have played an important role in the validation of non-adiabatic molecular dynamics approaches which have since then been proposed.

These models are simple enough to be amenable to an exact numerical solution. Without undermining the importance of a direct comparison against the experiments, exactly solvable models provide a controllable way to test the approximations of a newly proposed numerical approach. They furthermore illustrate three prototypical non-adiabatic behaviors, also found in more realistic molecular models [IC20]. The first Tully's model (Model I) is an example of nuclear wave-packet splitting at a single avoided crossing between two adiabatic potential energy surfaces. It is then similar to the well-known Landau-Zener model. Model II is an example of a double avoided crossing and subsequent interference between nuclear wave-packets. Model III is an example of an extended non-adiabatic coupling over two nearly degenerate Adiabatic Potential Energy Surfaces (APES), which also yield quantum interference. These models were designed to benchmark simple scattering problems in terms of probability of reflection or transmission over different APES. They are not meant to test the ability to model quantum tunneling or non-radiative relaxation. Benchmarking non-adiabatic dynamics across so-called conical intersections [Tan07, BMK+03]—which are ubiquitous in multidimensional APES—is also beyond the scope of these simple one-dimensional models.

Relatively simple, yet fully *ab initio*, molecular benchmarks, analogue to the original Tully's models were recently proposed in [IC20]. Namely, the features of Tully's Model I, II, and III are illustrated by ethylene, $H_2C=CH_2$, 4-N,N'-dimethylaminobenzonitrile (or DMABN), $(CH_3)_2NC_6H_4CN$, and fulvene $(CH=CH)_2C=CH_2$, respectively.

Ethylene is a simple molecule which has been intensively modeled because of the two conical intersections between its ground-state, S_0, and first excited singlet, S_1, APES. In fact, the non-radiative decay from S_1 following a Franck-Condon excitation proceeds through only one of the two conical intersections. To this extent, ethylene is an analogue of Tully's Model I. The first, S_1 and second excited singlet, S_2, APES of DMABN are close enough to observe repeated non-adiabatic transitions following an initial photo-excitation, so that DMABN is, up to details, an analogue of Tully's Model II. Finally, fulvene presents a sloped conical intersection between S_1 APES and the ground state driven by the stretching of the double bond of the methylidene group, $=CH_2$. This makes it an analogue of Tully's Model III.

There are no exact numerical benchmarks available for these molecular analogues. Yet, some valuable insights come from the comparison of the non-adiabatic molecular dynamics obtained using different approaches. For instance, the decay of the S_1 population of ethylene or the S_2 population of DMABN shows a substantial agreement between Ab Initio Multiple Spawning (AIMS), TSH, and Decoherence Corrected Surface Hopping [GPZ10] (DCSH, section 2.5.4.2).

In the case of DMABN, the DCSH generates a dynamics which is systematically closer to that given by AIMS (section 2.5.4.3). The finding is somehow expected as the original formulation of the TSH is known to generate over-coherent non-adiabatic dynamics, and several ways to fix the issue has been advanced [SJL+16]. The non-adiabatic dynamics given by the AIMS algorithm is not exact [IC20, MC18], but expected to follow more closely the exact Born-Huang expansion (Eq. (2.83)) of the electron-nuclear wave function. It also describes the decoherence due to components of the nuclear wave function which evolve on very different APES.

In the case of DMABN, the DCSH achieves the same accuracy as AIMS, with rather reduced computational requirements. Still, the molecular model provided by DMABN lacks an important feature [IC20] of Tully's model II, i.e., approximate coherence and Stckelberg oscillations [Tul90]. As a consequence, the agreement between DCSH and AIMS may not extend to molecular models which display nuclear wave function re-coherence after a multiple avoided crossing.

As shown in Figure 4.27, the relaxation of fulvene from its S_1 APES displays some notable discrepancy between AIMS, TSH, and DCSH. Although the initial decay of the population presents a substantial agreement between the three methods, the subsequent non-adiabatic dynamics develops quantitatively different. Due to the sloped conical intersection, part of the S_1 population which has quickly transferred to the ground state is temporarily transferred back through a recrossing of the potential energy surfaces, as in Tully's Model III. This dynamics is qualitatively described by all the methods considered, but both variants of Surface Hopping give a S_1 population almost three times larger than what returned by AIMS. In fact, DCSH is in slightly better agreement with AIMS than TSH.

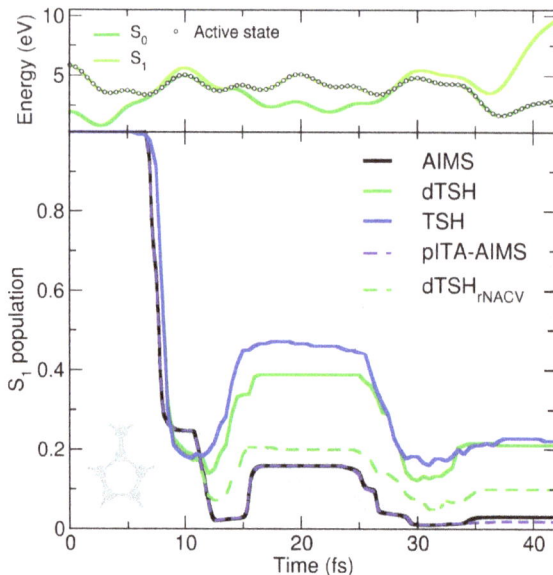

Figure 4.27   Top panel: Energy paths as sampled by one Surface Hopping trajectory. Note the multiple recrossing. Bottom panel: Population decay from the S_1 potential energy surface generated by different non-adiabatic molecular-dynamics algorithms, namely *Ab initio* Multiple Spawning (AIMS), Trajectory Surface Hopping (TSH), Decoherence-Corrected TSH (DCSH), pseudo-Independent Trajectory Approximation in AIMS (pITA-AIMS), and DCSH with nuclear velocities rescaled along the nonadiabatic coupling vectors (dTSH$_{rNACV}$). The shaded area gives the estimated error of the AIMS propagation. Adapted from [IC20]

The authors of [IC20] carefully investigate the origin of these notable discrepancies by considering small modifications of the non-adiabatic molecular dynamics algorithms. They find that, in this example, the implementation of velocity rescaling which follows a trajectory surface hop has a very strong influence on the overall dynamics. Velocity rescaling is required to conserve the total energy, but its implementation is not unique. In the case of fulvene, when only the speed is rescaled (i.e., an isotropic rescaling), the above mentioned discrepancy is observed. However, if the velocities are rescaled anisotropically along the direction of the instantaneous non-adiabatic couplings, an improved agreement between TSH and AIMS. is observed.

Since TSH is a non-adiabatic molecular dynamics algorithm with many practical advantages, one can highlight a few practical lessons learnt from the careful analysis of [IC20]. As noticed by other authors [SJL+16], adding corrections to the original FSSH of TSH (section 2.5.4.2) can improve its predictability, without compromising its numerically efficiency. In fact, there is still no universal consensus about the best implementation of the TSH, although recent advances have put the underlying theory on a much solid ground [SJL+16, SOL13]. A shared consensus may grow in the near future as different flavors of TSH are getting more widely available in general purpose and open source electronic structure codes [NWB+20, COB18]. There is growing evidence that DCSH can be confidently employed to model nonadiabatic molecular dynamics not showing strong re-coherence after multiple APES recrossing. In practice, the impact of re-coherence may not always be clear *a priori*, and benchmarking and validation of TSH on simple molecular models will still play a role in the future.

4.2.3 INTERACTION WITH THE ENVIRONMENT

Generally speaking the notion of environment covers a wide range of possibilities ranging from contact to a material, such as in embedding or deposition, to inclusion into a set of confining external potential, such as in a trap, to a chromophore in an else-wise inert molecule (see also the introduction to section 2.5.5). In relation to the topic of this book on irradiation we focus on dynamics and restrict the discussion to the presence of a material. We first illustrate the impact of such an environment on electron response (section 4.2.3.1) in the case of a deposited cluster. Of course, the first impact of an environment is on the shape of the system, which will also be considered in section 4.2.3.1. As a second example we illustrate the impact of an environment on the ionic response of a chromophore (section 4.2.3.2).

4.2.3.1 An Example of Impact on Shape and Electronic Emission

We first consider an example of the impact of a surface on a cluster. We analyze the impact of the surface on the shape of the deposited cluster by considering the optical response and show how the angular distribution is also affected, following [BDM+10]. Calculations have been performed at TDLDA level for Na_8, supplemented by a polarizable QM-MM modeling of the MgO(001) surface for deposited Na_8 [DRS10].

The optical absorption spectrum (section 3.4.2) of a free Na_8 and of Na_8 deposited on an MgO(001) surface are shown in the left panel of Figure 4.28. In free Na_8 there is one dominant clean peak reflecting the shell closure of the electronic cloud in this case [Bra93]. Optical responses are almost degenerate along the three principal axis in this case. The highly polarizable MgO (001) surface leads to some spectral fragmentation, about 1.4eV for MgO(001). It is especially large along the direction perpendicular to the surface, because of symmetry breaking. As is well known the optical response thus directly reflects the shape of the cluster, almost spherical when free, deformed when deposited.

The presence of the surface thus impacts the response of the system to a laser as it modifies eigenfrequencies. But the impact of the surface covers

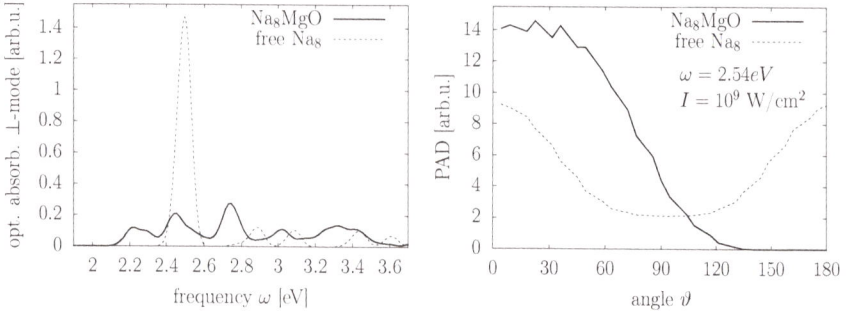

Figure 4.28 Comparison of free Na_8 with Na_8 on MgO. Left panel: Optical absorption spectra for the mode orthogonal to the MgO surface. Right panel: PAD after excitation by a laser pulse with frequency 2.54 eV, intensity 10^9 W/cm^2, and overall pulse length of 60 fs. The angle $\vartheta = 0$ points in away from the MgO surface in perpendicular direction. From [BDM$^+$10].

much more than this. We illustrate the point in the right panel of Figure 4.28 where are plotted angular distributions (PAD) following a laser irradiation, for free Na_8 and Na_8 on MgO. Before discussing the comparison between the two cases, let us briefly discuss the geometry. In the case of the deposited Na_8 one considers emission with respect to polar angle θ, measured with respect to the direction z perpendicular to the surface, which is also the laser polarization axis. The signal is integrated over the azimuthal angle ϕ along the surface. For free clusters the PAD is again measured along θ and integrated over ϕ but one has to further average signals over all possible orientations of the cluster, which delivers an Orientation Averaged PAD (section 3.4.4.4). The comparison of the two cases is quite telling. The OAPAD of the free cluster displays a clear alignment along laser polarization axis, while the PAD of the Na_8 on MgO shows a strong suppression of backward emission through the substrate.

The shape of the PAD is both laser frequency and intensity dependent. Depending on the intensity one may observe a maximum of emission around $\pi/4$, possibly accompanied by a sizable emission under rather flat angle ($\theta \simeq \pi/2$). The pattern is thus more involved than reflecting a mere re-scattering of the reflected electrons $\theta \to \pi - \theta$. At sufficiently high laser frequency and intensity one even observes a progressive restoring of the symmetric shape characteristic of the free cluster [BDM$^+$10].

4.2.3.2 An Example of Impact on Ionic Response : Photo-Isomerization of A Retinal Protonated Schiff Base

The conversion of light into mechanical energy, central for the vision process, is carried out by a retinal protonated Schiff base, a chromophore present in the rhodopsin family of proteins located in the eye. The retinal is simply a conjugated carbon chain terminated in a β-ionone aromatic ring on one end,

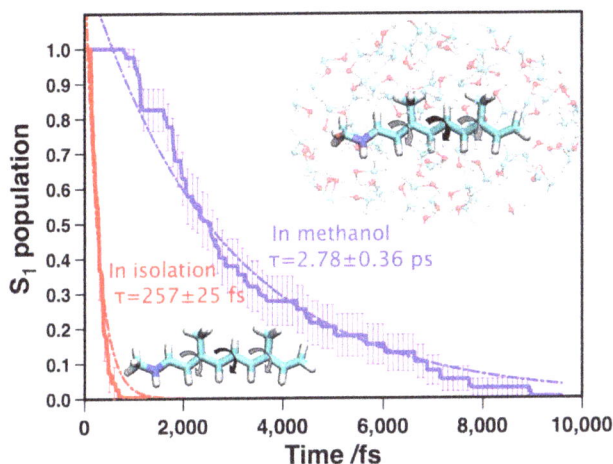

Figure 4.29 Excited state (S_1) population of the retinal model in gas-phase (red) and in methanol (purple). The dashed lines correspond to exponential decay with the fitted lifetimes. The insets show the systems studied. Figure reproduced from [POM15].

and a lysine on the other, which forms part of the scaffold of rhodopsin. Photon absorption promotes the electronic system from the ground state S_0 (APES) to the first excited state S_1. Torsional motion around C=C double bonds (see arrows along the molecular chain in Figure 4.29) leads to a crossing between S_0 and S_1, when the torsional angle twists to about 90°. Due to the different symmetry of the two APES, the crossing is a conical intersection. Therefore, the electronic system should be described within a multi-reference framework. When the torsional angle goes through the conical intersection, the electronic system experiences a non-adiabatic transition from S_1 to S_0, ending up in a *cis* isomer in the ground electronic state. There are several double bonds in the retinal that can undergo photo-isomerization. Which one will happen depends on the initial configuration of the molecule, which is statistically drawn from a thermal distribution. Of course, it is also possible that the retinal remains in the all-*trans* configuration and the absorbed energy is distributed among molecular vibrations without isomerization. The quantum yield, i.e., the probability that the retinal undergoes isomerization, can be computed by averaging the outcome over a statistically significant number of independent trajectories. One can also obtain the branching fraction, i.e., the fraction of trajectories undergoing isomerization around a given double bond. Similarly, it is possible to estimate the lifetime of the electronic excitation, i.e. the time required for it to decay back to the ground state.

Experimentally, lifetimes and quantum yields can be measured by femtosecond spectroscopy and liquid chromatography. This has been done initially in the protein environment, obtaining an extremely short lifetime of 200 fs [SPMS91], which leads to a very efficient process with a large quantum yield of 0.65 [KMY$^+$05]. Notice that in the protein environment, rhodopsin is in the 11-*cis* (bent) configuration, and isomerizes to 11-*trans* upon photon absorption. However, the opposite is observed in a liquid environment. Experimental studies in n-hexane, acetonitrile, methanol and 1-butanol, showed that the quantum yield increases with the polarity of the solvent [KKK$^+$91]. More importantly, they also show a significant increase in the lifetime of the excited state, from 200 fs to more than 2 ps, thus indicating the appearance of a barrier to isomerization [HZR$^+$96, LSES96], contrary to the barrier-less event in the protein environment.

The effect of the solvent has been studied by means of QM-MM ab initio multiple spawning (AIMS) simulations [POM15, CM18] as discussed in section 2.5.4.3, by comparing the behavior in the gas phase and solvated in methanol, of a retinal model in which the aromatic ring and the lysine on the two ends have been replaced with methyl groups. The systems studied are shown in the insets to Figure 4.29, lower left for gas-phase, and upper right for solvated. In these simulations, the methanol is treated using the all-atom classical force field OPLS/AA [JMTR96], see section 2.5.2. These simulations are, computationally, very demanding because of the size of retinal, the simulation time, and the number of trajectories. For these reasons, the only viable solution is to describe the electronic structure of the retinal at a simplified level. In this work, firstly, only the S_0 and S_1 APES are considered. Secondly, the electronic structure is described through a re-parameterized multi-reference semi-empirical method [TTM04] fitted to CASCI calculations [SM10] (see also section 2.3.4.3). The semi-empirical approach, in which the electronic integrals are parameterized, include both static and dynamical correlation, and is quite a reasonable, affordable approximation to CASPT2, which would be required for multi-reference excited state dynamics (see section 2.3.4.3). The full multiple spawning approach is suitable to describe nuclear quantum dynamics on multiple electronic states with non-adiabatic transitions (section 2.5.4.3). The results of these simulations are shown in Figure 4.29, where one sees that, in the gas phase, the excited retinal decays to the ground state in a time scale of 257 fs, while in methanol the time scale increases tenfold, to 2.78 ps. The simulations also reproduce qualitatively the experimental two-dimensional fluorescence spectra in methanol. The two-time scales obtained from simulations, 93 fs and 2.17 ps, are somewhat smaller than experimental ones, at 205 fs and 3 ps. A statistical analysis of the trajectories shows that, in both cases the dominant isomer is the 11-*cis*, but in methanol the process is more selective toward this isomer in comparison to 9-*cis* and 13-*cis*. The quantum yield is about 40% in both cases.

To understand the reason for the delay in isomerization in methanol, the energy profile in the electronic excited state S_1 has been calculated as a function of the torsion angle for the several relevant double bonds [POM15]. This

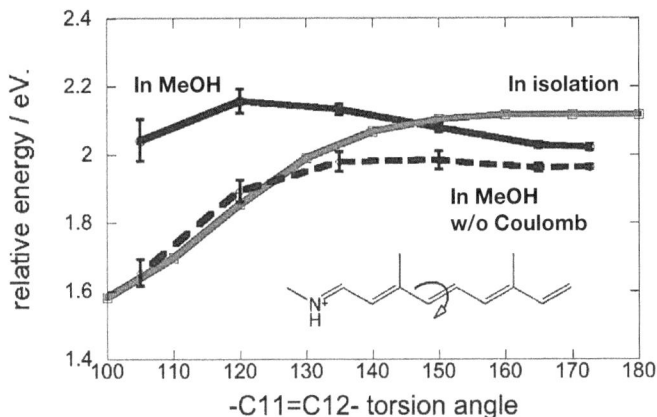

Figure 4.30 Example of an energy profile in the excited S_1 state for one most relevant torsion angle, in the gas phase (red line), solvated in methanol (full blue line), and solvated in methanol with the partial charges switched off (dashed blue line). From [POM15].

was done via constrained minimization, by fixing the torsion angle and the heavy atoms (C and N) in the retinal model, while optimizing the remaining degrees of freedom. In the solvated case they also optimized the degrees of freedom of the methanol molecules and averaged the energy profiles over a few independent solvent configurations, to account for configurational disorder. An example of a profile is shown in Figure 4.30. The grey line is for the gas-phase retinal model, while the black one is for the solvated case. Similarly to the protein environment, the gas-phase profiles are barrier-less toward isomerization barrier while, in agreement with experiment, the solvated ones exhibit a small barrier of 0.1-0.2 eV, thus explaining the delay. Setting the methanol partial charges to zero (dashed blue lines in Figure 4.30) suggests that the origin of the barriers are electrostatic rather than steric interactions.

4.3   BEYOND CLUSTERS: METALLIC NANOPARTICLES AND PLASMONICS

We discuss here some aspects of the plasmonic response of metals in relation to recent applications in photo-catalysis. Plasmons quickly decay into energetic (or "hot") electron-hole pairs which, in principle, can be harnessed to catalyze chemical reactions. In practice, several mechanisms must be taken into account to estimate the efficiency and selectivity of "hot" carrier photocatalysis, as discussed in section 2.4.3. State-of-the-art first principles calculations which account for the different scattering mechanisms can achieve a remarkable agreement with the experiments. In the case of metallic nanoparticles and nanostructures, the presence of surfaces must also be accounted

for. Along with accurate estimates of the plasmon decay rate, thermalization and transport of the "hot" carriers must be modeled to predict their
photo-catalytic action. Achieving good energy and spacial resolution requires
multiscale approaches, which are currently developed.

4.3.1 PLASMONIC RESPONSE OF SIMPLE METALS

Both the jellium-based DFT and hydrodynamic approaches can model the
plasmonic resonances of metallic nanostructures (see section 2.4.3). The domain of applicability of the jellium-based DFT approach is limited to the
smallest size nanostructures of about a few tens of Å [Bec87, MKN$^+$12,
SZGV$^+$13]. But much larger spherical nanoparticles can be modeled because
of their symmetry [VGGF$^+$16, MLKN14].

The "spill-out" of the electrons outside the uniform ionic background is due
to the quantum delocalization of the electrons and extends just for a few Å
outside the metallic surface [Bra93, Ler11, TSK$^+$15]. The electron "spill-out"
is missed by simple hydrodynamic approaches, which characterize the electron liquid just by its dielectric function [Mor21, Cir17, SZGV$^+$13]. Corrections to these simple approaches have been devised and compare well against
the jellium-based DFT models. Corrected hydrodynamic models offer a compromise between scalability and transferability to larger and more complex
nanostructures.

A challenging application for both the jellium-based DFT and hydrodynamic approaches is the "gap plasmonics", in which two metallic nanostructures, e.g., two nanospheres are separated by a few Å dielectric gap. Strong
hybridization of the plasmonic modes of the two nanostructures yields a very
localized field enhancement (section 4.1.1.4). The configuration bears similarity to an optical cavity, with a very large Purcell's factor, i.e. the ratio between
the spontaneous emission in the cavity and in vacuum. Because of the electron "spill-out", charge-transfer plasmons can also appear for very small gaps.
The challenge in the modeling comes from the different scales of the problem:
the smallest scale of the gap and electron "spill-out" and the relatively larger
scale of the metallic nanostructures.

Plasmonic nanocavities constrained by gaps between metallic nanostructures differ from conventional Fabry-Perrot nanocavities in a fundamental
way. Plasmons are inherently lossy excitation modes, which limits the quality
factor of plasmonic nano-cavities. Eventually, the energy stored in the plasmonic mode is dissipated to the lattice by Joule heating. The microscopic
kinetic mechanisms which eventually yield the Joule heating are involved and
span several time- and length-scales. It is still challenging to model all these
kinetic mechanisms within the same numerical framework.

The initial decay of the plasmonic excitation stems from the coupling between the plasmon with incoherent electron-hole excitations, a mechanism
referred to as the "Landau damping" (section 2.4.3.1), in analogy, up to finite
size effects, with the well-known damping process in plasma physics.

4.3.2 "HOT" CHARGE CARRIERS AND PHOTO-CATALYSIS

Plasmonic excitations are inherently "lossy" because of the scattering processes always present in metals. The presence of surfaces adds a new source of scattering, which can be prevalent in nanostructures. Efforts have been dedicated to the reduction of these loss mechanisms [BCZ+17], but there are inherent limitations—especially in the optical range of frequencies—dictated by the current plasmonics materials [KS12].

Along with the ongoing research of low-loss negative permittivity materials, benefits of the plasmonics loss mechanics have been realized [BCZ+17]. The electromagnetic energy initially absorbed and stored in a plasmon mode is ultimately dissipated into the local environment as heat. Hence, plasmonics in nanostructures can be used as controllable heat generators at the nanoscale. Thermo-plasmonics, i.e. the application of plasmonics to heat generation, is currently a growing area of research with potential impact in many sectors, including chemical engineering (e.g., photothermal assisted catalysis) and medicine (e.g., photothermal treatment of cancer [BQ14, BQ13]). The photothermal effect is typically broad-band or non-resonant and part of the energy is likely to be dissipated without assisting a target chemical process, e.g., the catalysis of a chemical reaction.

"Hot" charge carriers (electrons and holes) directly generated upon plasmon decay can be harnessed to drive reactions more efficiently in a tuned or resonant way [MZG+14, MLL+13]. Here, the adjective "hot" is generally extended to both: i) the transient (≈ 0.1 ps) non-equilibrium (or non-thermal) distribution of charge carriers from plasmon decay (plasmons decay, i.e. dephase, in ≈ 10 fs [Har11]) and ii) the (≈ 1 ps) Fermi-Dirac distribution of charge carriers reached upon thermalization at a temperature, T_e, which will be generally larger than the local lattice temperature, T_l [BYW+19]. In other cases, only thermalized carriers at $T_e > T_l$ are referred to as "hot" [Har11].

Non-equilibrium "hot" carriers will eventually thermalize to the local lattice temperature, T_l, mainly through electron-phonon scattering [Har11]. This electron-lattice thermalization process is often modeled with a phenomenological Two-Temperature Model (TTM) [All87, KLT57],

$$\begin{cases} C_e\left(T_e\right) \frac{dT_e}{dt} &= -g_{eph}\left(T_e - T_l\right) , \\ C_l\left(T_l\right) \frac{dT_l}{dt} &= g_{eph}\left(T_e - T_l\right) , \end{cases} \tag{4.8}$$

where $C_e\left(T_e\right)$ is the heat capacity of the charge carriers, $C_l\left(T_l\right)$ is the heat capacity of the lattice and g_{eph} is the effective electron-phonon coupling constant. Note that the r.h.s. of Eq. (4.8) is a linearization of a more complicated expression, in principle not applicable to far off equilibrium relaxation. The use of the equilibrium expressions for the heat capacities $C_e\left(T_e\right)$ and $C_l\left(T_l\right)$ is also questionable, although most often used in practice.

The experimental evidences of the catalytic action of "hot" charge carriers has been debated [ZSZ+18, SBUD19, ZSR+19], since it is not always straightforward to rule out competing photothermal catalytic mechanisms [CBA+20].

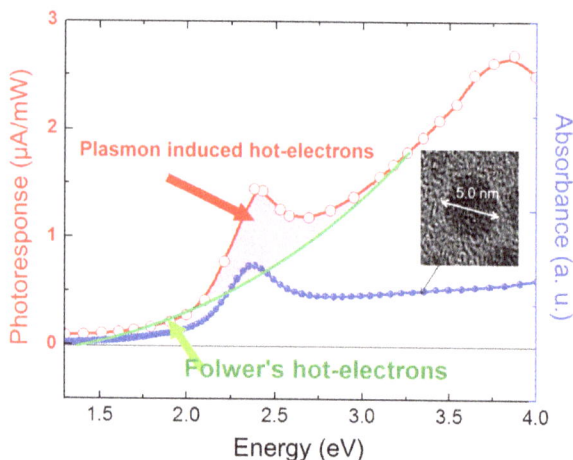

Figure 4.31 The photo-response of a chemical reaction catalysed by ZnO nanorods decorated with Au nanoparticles (red curve) is compared to the absorption spectrum of the nanoparticles (blue curve). The excess photo-response with respect to the expected background (Fowler's theory) aligns well with the peak of the nanoparticles absorption. This alignment is taken as an evidence of the contribution of "hot" carriers to the photo-catalytic process. From [CCC+12].

As shown in Figure 4.31, the alignment of the energy dependencies of the plasmonic absorption (e.g., of an Au nanoparticle) and the excess photoreaction yield provide an evidence of the catalytic action of "hot" carriers. The excess is defined with respect to a background expected without the plasmonic excitation. As shown in [CCC+12], this background is well described by the Fowler's theory of internal photoemission [Dal71]. Internal photoemission occurs when an electron is transferred from the Au nanoparticle to the ZnO nanorod, leaving a hole behind. The extra hole promotes the water oxidation reaction, while the transferred electron promotes the hydrogen reduction. An unambiguous measurement of the temperature of the catalyst, especially at the nanoscale [MN20, BBBQ20] nevertheless remains challenging.

Modeling of "hot" charge carriers has focused on two main processes: generation of "hot" carriers upon plasmon decay and the subsequent transfer mechanism into the attached semiconductor or molecule. Figure 4.32 shows a comparison between first principles modeling (section 2.1.5) and experimentally determined values of the plasmon decay rates in Al, Cu, Ag and Au at room temperature [BSN+16b]. The decay rate is proportional to the imaginary part of the dielectric function [BSN+16b]. The experimental values are obtained from ellipsometry measurements [Pal85]. The mechanism considered for plasmon decay is the generation of a single electron-hole pair, with an energy equal to the "plasmon energy". Decay through the generation of multiple electron-hole pairs of phonons has not been considered.

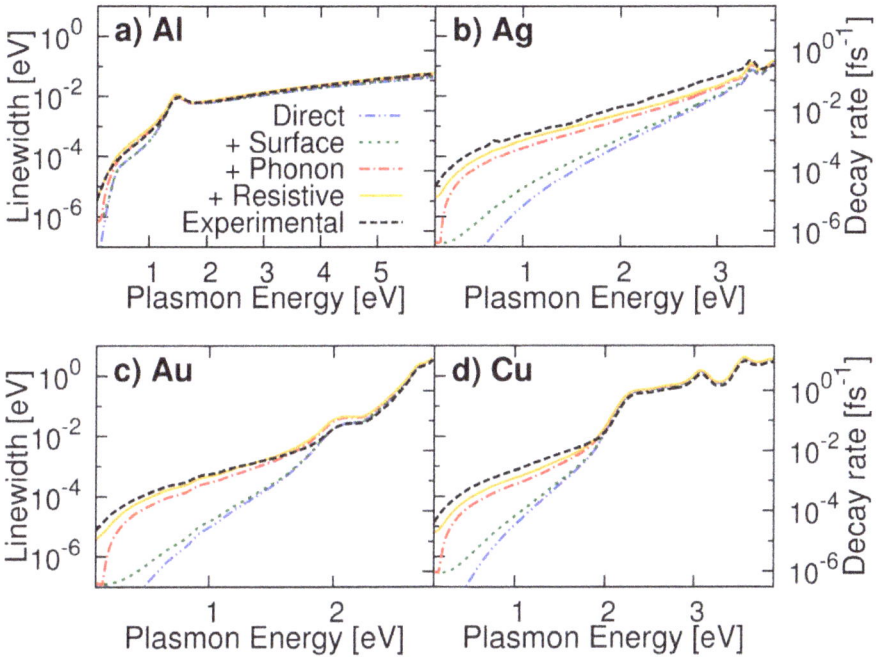

Figure 4.32 FWHM (or linewidth), along with the corresponding decay rate, of a plasmonic excitation in four representative metals. The experimental values are obtained from ellipsometry measurements [Pal85]. The results of the first principles modeling show the contributions from direct intraband transitions, surface and phonon scattering, and the resistive dissipation due to the conduction electrons. In the case of the noble metals, the inclusion of all contributions is required to reach a good agreement with the experiments. From [BSN+16b].

The plasmon decay rate is reproduced accurately if the contributions from direct interband, surface-assisted interband, phonon-assisted interband and intraband transitions are all accounted (see section 2.4.3.2). Direct interband transitions account for most of the decay rate above the $d \rightarrow sp$ absorption edge, which for noble metals is in the visible range of the spectrum. Intraband transitions originate the resistive losses, which dominate at low energy in the case of surface plasmon polaritons. Brown *et al.* take the resistive losses proportional to the imaginary part of the Drude dielectric function, Eq. (2.73), where the plasmon frequency is obtained by integrating Eq. (2.78) numerically, and the scattering rate computed in a relaxation time approximation (RTA, see section 2.3.7.2).

In the case of localized plasmons in nanospheres, the decay rate is determined by surface-assisted transitions [BSN+16b]. A good agreement between theory and experiments is found for the real and imaginary parts of the dielectric function of plasmonic metals [BSN+16a]. Numerical investigations reported in [BSN+16a, BSN+16b] made use of MBPT (section 2.3.6) based on accurate DFT+U calculations (section 2.2.3) of the bulk plasmonic metal band structure.

High level DFT [ADN+20, SNJ+14] or GW [MBNSGL16, RKT+12] (section 2.3.6) electronic structure calculations are crucial to reproduce the correct $d \rightarrow sp$ absorption edge. This is especially relevant for Cu and Au, as the bulk plasmon frequency is close to that absorption edge. The colour of these noble metals is strongly influenced by the position of the $d \rightarrow sp$ absorption edge relative to the bulk plasmon frequency [PRM19, Poo83].

In an alternative approach by Govorov and co-workers, "hot" charge carriers generation and decay has been modeled using a quantum kinetic formalism (section 2.3.7.1) based on DFT[GZ15]. This approach is similar to the RTA scheme, see section 2.3.7.2 for more details. The relaxation operator is a collision term (see section 2.3.7.1) which accounts for the electron-electron interactions neglected at the given DFT approximation, e.g., LDA. The relaxation channels modeled by $\Gamma \{\rho\}$ (Eq. (2.64)) include both energy relaxation and dephasing. The relaxation operator is built so that it vanishes for the density matrix of a thermal state at a given temperature, $\Gamma \{\hat{\rho}\} = 0$. To this extent, this semi-empirical approach is also analogue to the Bhatnagar-Gross-Krook (BGK) approximation for the Boltzmann equation [BGK54]. The relaxation times are typically derived from experimental data [GZ15]. A quantum kinetic approach can also be coupled to a finite-element method description of the electrostatic field to model general metallic nanoparticles geometries, as shown in Figure 4.33 [SBK+20a, BKW+17].

From the knowledge of the plasmon decay rate into electron-hole pairs, it is possible to infer the number of "hot" charge carriers generated upon plasmon decay. However, not all of them will be available to assist a catalytic reaction, as their number and energy decrease through recombination and thermalization, respectively. To estimate the efficiency of "hot" carrier photo-catalysis, it is important to assess the thermalization time of the "hot" electrons and holes. To this end, the electronic heat capacity, C_e, and the effective electron-phonon coupling constant g_{eph}, see Eq. (4.8), can be obtained from first principle calculations [WBEV16, BSN+16a, LZC08]. It should be noted that those calculations are based on rigid lattice approximation and assuming an equilibrium (i.e., Bose-Einstein) phonon distribution at the lattice temperature, T_l. The accuracy of these approximations is expected to degrade if the electron-phonon interaction drives the phonons far off equilibrium, and a single lattice temperature can no longer be defined [MCFO17, SCD17, WBEV16]. The accuracy of all methods based on phonon dynamics is also questionable when the lattice temperature gets closer to the melting temperature. In this regime, large lattice anharmonicities and atomic displacements require a proper molecular dynamics description [WDTZ15, DR06, IZ03].

Before thermalizing, "hot" charge carriers generated upon plasmon decay can reach the surface of the plasmonic metal and be injected into the attached

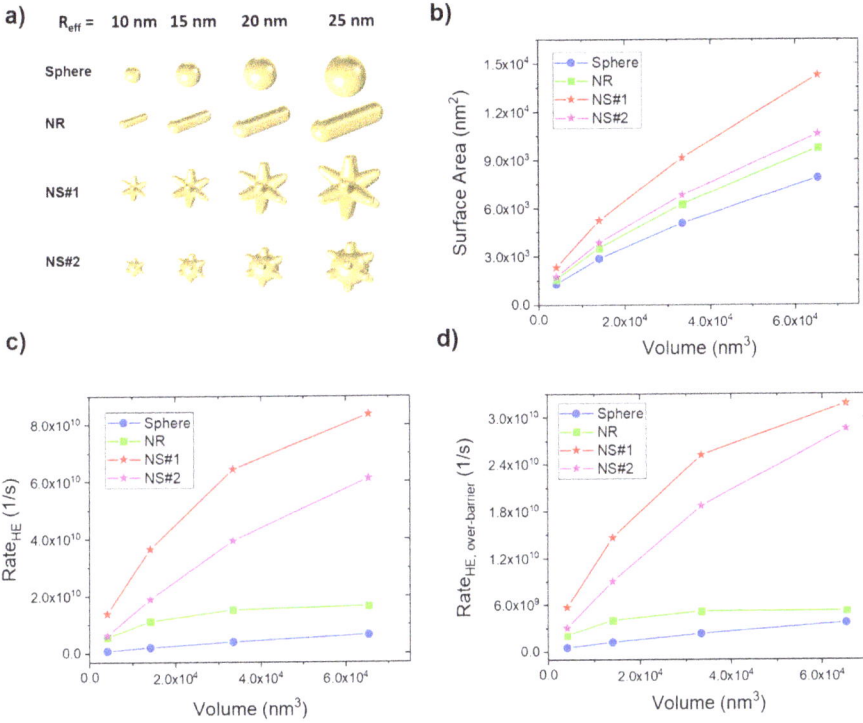

Figure 4.33 Rate of "hot" electron generation for nanoparticles of different shape and size. Panel (a) illustrates the geometries of the gold nanoparticles modeled in Ref. [SBK⁺20a], nanospheres, nanorods, and two types of nanostars. Panel (b) shows the surface area of the nanoparticles as a function of their volumes. The surface-to-volume ratio is larger for nanostars, which are also more efficient for "hot" electron generation (see below). Panel (c) reports the "hot" electron generation rates as a function of the nanoparticle volume. Panel (d) differs from Panel (c) as only the "hot" electron available for photo-catalysis ('over-barrier') are accounted for in the rate. The trends are comparable. From [SBK⁺20a].

semiconductor or molecule, e.g., via internal photo-emission. A multiscale scheme based on first principles Boltzmann transport equation along with classical modeling of the field enhancement in a plasmonic metallic nanostructure has been demonstrated [Jer19, TJS⁺18]. In this way, both the spatial variation needed to resolve the "hot" carrier transport and the energy variation needed to resolve the "hot" carrier relaxation are included in the same numerical scheme [Jer19]. As the Boltzmann transport equation is linearized, this numerical approach may not be adequate to model relaxation far off equilibrium modeled through the TTM [BSN⁺16a].

5 Conclusions and Future Directions

The aim of this book is to offer the reader a critical view of practical tools to analyze irradiation dynamics. We focus on the highly demanding case of electrons dynamics even if coupling of electrons to ions and possibly an environment are an irreducible part of the problem. It has indeed to be considered as well, but in the perspective with respect to electrons. At the present stage we thus need to gather and summarize all we have seen in the former chapters in terms of formal and practical developments. The point is to provide an easy to use access to what one should know on the various theories, what they can achieve and what remains to be done. Being aware of both the successes and limitations of these theories is certainly a major issue.

In Chapter 4 we have seen numerous examples of applications of the formalisms introduced in Chapter 2 using the analysis tools described in Chapter 3. We have seen in several examples that there is yet no theory allowing to cover the many spatial and temporal scales encountered during an irradiation scenario. There nevertheless exist numerous approaches which are able to deal, even sometimes at a high level of sophistication, with some aspects of the process. We thus somewhat have at hand a puzzle of theories that cover complementary aspects.

The first aspect to be addressed is thus to summarize what can be achieved with today's theories both formally and practically, and in relation to given physical situations. Stated in simple terms the question we shall try to answer to is: what is the theory to be used in a given context, what it can do and what it cannot do? This aspect will be extensively discussed in section 5.1.

Because the problem of irradiation remains an open challenge, once one has become aware of the limitations of today's theories, the next step is to point out major open theoretical challenges. This aspect is not trivial and does not refer to "simple" computational limitations, by far. We have already indirectly addressed such questions in particular in Chapter 2. But with the examples of Chapter 4 these open questions have taken a more solid flavor. We shall thus address these challenging, and often very open, questions in the second part of the present chapter (section 5.2). The list covers both formal and computational issues, some of them known for decades but still resisting recurrent efforts. It is by no means complete but gives a flavor of where to invest most efforts, at least form our point of view.

DOI: 10.1201/9781003127949-5

5.1   A SUMMARY VIEW: WHAT CAN THEORIES OFFER?

This section tries to summarize what we learned from the capabilities of the theoretical tools, presented in Chapter 2, to describe the various applications, discussed in Chapter 4. The workhorse for describing electronic dynamics over a wide range of dynamical regimes is TDLDA (section 2.2.3.4) and, for higher excitation energies semi-classical methods as VUU and effective electronic MD. It does a surprisingly good job as a robust, reliable, and inexpensive approach. And yet, TDLDA has limits and we have seen them in several places. This opens the wide field of correlations which are everything beyond TDLDA (section 2.3).

The importance of correlations depends not only on the actual process and its dynamical regime, but also on the observable of interest: To give a few examples: Genuine two-body observables, as coincidence measurement of two-electron emission [WGW$^+$00, UMD$^+$03], require an explicit description of correlations already in the ground state. Photo-absorption spectra are computed in the linear regime of very weak excitations (section 3.4.2), but measuring them with extremely high resolution reveals the impact of complex configurations. Electron TDLDA following strong irradiation is initially well described by TDLDA while dynamical correlations must be taken care of at later stages of the evolution to account properly for the attenuation of emission due to internal heating (section 4.1.1.3). A summary description of the kinetic distributions of fragmentation after violent irradiation can still be given by TDLDA while detailed fragment probabilities require correlations. All these examples together indicate that correlations cannot be easily quantified in general.

5.1.1   TIME AND CORRELATIONS

It is useful to provide a sketch of how correlations develop in the course of time in continuation of the discussion in section 1.2.2 with the schematic Figure 1.5. To that end, we restrict considerations to the correlation content of the actual quantum state. We can derive a measure from the one-body density matrix in terms of natural orbitals Eq. (2.20). A pure mean-field state is characterized by occupations $n_i = 1$ or 0 whereas a correlated state produces non-integer occupations $0 \leq n_i \leq 1$. A convenient measure for the amount of correlations is thus the one-body entropy

$$S = -\sum_i \left[n_i \log(n_i) + (1 - n_i) \log(1 - n_i) \right] . \tag{5.1}$$

Note that this is a purely theoretical analyzing tool, not to be confused with the entropy as a physical, thermodynamical quantity.

A typical time evolution of the one-body entropy S in a strong irradiation process is sketched in Figure 5.1 where the red line indicates the pursuit of S for the hypothetical exact solution. Dominant physical process are indicated

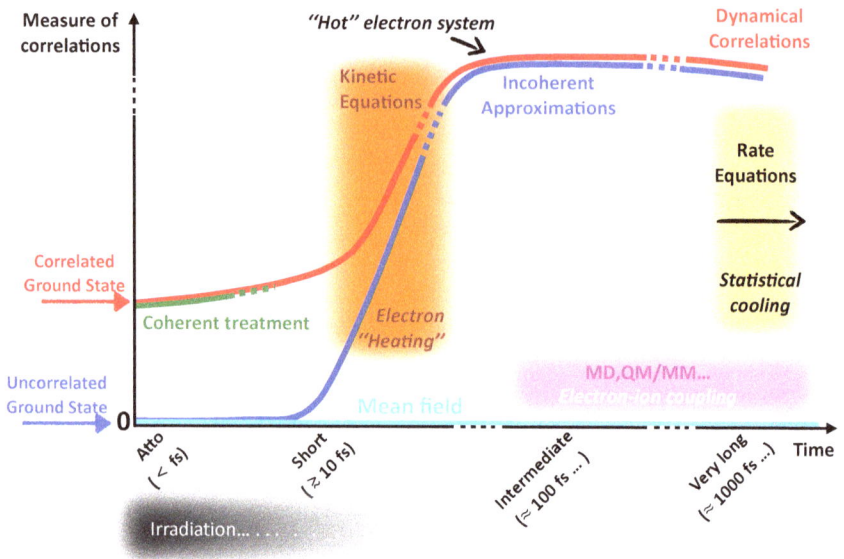

Figure 5.1 Schematic illustration of the time evolution of correlations, quantified in terms of a one-body entropy Eq. (5.1), in a strong irradiation process. The red line shows the correlations from a (hypothetical) exact solution, the dark blue line stands for mean-field theory (as TDLDA) plus dynamical correlations (as RTA), and the cyan line for pure mean-field theory. Typical dynamical regimes, in their relevant times, are indicated in italics, together with the appropriate theoretical approaches for each regime.

in italics. The abscissa comments the time spans and a shaded area at early times symbolizes the duration of the irradiation. Already before the start, possible ground state correlations set a finite initial value for S. The subsequent dynamics adds further "dynamical" correlations which enhance steadily the entropy. On the longer term this reflects the heating up of the electron cloud via electron-electron collisions, a phase which is is dominated by increasing incoherent dynamical correlations. Coupling to ionic motion comes into play at the typical ionic time scale. At some later stage the electron cloud becomes thermalized and slowly starts to cool down by thermal electron emission and ongoing energy transfer to ionic motion.

Major physical phases are associated with typical theoretical modeling and also indicated in Figure 5.1. The initial phase is well described by coherent propagation and we have seen that even the robust TDLDA does a good job here. Dynamical correlations become increasingly important as time evolves. This is the regime of kinetic equations (see section 2.3.7). At about the same time, coupling to ions, and possibly environment, comes into play. Ions can be handled by MD and environment by QM/MM (see section 2.5). Later on comes the thermalization phase whose extremely long time spans can only be handled with rate equations.

The dark blue line in the figure indicates the one-body entropy for a mean-field approximation augmented by dynamical correlations (e.g., RTA). That starts at $S = 0$ because ground state correlations are absent by construction. To be precise, they are embodied effectively by the energy-density functional in DFT. But a mere mean-field theory as TDLDA produces $S(t) = 0$ all times (cyan line). We have seen that this is insufficient, at least for TDLDA, which calls for inclusion of dynamical correlations, e.g., still coherent ones by the GW approach (section 2.3.6) at lower energies or incoherent ones (see section 2.3.7). Dynamical correlations dominate sooner or later for strong excitations and the two curves for the exact state (red) and for mean-field plus dynamical correlations (blue) come close while a mere mean-field description (cyan) stays far off. This sketch serves to communicate that dynamical correlations must be accounted for at longer times and that kinetic equations are a sufficient means to do that in an affordable manner. Of course, this is a first-order orientation through the diffuse landscape of correlations. As said above, the final judgment must also take into account the actual observable under consideration.

5.1.2   TIME, SIZE AND EXCITATION ENERGIES

Time is probably the most important sorting parameter and Figure 5.1 has discussed the appearance of correlations in terms of evolution time. But there are other "parameters" which can be also decisive. As the next ones, we discuss the impact of system size and deposited energy. This is done in Figure 5.2 where we have sorted available approaches, and range of applicability thereof, according to time again, for a direct connection with Figure 5.1, but also according to system's size and deposited excitation energy. It should be noted that the ranges of applicability discussed here are mostly indicative and qualitative. They do not imply strict limits. This is even more true as such limits may depend on the actual observables under consideration.

Quantum mean-field approaches such as real-time TDDFT/TDLDA (section 2.2.3) have a rather wide range of applicability in terms of sizes and times. But they are restricted to small excitation energies and even in that energy domain, we know that they miss part of coherent correlations, namely those which cannot be effectively incorporated into exchange-correlation functionals, section 2.2.3. These correlations are accounted for in the few available quantum approaches accounting for coherent correlations in dynamics (section 2.3). Unfortunately, these approximations are very involved, thus bound to rather small systems and short times. Extensions of these approaches beyond these limits are presently out of reach.

At larger excitation energies one may switch to quantum approaches incorporating dynamical correlations in an incoherent manner. This way one misses

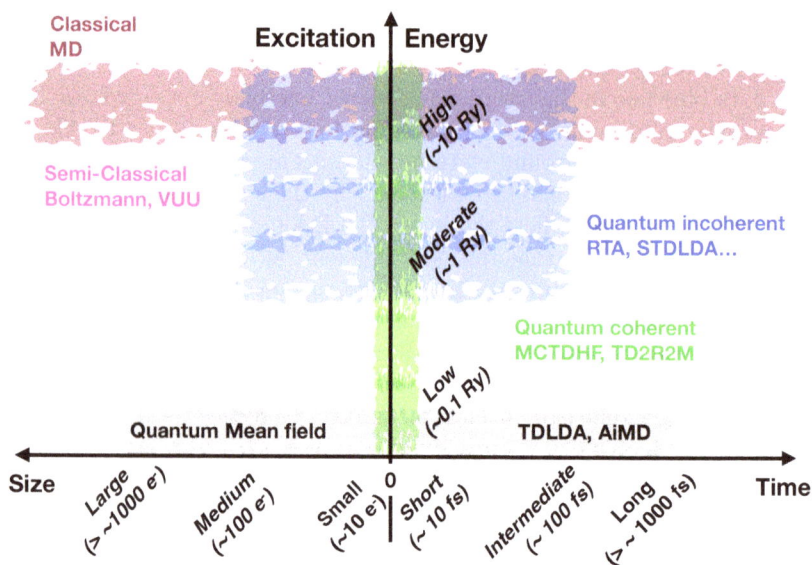

Figure 5.2 Capabilities of various theories to address time, sizes, and excitation energies. Ranges of applicability are to be seen with a grain of salt as no strict limits exist, even more so as applicability may depend on considered observables.

coherent correlations, but these are expected to be relevant only at smaller excitation energies. For larger excitations, incoherent effects dominate and lead, e.g., to dissipative effects at electronic level (see for example section 2.3.7.2). The range of applicability for incoherent approaches in times and system sizes is very much larger than for coherent correlations. This allows us to address numerous moderate size systems on physically relevant longer times. Still, these dissipative theories are more expensive than a pure mean-field treatment. This poses a challenge for computational developments to extend their range of applicability in the near future.

At even larger excitation energies quantum features tend to be wiped out and semi-classical or even classical approaches become acceptable. Semi-classical approaches based on kinetic equations (VUU, section 2.4.1) are a good compromise here. Their range of applicability with respect to mass and time is still wider than comparable quantum approaches with incoherent correlations. A problem is that there is no simple criterion for the quality of semi-classical approaches because it depends not only on energy, but also on the type of binding. Simple metals are usually well described but covalent molecules are often poorly treated. This limits semi-classical approaches in what concerns their flexibility.

Finally, at even higher excitation energies, fully classical approaches can be used (section 2.4.2). We refer here to the Molecular Dynamics (MD) approaches which are widely used to describe electrons in high intensity laser irradiation, though being limited to very violent excitation scenarios [SSR06]. Again, not only energy but also time plays a role to allow MD approaches. At late stages of an irradiation process, if the electron cloud has thermalized, electronic MD might also become a viable choice for describing long time evolution. After all, classical MD approaches allow us to address very large systems on very long times and are thus quite useful.

5.1.3   WHICH THEORY FOR WHICH PROPERTIES?

We have seen in section 5.1.1 which theories are applicable to which time scale. We have complemented this analysis in section 5.1.2 by taking into account the accessible numbers of active electrons and the excitation energy in the system. Figure 5.3 tries to provide a complementing summary by adding some technical aspects in terms of observables. We consider here a few representative theories at various levels of sophistication, focusing on the ones mostly used along this book. Some references to the sections in which they are described are provided in the last column of the upper panel. The lower panel displays a list of characteristics in terms of number of electrons, accessible physical times and physical properties. Finally, each property, for each theory, is attributed a "degree of feasibility" ranging from full access (black) to no access (white). Three complementing fillings are considered in between these two extreme cases and try to reflect the more or less complete capability of a theory to fulfill a certain characteristic or access to a given quantity. These indications are qualitative and should thus be taken with a grain of salt, but they give a flavor of what can be safely done and what cannot. This last statement covers two aspects. First it is a status as of today so that it can very well evolve with time, especially toward more feasibility. Second, and most important, feasibility does not mean relevance. A given theory like classical MD for electrons is so simple that it works formally in every regime, but it is not valid everywhere because it misses quantum mechanical features. The question where it is reasonable to apply theory, i.e. the area of applicability, is addressed in Figs. 5.1 and 5.2 discussed in sections 5.1.1 and 5.1.2 respectively. They need to be read together with the present figure.

Ranges of size N (number of electrons) and times t mean preferably feasibility. The trend is clear, the more refined a theory the more restricted to small systems and short times while more becomes possible the more approximations are involved. They imply also, to some extent, ranges of validity. Approximations are usually better validated at their upper end of feasibility (concerning size and time). For example, electronic MD becomes more and more valid for longer times and larger systems while it is not a good idea to use it for short times and small systems.

Approach	Section
MD (elect.)	2.4.2
AIMD	2.5.2
GW	2.3.6
TDLDA(-MD)	2.2.3.4
VUU(-MD)	2.4.1
RTA	2.3.7.2
STDLDA	2.3.8
MCTDHF	2.3.4.3
TDCC/TDCI	2.3.4
TD2RDM	2.3.5
TSH	2.5.4.2
MS	2.5.4.3

Property columns — Class: Quant./Class. (Classical, Semi-class., Quantum); Size (N < 10, N < 1000, N > 1000); Time span (t < 10 fs, t < 100 fs, t < 1 ps, t < 1 ns, Ensemble); Electrons (Time Res., Correl., Fluctuat., Temp.); Coupling; Ions (Time Res., Correl., Temp.); Section.

Legend: no / hardly / partly / mostly / yes

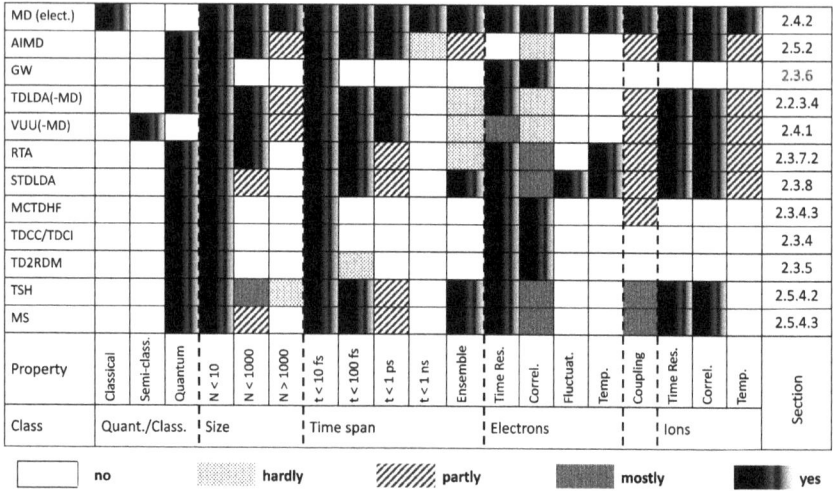

Size	N < …	"Maximum" accessible electron number
Time span	t < … fs	"Maximum" accessible physical times in fs
	Ensemble	Description of dynamics from an ensemble of trajectories
Electrons	Time Res.	Time resolved (real time) dynamics
	Correl.	Electron-electron correlations (coherent and/or incoherent), section 2.3
	Fluctuat.	Electronic fluctuations
	Temp.	Access to electronic temperature
	Coupling	Level of electron-ion coupling in dynamics (section 2.5)
Ions	Time Res.	Time resolved (real time) dynamics
	Correl.	Ion-Ion correlations (coherent and/or incoherent)
	Temp.	Access to ionic temperature

Figure 5.3 Upper panel : The various approaches and their range of feasibility as "gray"-scale matrix. Black signals feasibility and white unfeasibility. Intermediate cases are indicated by various fillings as indicated. Abbreviations needing explanations are detailed in the lower panel. See text for more details.

The entry "Ensemble" (in "Time Span" class) is a technical characteristic of the dynamics. It refers to the nature of the description: whether it relies on an ensemble of trajectories (typically 100-1000) or whether it is attained from a single trajectory.

The entry concerning electron-ion coupling has to be taken with a grain of salt. Most theories combining electron motion with classical ionic MD are doing well for the dynamics of direct coupling (see section 2.5.3), but fail in thermalizing properly the ionic ensemble, which requires fluctuations in the coupling 2.5.4. Theories that contain by construction electronic fluctuations may also solve the thermalization problem.

5.2 MAJOR OPEN THEORETICAL QUESTIONS

5.2.1 ELECTRONIC CORRELATIONS IN THE FAR OFF EQUILIBRIUM REGIME

5.2.1.1 Coherent and Incoherent Dynamical Correlations

Looking back at the application examples in Chapter 4, we realize that the majority of them has been described on the basis of TDLDA (section 2.2.3.4) which is the workhorse for practical calculations. Some examples roughly account for dynamical correlations using RTA (see section 2.3.7.2) to cope with longer time scales or higher excitations. If system size and excitation energy allows, (semi-)classical methods (section 2.4) become a preferred choice because they are simpler to use in those demanding scenarios and they easily allow us to include correlations. In the dynamics of finite quantum systems a widely open problem remains the description of correlations in intermediate dynamical regimes still dominated by quantum effects. While low energy phenomena are dominated by coherent correlations, regimes corresponding to moderate excitation energies can be addressed using simple treatments of dynamical correlations, even if they may require elaborate techniques such as stochastic approaches (section 2.3.8).

The problem is, in fact, more general. Non-equilibrium (possibly far off equilibrium) many-particle dynamics has been studied for decades in several fields of physics ranging from ultracold atoms to hot nuclear matter, from condensed matter to dense plasmas. Its long history can be traced all the way back to Bohrs early pioneering work on charged-particle penetration and stopping in matter, that preceded even his early quantum theory and later on nuclear collision dynamics [BW39]. The nuclear physics domain addresses dynamical correlations particularly in energetic heavy ion collisions, but mostly with classical or semi-classical theories, and only occasionally with approximate quantum treatments [AARS96, DST00, BGG+12]. Non-equilibrium dynamics is also an issue in transport processes in solid state physics [BD19] and quantum fluids [KB62, PN66]. Recently it has become an important issue in ultracold bosonic and fermionic gases [Bon98, LLM+20].

Finite fermion systems as, e.g., molecules are much more demanding and, as we have seen, correlations cannot be ignored in many cases. They have an impact on key observables such as dipole response (Figure 4.3) and ionization (Figure 4.6). They become crucial on medium to long times for a correct distribution of energy share (Figure 4.4) with an effect on electron emission PAD (Figure 4.13). They become the more important the higher the excitation energy (section 4.1.3). The importance of correlations depends not only on time or energy, but also on the observable. For example, density correlation functions or coincidence measurements of two-electron emission (see section 3.4.5) require correlations from the onset. One successful example will be discussed later in Figure 5.7, where electrons are described at the highest possible level of sophistication with full many-body correlations. But this was possible only because the system and the number of active electrons were very small. Already for moderate-size systems manageable schemes for time-dependent correlations are rare although there exist many promising lines of development as outlined in section 2.3. They build upon successful methods to treat ground-state correlations such as Complete-Active Space Self-Consistent Field (CASSCF), multi-reference configuration interaction or coupled-cluster methods [GL20, LSM+20, BBC+20], as well as alternative routes through Reduced Density Matrix (RDM) functional theory [PG15], Greens functions and many-body perturbation theory [BDJL20].

Time dependent extensions of such quantum chemistry methods can now provide dynamical correlations at short times, low excitation energies, and small to moderate system sizes [LLM+20], but with little hope to extend them directly to the dissipation-dominated regime at higher energies and longer times. The Multi-Configuration TDHF (MCTDHF, section 2.3.4.3) method with the dynamically correlated wave function expanded over a coherent set of TDHF (or pure) states [LKY17] is conceptually simple and can exactly account for dynamical correlations. But the unfavorable factorial scaling with particle number and the explosive growth of relevant configurations with excitation energy sets a severe limit to its applicability. TD-CASSCF reduces the number of dynamically active particles but still leads to a factorial scaling. First implementations of the TD Coupled-Cluster theory for simple model systems have been reported [Kva12]. Alternative routes include explorations beyond standard TDDFT [EM16], trajectory surface hopping approaches (section 2.5.4.2), TD density matrix functional theory [PG15] (section 2.3.5), non-equilibrium Greens function approaches [SvL13], many-body perturbation theory [SFM+19], TD quantum Monte Carlo methods [GDGA18], and TD Density matrix renormalization [RLJHC17]. Although much progress has been achieved with these developments, they remain still limited concerning system size, time span, and excitation energy.

Large systems and high excitation energies render approximate semi-classical and even fully classical approaches applicable (section 2.4). But between these complementary approaches (fully quantum and semi classical ones), there remains a large gap into which many laser irradiation scenarios fall, in particular in the regime of moderate excitation energies (deposited energy a few times the first ionization potential). In those scenarios, quantum effects remain still relevant, but the duration and energy deposition already lead to dissipative effects and thermalization of the electron cloud. That cannot be accessed in practice by a fully coherent description of correlations within any one of the above mentioned methods. On the other hand, classical and semi-classical approaches (section 2.4) describe such dissipative features but ignore any quantum effects, as observed, e.g., in PES [KJJ+10, HRA+17]. There is thus clear need to develop approximate methods in between these regimes.

5.2.1.2 A Few Promising Examples

As we have seen, the quest for a quantum theory of dynamical correlations accounting for both coherent effects at short times and incoherent ones at long times remains open. Presently, one thus has to rely on theories covering only partial aspects thereof. We have for example already seen along this book a few applications within the RTA (section 2.3.7.2). Let us recall the impact of such incoherent correlations on dipole damping in Figure 4.3 or on ionization in Figure 4.4. But RTA is a rough approximation even at the level of incoherent correlations and one often needs more elaborate approaches. We present here two examples addressing the problem with two complementary quantum methods. The first one is the Time Dependent second Reduced Density Matrix (TD-2RDM) method, which includes coherent correlations at the second order of the quantum BBGKY hierarchy (section 2.3.5) by a truncation of the hierarchy [Bon98]. The second one is STDLDA treating incoherent correlations in a stochastic manner (section 2.3.8) after approximating the correlated state of a system by an incoherent ensemble of pure states.

The TD-2RDM is a quantum kinetic equation obtained by closing the BBGKY hierarchy through an approximate reconstruction functional for the higher-order density matrix in terms of the 2RDM with help of dedicated functionals [LBS+17]. It allows for an explicit and accurate account of 2-particle correlations and extraction of 2-body observables. Numerical implementation of TD-2RDM scales polynomially rather than factorially with systems size, thus promising the accurate treatment of larger systems. Quantitative comparisons with MCTDHF of non-linear dynamics in Be and Ne atoms exhibit an excellent agreement [LBS+17]. TD-2RDM thus offers an affordable alternative to MCTDHF and constitutes an exact benchmark approach on short times and for small systems. An example of application of TD-2RDM is shown in Figure 5.4. The figure displays the time evolution of the dipole moment (left panel)

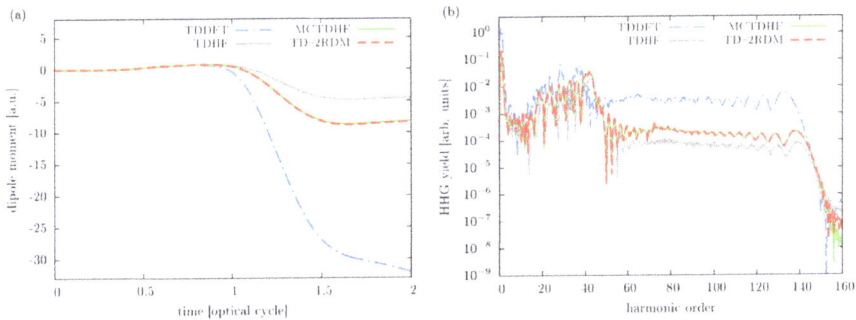

Figure 5.4 Time evolution of the dipole moment (left panel) and HHG spectrum (right panel) in the Ne atom, following a two cycle laser pulse of intensity 10^{15}W.cm^{-2}, frequency 1.5 eV and Carrier Envelop Phase 0 (section 4.1.3.1). From [LBS+17].

and the higher harmonic spectrum (right panel, see section 4.1.3.3), and compares various levels of theory: TDHF (section 2.2.2), TDDFT at TDLDA level (section 2.2.3), MCTDHF (section 2.3.4.3), and TD-2RDM (section 2.3.5). Several aspects need to be pointed out. First, TD-2RDM and MCTDHF lead to exactly the same results. Furthermore, these results significantly differ from TDDFT ones while differences with TDHF are in this case much smaller. The effect is all the more striking that differences appear very quickly and do not seem to level off with time. Note that the actual time span is still very small in this example: it reaches a few fs at best and this for a rather small system (Ne atom). We are hitting here a practical difficulty in solving such an involved equation in which the reconstruction functional turns out to be extremely sensitive [LBS+17]. Extension to much larger times is thus delicate.

An alternative route is the stochastic treatment of incoherent correlations via STDLDA (section 2.3.8). The capabilities of this approach are illustrated on an example in Figure 5.5 in a 1D model system. The system is excited by an instantaneous particle-hole excitation at initial time which simulates an ultra-fast laser excitation (section 4.1.3.2). The figure compares TDLDA and STDLDA (left panels). Correlations build up in the course of time out of energy conserving 2 particles - 2 holes transitions which are collected in an ensemble of TDLDA trajectories (section 2.3.8). The transitions can be seen

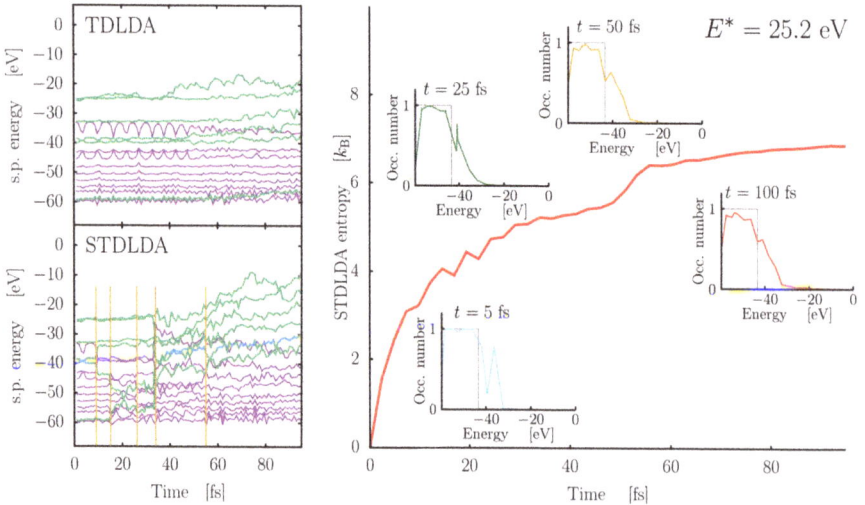

Figure 5.5 Time evolution of the single-particle energies (left panels) in TDLDA (upper left) and STDLDA (lower left) following an instantaneous initial time particle hole excitation which delivers about 25 eV excitation energy. Occupied states are in purple, unoccupied states in green. Jumps ($2p2h$ transitions) are indicated by brown vertical lines (lower left panel). Right panels display the time evolution of STDLDA one-body entropy Eq. (5.1) with snapshots of occupation numbers distributions as insets. Adapted from [DLR+18].

from the jumps in single particle energies (lower left panel). This amounts to treat electron-electron correlations at the level of a quantum Boltzmann collision term realized as a stochastic process. In the long term, this leads to a thermalization of the electron cloud as illustrated in the right panel of the figure in terms of occupation numbers distributions (insets) and in terms of the single particle entropy S Eq. (5.1). The drive toward thermalization is very clear both on S and on snapshots of occupation numbers (insets).

A few remarks are in place here. First, the time evolution runs over long times which allows one to properly treat thermalization effects, as done also, but in a more approximate way, in RTA (section 2.3.7.2). At the price of overlooking short time coherent correlations, such an approach thus looks quite appropriate for long-term evolution, where incoherent correlations are anyway expected to take the lead (see the discussion around Figure 5.1). As a second point, we recall that this example has been computed in a 1D model system as a proof of principle. STDLDA scales with system size as TDLDA. A limitation is the size of the ensemble. The larger the ensemble, the higher the resolution with respect to less dominant reaction channels. The proper balance between ensemble size and resolution depends on the application. An important aspect, as demonstrated by the present example, is that no instability is occurring in such calculations which gives way to long-time simulations. Another advantage is that the ensemble of mean field states in STDLDA allows one to access all density matrices at any order. Of course, these density matrices are approximate, as the whole STDLDA is, but they explicitly fulfill consistency relationships between the many-body levels, as required by the BBGKY hierarchy, something which is hard to reach with methods relying on a truncation of the hierarchy [GKRT93].

5.2.1.3   An Example of A Yet Unresolved Problem

Figure 5.6 provides an interesting illustration of what can be achieved and what is left yet to be achieved in a realistic case of irradiation of a standard molecule, namely C_{60}. It presents a comparison between experimental and TDLDA computed PES at laser frequency $\hbar\omega_{las} = 20$ eV, which lies well within the one-photon regime (section 4.1.2.2). For reasons of computational cost, the theoretical pulse has a duration of 60 fs, while the experimental pulse duration is on the order of several picoseconds but one may hope that in the one-photon regime the difference in pulse lengths does not impact too much the results. However, the long duration of the experimental pulse may explore effects of thermalization which slightly soften the PES. Ionic motion as such is included in the calculations because that comes into play already for the time span considered. Note that the experimental and the theoretical PES in the figure have been normalized independently to deliver a fair comparison of the heights of the peaks from the HOMO and HOMO-1 orbitals (higher energy peaks in the 11-13 eV range). This is justified by the fact that the final ionization is unknown experimentally.

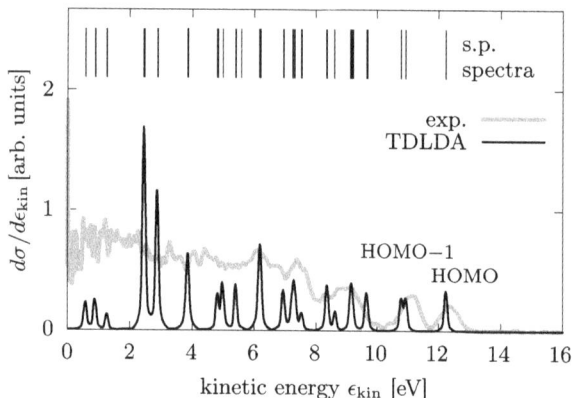

Figure 5.6 Theoretical (TDLDA, red line) and experimental (green line) PES in a one-photon regime irradiation ($\hbar\omega_{las} = 20$ eV, intensity $7.8 \ 10^9$ W.cm^{-2}). The vertical bars show the static single-particle density of states upshifted by the laser frequency and downshifted by 0.3 eV to account for the Coulomb shift due to the final ionization of 0.1. Pulse duration is 60 fs for the TDLDA computation and several ps for the experiment. Adapted from [WGB+15].

Figure 5.6 exhibits interesting similarities and differences between experimental and computed PES. The theoretical PES is highly structured and shows well-separated s.p. transitions down to the deepest states. To demonstrate that we see, indeed, the s.p. transitions, the (static) s.p. spectrum upshifted by the laser frequency is indicated near the upper border of the panel, downshifted by 0.3 eV to account for the Coulomb shift stemming from the residual positive charge associated with the net ionization (0.1 in that case).

The comparison with the experimental PES is fairly good for HOMO and HOMO-1 levels. Still, the experimental peaks are broadened due to vibrational energy that is not taken into account in the theory. This stems from the finite temperature of around 500 K of C_{60} in experiments, the effect being enhanced by the picosecond duration of the laser pulse. The impact of a thermal ensemble can be accounted for with some extra expense, as outlined in section 4.2.1.1. But this is not the issue here as it would most probably not change peak locations.

Truly qualitative differences develop toward lower energies with the experimental PES exhibiting a broad, unstructured contribution. This shape is attributed to electron-electron collisions which partly hinder direct electron emission turning to involved reaction paths as multi-step emission or auto-ionization mechanisms. A proper description of such processes requires to go beyond TDLDA by including correlations (incoherent and possibly coherent ones, see section 2.3). In the above case the comparison between data and theory is only relevant for the highest PES energies where direct electron emission prevails which still allows to associate the peaks with s.p. energies in the mother system.

5.2.2 ION-ELECTRON CORRELATIONS

5.2.2.1 General Considerations and Context

A finite electron cloud alone poses already a couple of intriguing problems as we have discussed in the previous section. The coupling to ionic motion adds another layer of complexity. Ionic motion can more easily grow to large amplitude which amplifies electronic fluctuations. On the one hand, this is beneficial as it makes them better accessible to measurement. On the other hand, this renders the outgoing reaction channels too different to be tractable by one common mean field. Thus ion-electron correlations are a central issue. We already discussed that from an experimental viewpoint in section 1.3.1 (see in particular Figures 1.8 and 1.9). These correlations are present all over an irradiation process. At low to moderate excitation and/or net charge, ions will, for example, impact the optical response (see for example section 4.2) by coupling electronic eigenfrequencies to ionic vibrations. At higher excitation energies, ionic motion leads to dissociation and fragmentation processes, typical of the late evolution of ionized species (see the examples of Figures 1.1 and 1.3). Therefore, a purely electronic dynamics is relevant only at early stages of a reaction. Sooner or later one must consider coupling to ionic dynamics.

A first difficulty then immediately becomes apparent: the time scales spread over several orders of magnitude from sub-fs, to cover properly the doorway of electronic excitations, to hundreds of fs or even ps, for the consequences of ionic motion. This means that modeling of such processes will require access to several complementing theories (see section 5.1). These have to be properly interfaced as no approach today allows one to deal with these different time scales within one single theoretical framework.

The second, important, difficulty concerns the large amount of possible exit channels leading eventually to fragmentation for sufficiently high excitation energies. We will discuss that extreme case separately in section 5.2.3, but it is already present in "simple" dissociation processes (see the example of section 5.2.2.2 below). Even before it comes to high excitation energy regimes, one has to deal with a multitude of ionic trajectories for which one needs involved theories beyond Ehrenfest dynamics, see sections 2.5.4 and 4.2.2. And well manageable theories in that regime are presently still under development. In the case of the strong perturbations typical of many irradiation scenarios, today's most elaborate computations are thus still restricted to rather simple cases in which only a few exit channels (often only one or two) are considered. Extension to larger systems represents a rather big step and is not yet directly accessible.

5.2.2.2 An Example of A Full Treatment

An example of an experiment which can be described theoretically in full detail is displayed in Figure 5.7. The case concerns the photo induced dissociation of H_2 into a proton, an H atom and an electron: $H_2 + h\nu \rightarrow p + H + e^-$. The study focuses on symmetry breaking effects in this reaction and presents a direct comparison between complete experimental and theoretical results. The term "complete" refers here to the fact that the full kinematics of electrons and ions is measured. Conservation laws in this small system immediately provide information on the third partner, the hydrogen atom. On the theory side, both electrons and ions, have been treated fully quantum mechanically and all possible reaction channels implying various ionic and electronic pathways have been accounted for. The interesting aspect here is that one needs a full quantum treatment of both electrons and ions because the symmetry breaking mechanism requires an entangled electronic-ionic wavefunction. Figure 5.7 illustrates both the complexity of such an approach (left panel) and its success (right panels). The left panel shows the APES of the system including doubly excited states of the molecule. The right panel displays a direct comparison

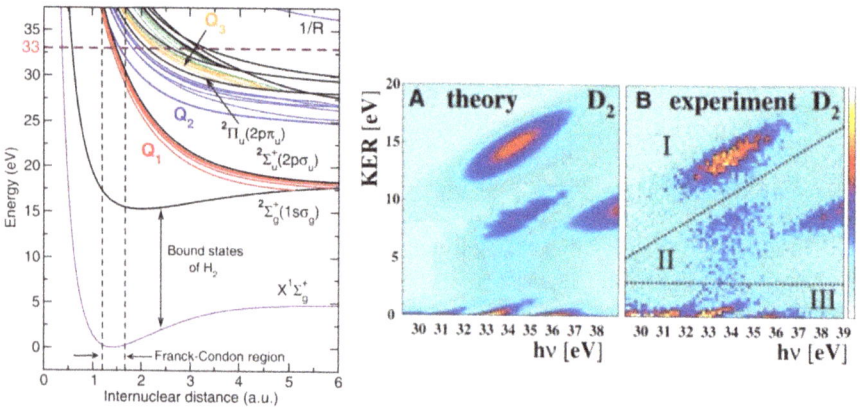

Figure 5.7 Symmetry breaking in H_2 dissociation following single photon absorption. The left panel displays the energy-level diagram and pathways to dissociative ionization. The total energy of H_2 and H_2^+ are shown as a function of inter-nuclear distance (a.u., atomic units). $^2\Sigma_g^+$ and $^2\Sigma_u^+$ represent the lowest gerade and ungerade states in H_2^+ while red and blue curves are the lowest two series of doubly excited states of H_2. The right panel provides a comparison of experimental and theoretical ionic Kinetic Energy Releases in D_2 as a function of photon energy $h\nu$. Computations have been performed treating both electrons and ions at quantum level. Note the remarkable agreement between theory and experiment as well as the complexity of the excited states landscape in such a simple molecule as H_2. Adapted from [MFH+07]. Courtesy of F. Martin.

of experiment and theory in terms of the ionic Kinetic Energy Release (KER) performed in the slightly different case of D_2 rather than H_2 but without changing symmetry properties of the molecule. It displays several structures which can all be reproduced by theory.

However, the complexity of the energy diagram in the left panel also points out the limitations of such a comprehensive approach. It is hardly conceivable to envision such a complete treatment of electrons and ions for much larger systems, not only in terms of computational effort but also because of its intrinsic complexity. Larger systems need a reduction on the experimental side and they are more forgiving on the theory side because of the larger number of degrees of freedom which may justify average treatments, such as mean field, semi-classical, or statistical approaches. Demanding remain, again, the intermediate cases where some quantum effects or correlations are required for a pertinent description.

5.2.2.3 Beyond Time-Local TDLDA in Ehrenfest Dynamics

Ehrenfest Dynamics (ED) remains the basic tool for treating ion-electrons correlations (section 4.2). But ED is fully coherent because both, the ions are described classically and the electrons normally evolve via adiabatic interactions. Indeed, in the majority of cases the exchange-correlation potential in TDDFT (usually at TDLDA level) is evaluated within the adiabatic approximation (time local), despite the fact that memory of the time evolution (or frequency dependence) of the electronic density plays an important role, as can be assessed from the initial-state dependence that is a necessary ingredient for Runge-Gross theorem [RG84].

The properties of the exact exchange-correlation potential in TDDFT have been discussed in [MBW02]. Memory-dependent exchange-correlation kernels are the subject of much study at the present time [KB04, RRM10, DLM22], but real-life applications are rare, not only because there is no general consensus about them, but also because of the high computational cost of carrying and integrating the time dependence in the exchange-correlation potential.

An interesting alternative approach is Time Dependent Current Density Functional Theory (TDCDFT) [VR87, VUC97, EVR$^+$15] among other works. Unlike TDDFT, a spatially local formulation of TDCDFT turns sufficiently accurate because non-locality in time can be taken care of by means of non-locality in space [Ull12, DBG97]. Said this, today, none of these two approaches seems mature enough to make it into the main stream in the near future, let alone to be combined with Ehrenfest dynamics for the nuclei.

On the other hand, there is no fundamental reason why one could not go beyond mean-field approaches (TDHF or TDLDA) for the electron-electron interaction. One possible way could be to go beyond TDLDA and couple the nuclear dynamics, i.e. Eq. (2.86), with any of the many-body methods of section 2.3 that allow for the construction of the electronic density from the density matrix $\varrho\,(\mathbf{r}, t) = \rho(\mathbf{r}, \mathbf{r}, t)$, as this is the only quantity that enters the calculation of the force in ED. Such approaches are rare in the literature, in part because the implementation of forces is not straightforward, particularly for basis sets that move with the atoms, and also because the computational cost of the electronic part makes it prohibitive to run meaningful molecular dynamics simulations at that level. For example, attempts to define total energies and forces within a GW scheme can be found in the literature [IBL03, LSD$^+$20], but so far these have not been extended to the time domain.

5.2.3 MULTICHANNEL DYNAMICS

5.2.3.1 How to Access Multichannel Dynamics?

The term multichannel dynamics covers irradiation scenarios in which a multitude of exit channels are activated. This may be the dissociation of a molecule along several different pathways leading to various smaller molecules, atoms, ions, radicals etc. Examples thereof will be discussed in sections 5.2.3.2 and 5.2.3.3 and we already discussed such a fragmentation example, from a global viewpoint, in section 1.1.3. Modeling such situations can become extremely cumbersome as soon as the number of accessible exit channels grows. This is especially true for a fully dynamical description of the process. When considering, for example, a fragmentation into various products a first approach is to explore from a static point of view energetically favorable products as illustrated in Figure 5.9. But such a picture remains far away from a dynamical description. A step forward consists in computing the APES with corresponding reaction barriers associated with the various fragmentation pattern. This suffices for a statistical evaluation of the weight of each exit channel, based on relative energies. But such a statistical description still misses dynamical aspects. The step toward dynamics would require to develop a dynamics along the various relevant APES. This is well beyond a standard mean-field dynamics, even including ionic motion, which can only provide an average description of the various possible dynamical trajectories.

The feasibility of a theoretical treatment of such processes beyond mean field depends on the actual setup, the amount of excitation energy and the charge state acquired during the process. For moderate perturbations, the dynamics of the system remains within a "small" phase space. In molecular physics terms this means that only a small number of APES, associated to excited states, are accessible. Dynamics can then be confined to these surfaces and described in terms of methods such as TSH (section 2.5.4.2) in which the various evolution channels are explored stochastically. But that implies to know in advance on which APES dynamics will take place. This is not a severe constraint at low energies where accessible states are not numerous. But it obviously becomes a strong limitation as soon as the number of accessible channels grows large. A solution could be to compute the reaction surfaces on the fly together with the dynamical evolution. The exact factorization (section 2.5.4.5) can help here as a sound starting point for developing appropriate approaches along that line which, however, has yet to be done. Even then, also the handling of dynamically generated trajectories will reach its limits if the phase space grows further. Here remain presently only average, statistical, or classical approaches.

5.2.3.2 An Example of Fragmentation Dynamics

Figure 5.8 provides an example from a calculation of fragmentation of Uracil (Ura, $C_4H_4N_2O_2$, one of the four nucleobases of RNA) following irradiation by a 100 keV proton beam [LTdPV$^+$11]. The results of the calculation are compared to experimental data. The Coulomb pulse from the bypassing proton leads to a doubly charged Uracil which in the long run dissociates into several fragments. The computations are performed with TDLDA-MD (section 2.5.3.1) for short times followed by an evolution using "ab initio" MD

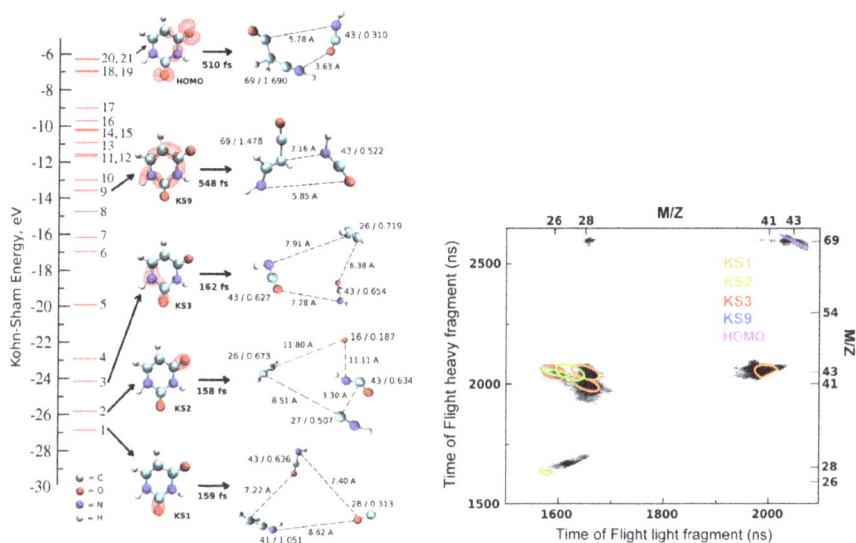

Figure 5.8 TDLDA + AIMD analysis of fragmentation. Left panel: Fragmentation resulting from two- electron removal from the HOMO, KS9, KS3, KS2, and KS1 orbitals (top to bottom). Energies of the occupied Kohn-Sham orbitals of neutral uracil are indicated in the left part and electron densities in the middle part of the panel. On the right of the panel are shown the fragments observed at the end of the TDLDA + AIMD propagation, and their corresponding mass over charge ratios. Finally fragment-fragment distances (dashed lines) are given in Å. Final times (end of the dynamics) used for fragment analysis are also indicated. Right panel: Two dimensional Time Of Flight coincidence spectrum of charged uracil fragments for mass to charge ratio larger than 22. Theoretical results for KS1, KS2, KS3, KS9, and HOMO induced fragments are also indicated for comparison. From [LTdPV$^+$11].

(AIMD, section 2.5.2), which allows to track the dynamics on a sufficiently long time span to explore the fragmentation in detail.

As the collision is fast, one can assume that the two electrons are removed almost instantaneously from the initially neutral molecule. Thus the excitation is modeled theoretically by instantaneously removing two electrons from a given level of the ground state configuration. These are denoted KSn with n labeling the level from which the electrons are stripped, see the sequence of levels in the left panel of Figure 5.8. The dynamics is then computed for the remaining Ura^{2+}. Varying the initially depleted level induces variations of deposited excitation energy (see the sequence of energies in the left panel of Figure 5.8) which leads to different fragmentation paths. The cases illustrated in Figure 5.8 are associated to more or less de-localized electron distributions. They lead to various fragmentation patterns. They are analyzed by extracting mass over charge ratios (up to the intrinsic limitations of fractional charges inherent to TDLDA), which allows for a direct comparison to experimental data as can be seen in the right panel of the figure. The calculations reproduce the most intense coincidence signals and, to some extent, their shape. This suggests that not only the fragments but also their relative velocities are reasonably described by theory. The actual comparison is thus very satisfying in spite of the much abbreviated excitation mechanisms in the computations.

Nonetheless, the success is surprising. A detailed TDLDA description of the excitation process would deliver at the end of proton collision a coherent superposition of all conceivable hole states. However, the above model treatment starts from an ensemble of well defined two-hole states. Beyond that stays the tacit assumption that decoherence proceeds extremely fast producing almost immediately the starting ensemble whose members can then be tracked each one by a separate TDLDA-MD calculation. This corresponds to STDLDA (section 2.3.8) simplified by sampling only once at the beginning and that seems to work. The case is even more surprising as the initial sampling picks only states with the two holes in the same level (spin up, spin down) and not all combinations of holes in different levels. It is not yet clear whether this is a lucky accident or whether it points out principle simplifications which could be applied in other systems too. Altogether, a promising example which prompts further investigations.

5.2.3.3   Rare Events

We have evoked in section 5.2.2 the difficulties raised by the occurrence of different reaction channels, which can hardly be treated by time-dependent coherent states. And we have discussed similar problems in the case of fragmentation dynamics (section 5.2.3.2). Things are getting even more intriguing for rare reaction channels. Even if a stochastic treatment applies, it would require huge ensembles to have a chance to include a rare channel. On the other hand, a rare channel can become important. A typical example is provided by collisions with electron beams. The scattering cross section is strongly dominated by the elastic channel in which the electron elastically bounces on the molecular target. However, the interest lies often in the inelastic channels, whose weights are smaller than the elastic channel and particular outcomes often have particularly small weights.

Electron attachment and ensuing molecular dissociation is a typical example of such a situation in which a rare reaction path is physically essential. This mechanism plays a crucial role in the understanding of irradiation processes in a biological context, in spite of the fact that this reaction channel is rare as compared to other more trivial ones. We shall thus take electron attachment and ensuing dissociation, coined Dissociative Electron Attachment (DEA), as a test case for a dynamical treatment of a rare channel.

A physically relevant example for medical applications is the case of water coating of molecules, because of the important role of water in irradiation of biological material. We have already addressed a somewhat similar situation in section 1.1.3, but there for the collisions of ions on water coated molecules and the impact of water molecules on fragmentation pattern. In the case of water, the electron attachment cross section in the physically relevant energy range (a few eV) is several orders of magnitude smaller than that of the elastic channel [IM05], but the attachment may lead to dissociation and formation of a strongly ionizing radical such as OH^-. Understanding and analyzing such processes thus represents a crucial issue, especially demanding for a dynamical description.

An example of a recent study on electron attachment is shown in Figure 5.9. The target molecule is Pyruvic Acid (PA, $CH_3COCOOH$) which is an

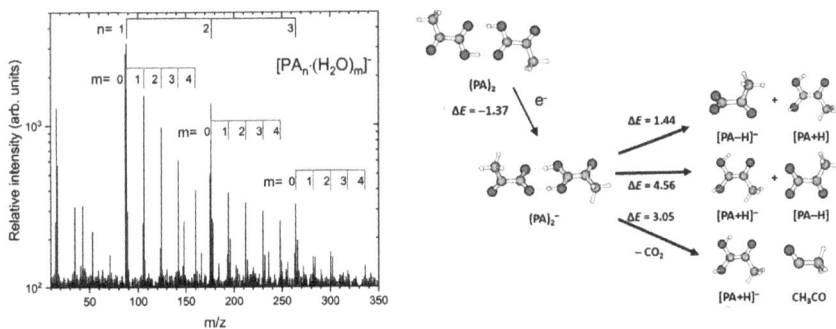

Figure 5.9 Dissociative Electron Attachment in microhydrated Pyruvic Acid (PA) clusters following an irradiation by low energy electrons (energies below 20 eV). Left panel displays the experimental cumulative mass spectrum of microhydrated PA clusters ($(PA)_n(H_2O)_m$). Fragmentation products can be identified by n and m values as indicated in the panel. Right panel displays theoretical structure calculations identifying possible fragmentation pathways in non hydrated PA_2. These calculations are purely static. From [PGK$^+$21].

organic acid playing an important role in cell biochemistry. It is considered as representing a fundamental building block in prebiotic chemistry and has been detected in carbonaceous meteorites. This has motivated studies on the response to radiation (from the interstellar medium) of PA clusters and PA clusters in ice. Experiments reported here have been performed on mixed clusters of water and pyruvic acid $(PA)_n(H_2O)_m$. They are irradiated by a beam of relatively low energy electrons (0 to 12 eV electrons) which are representative of the secondary electrons created in the interstellar dust grains irradiated by most types of high energy radiation (UV light, X-rays, high energy protons). In the considered electron energy range, below 20 eV, the most efficient fragmentation process is DEA through a resonant process visible via negative ion fragmentation pathways.

The left panel of Figure 5.9 displays the cumulative mass spectrum of microhydrated PA clusters $((PA)_n(H_2O)_m$, n = 1-3 and m up to 9). The right panel displays possible fragmentation pattern as suggested by structure calculations of PA dimers (no water in that case). Without entering details, the interesting point for our purpose is to note the richness of the fragmentation scenarios (left panel) in which both PA and water generate a multitude of fragmentation channels. Computations illustrate this complexity by displaying pathways in the simplified case of non hydrated PA dimer. It is also interesting to note that the theoretical investigation, although ab initio, in quantum chemistry terms, is restricted to geometry optimizations, which points out toward potential fragmentation yields. The step toward a fully dynamical description of such systems (possibly including water coating) is thus still quite big.

We have seen that multi-channel processes require different mean fields for each channel which can be managed with an incoherent ensemble of TDLDA trajectories, e.g., in STDLDA (section 2.3.8). We have argued that it will be awfully expensive to cover rare channels straightforwardly with that strategy. A solution can be to model the rare channel explicitly and to compute the transition probability to it specifically. A step in that direction has been explored in [LDR+15]. This is illustrated in Figure 5.10 for the case of electron collisions on water. The leading trajectory $|\Phi_{in}\rangle$ is computed by TDLDA. The initial state consists of the electron cloud of the molecule together with a Gaussian wavepacket modeling the incoming electron of energy E_{in}. The final states searched for are chosen as 2 particle-1 hole $\Psi_{pp'h}$ states built on the water ground state, which represent configurations with one electron added together with an excitation of the target. The attachment probability is then

$$\mathcal{P}_{att} = \frac{2\pi}{\hbar} \int_{-\infty}^{+\infty} dt \sum_{pp'h} |\langle \Psi_{pp'h} | \hat{V}_{coll} | \Phi_{in} \rangle|^2 \delta(E_{pp'h} - E_{in}) \qquad (5.2)$$

where E_0 is the ground state energy of the target, $\hat{V}_{coll}$ is the residual interaction for the collision channel and $E_{pp'h}$ the energy of the 2p-1h state. It is much similar to the jump probability in STDLDA, but evaluated only for the few intended reaction channels, and integrated over the whole time evolution.

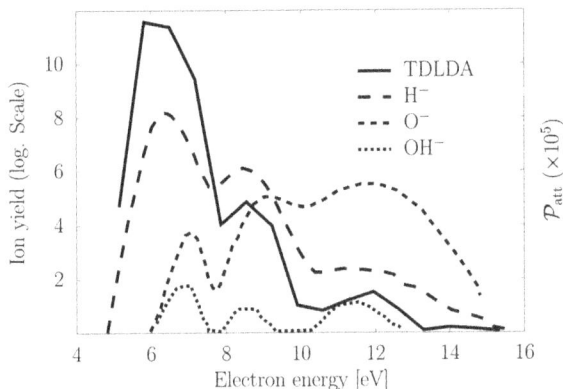

Figure 5.10 Example of a TDLDA analysis of electron attachment in water. Left ordinate axis: experimental ion yields; right ordinate axis: theoretical probabilities, Eq. (5.2). Both axes with arbitrary units adapted to match scales. See text for details. From [LDR$^+$15].

A further simplification is that the final states are still taken as stationary states. We thus deal with a mixed time-dependent and stationary picture.

Figure 5.10 shows the attachment probability as a function of the initial kinetic energy of the electron projectile. It is compared to available experimental data obtained from corresponding dissociation channels. The agreement is striking especially in terms of peak positions. This shows that one can access attachment probabilities within a perturbative treatment around a TDLDA trajectory. However, the (preliminary) modeling is limited to the attachment step. It does not provide a fully dynamical description of the attachment process itself nor of the long term evolution to dissociation. That may be achieved by continuing with TDLDA from the final states $\Psi_{pp'h}$ reached after the jumps. However, the DEA process involves the stretching of specific bonds while the electron is attached in a resonant state of the molecule, forming a transient negative ion. Therefore, a proper dynamical treatment of DEA would require a coupled electron-nuclear dynamics approach like TSH or MS (see sections 2.5.4.2 and 2.5.4.3), which, to the best of our knowledge, has not been realized up to now. This might be achievable for reasonably small molecules where the number of accessible states and transitions remains moderate. After dissociation, the electronic component can be assumed to be back in its ground state, and the process can be described through AIMD, as done for solvated thymine in [MKH$^+$19].

5.2.4 SOME COMPUTATIONAL ISSUES

In the previous sections we have discussed major challenges to a theoretical description of dynamical processes, identified the difficulties, and indicated possible solutions. We want here to address some open points mostly related to computational aspects.

5.2.4.1 The Ionization Problem

Ionization carries a subtle problem from the modeling viewpoint. Experiments measuring the kinematic characteristics of emitted electrons take place far away from the system (at macroscopic distance). But calculations do not cover such large scales. This ignores long-range Coulomb interactions between a departing electron and the system itself. The success of the many TDLDA simulations indicates that their spatial limitation is acceptable. However, one should keep in mind that it involves an approximation which may not always be justified, particularly for highly polarizable systems. Long-range effects may be included by a mixed description, either adding Volkov states [GVM+12] to the numerical basis or tracking escaping electrons far out with classical MD. Both are feasible, but add considerable expense to the calculations. Moreover, the extensions still require more testing and optimization. Thus one prefers to stay as far as possible with the well-established finite representations (grid or wavefunction basis).

Describing the dynamics of electrons in the continuum nevertheless remains a clear pending issue for which no good solution has yet been found. Approximations deliver an intermediate picture in which only part of polarization effects can be taken into account. Description of ionization, which is at the core of irradiation processes, thus remains only partially treated in today's theoretical approaches. This is clearly a limitation in our understanding of irradiation, both in terms of the underlying processes and mechanisms and in terms of access to relevant experimental observables such as energy- or angle-resolved distributions of emitted electrons (section 3.4.4).

5.2.4.2 On Numerical Simulations Themselves

Numerical simulations are inevitable for the description of irradiation processes because analytical approaches are rare if one aims at a sufficient degree of realism. Dedicated algorithms have thus been developed over decades and we now avail of well-validated numerical methods. This, nevertheless, does not mean that numerical simulations can always be taken for granted. Two aspects will be mentioned here.

The first concerns benchmarking. Of course, the ultimate goal is to compare theories with experiments to explore the relevance of theories. But comparison between theories is also an important issue to learn about the validity of a theory. There are two aspects here: i) comparison between competing theories and numerical realization thereof and ii) comparison between different numerical realizations of the same theory. We do not mention here comparison to "analytical" cases which are marginal in far-off equilibrium dynamics.

While a comparison between two different theories looks natural, a comparison between two implementations of the same theory may look more surprising. Numerical realizations of a theory must indeed be benchmarked with respect to each other. This may look trivial, but it is not. Building up computer codes for far off equilibrium dynamics implies numerous implicit numerical choices, whose final impact on the results is often hard to foresee. This may lead to unexpected surprises [OPP+20]. Numerical benchmarking is thus essential to validate computer implementations of a given theory and should ideally be done before discussing the capabilities of a given theory with respect to experimental results. There exist only few attempts along that line while the number of computer implementations tends to grow rapidly. More efforts in that direction are thus clearly needed.

A second aspect concerns the impact of irradiation on the computers. Cosmic rays, in particular charged particles, may hit an electronic component in a computer and spoil it. The computer continues to work but the hit can give rise to a wrong operation. Such errors, difficult to detect, are coined soft errors [AM15]. They may have dramatic consequences in automated devices such as autonomous cars and the problem is growing with the increasing miniaturization of electronic devices in which triggering thresholds become correspondingly smaller and smaller. There have been several physical and computational strategies developed to prevent or cure such problems. But cosmic rays, whose time and spatial occurrence is random, still remain an issue for electronic devices. And especially simulations using massive numbers of processors working on longer and longer times become more and more vulnerable to such soft errors. A simple way out would be to do computations twice, because the probability of double errors is marginally small. But this double check is desirable exactly for those long calculations which are already at the edge of feasibility. One is thus facing the paradox that addressing a computational description of irradiation is itself potentially affected by the type of processes it aims at describing [Bra17].

5.2.4.3   A Quick Glance at Quantum Computing

Quantum computing has been pondered since long [Ben80, Fey82]. Only since a few years the first practical realizations came up and have turned the topic into a hype. This is not surprising because quantum computing, once properly installed, opens the door to dramatic improvements in computing power. It will allow to attack numerical problems which are far out of reach for any conventional computer architecture. For a general overview see, e.g., [SS04, NC10]. Here, we are briefly discussing one aspect related to this book, the simulation of highly correlated many-body systems.

A quantum computer is an assembly of qubits and a qubit is formally a spin system, or equivalently a two-level system represented by ground state $|0\rangle$ and excited state $|1\rangle$. The most general state of a quantum computer consisting out of N qubits is then

$$|\Psi\rangle = \sum_{n_1...n_N} c_{n_1...n_N} |n_1...n_N\rangle \qquad (5.3)$$

where the $c_{n_1...n_N}$ are complex expansion coefficients and the basis $|n_1...n_N\rangle = |n_1\rangle \otimes ... \otimes |n_N\rangle$ is the tensor product of the qubits basis states. The Hamiltonian is a matrix in this basis space. The non-zero matrix elements of the Hamiltonian are defined by the topology of the qubit network. The setup of a quantum computer is flexible enough to accommodate a great variety of Hamiltonians for realistic applications.

The simple equation (5.3) demonstrates already the power of quantum computing. The left-hand side stands for a state of the quantum computer. It comprises at once the bulky expansion on the right-hand side, the form in which it would be treated in a conventional computer. Let us consider the example of a quantum computer with $N = 20$ qubits. The basis $\{|n_1...n_N\rangle\}$ then consists out of $2^N = 2^{20} \approx 10^6$ states which requires 10^6 complex coefficients $c_{n_1...n_N}$ for a numerical description of the state. This amounts to 16 MByte storage which may still be manageable in connection with a sparse Hamiltonian. However, the problem drives toward the limits of present days facilities if a full Hamiltonian matrix with about 10^{12} entries is required. The case of $N = 30$ qubits brings already the representation of the state as such with 16 TByte beyond efficient handling with conventional computers. It is the simple comparison of scaling, linear with N for quantum computers versus 2^N for conventional ones, which shows the immense potential of quantum computing.

There are a few bottlenecks nonetheless. One is the fact that a single qubit is prone to errors. This can be dealt with by assisting each qubit by a couple of control qubits which help to detect and counter-weight errors. This reduces the number of effective qubits by a constant factor, but does not alter the linear scaling of expense and preserves the superior scaling behavior of quantum computers. The other bottleneck lies at the side of the problem treated. It must have a structure which allows to map it into an ensemble of two-level (or spin) systems. The general mapping of spins into Fermion operators is too involved to be addressed here in detail, see the reviews [SS04, NC10, ?] or [SRL12] for a specific example. To estimate the expected feasibility, we discuss here superficially the connection with applications to the Configuration Interaction (CI) method (see section 2.3). There, the many-electron state $|\Psi\rangle$ is expanded about an independent electron state $|\Phi\rangle$ as

$$|\Psi\rangle = \left(1 + \sum_{ph} c_h^p \hat{a}_p^\dagger \hat{a}_h + \sum_{pqhj} c_{pq}^{hj} \hat{a}_p^\dagger \hat{a}_q^\dagger \hat{a}_h \hat{a}_j + \cdots \right) |\Phi\rangle \quad . \qquad (5.4)$$

The excitation modes in the CI expansion are independent of each other (formally speaking orthogonal) which allows to associate each excitation term

with one qubit. The number of terms in the CI expansion then relates immediately to the number N of qubits required. Today's quantum computers can
handle 30–50 qubits from which a part has to be put aside as control qubits
to reduce the errors still inherent in a qubit. That means that presently a de-
cent CI calculation is not yet possible. However, steady increase in N is to be
expected and with time going on realistic applications will become accessible.
This holds for correlated ground states of atoms and molecules. But, as argued
in section 2.3, the CI expansion is equally well applicable to time dependent
problems such that quantum computers could also deal with that. The limitation comes from the demands of dynamical simulations. The required size
of the CI expansion grows rapidly with excitation energy. Small excitations in
the linear regime will remain as feasible as ground-state calculations. However,
the high energy domain stays out of reach for a long while.

General Conclusion

This book addresses the theory of irradiation with a focus on the most demanding part, namely electron dynamics. Irradiation is a vast topic ranging from fundamental physics and chemistry questions to societal challenges with numerous applications, from health to industry. It concerns both finite systems and bulk material. We confined our discussions to finite systems namely from atoms to molecules and nanoclusters. Bulk materials do not imply the same constraints and thus may be addressed with other theoretical tools.

Because of their small mass, electrons are the first particles to respond to irradiation. This response may be close to instantaneous, often in the attosecond time scale. Coupling to ions and environment comes later and may extend up to picoseconds. Nonetheless, electrons remain active partners during the whole process. The focus of the book was thus on electron dynamics in finite systems which constitutes a complex and difficult problem for many reasons. Energetic irradiation processes involve a broad range of time and length scales which cannot be covered by one unique theoretical framework. One has thus to use different approaches each one specialized on a certain range of times and lengths, ideally complementing each other. We have presented in this book a great variety of approaches from quantum dynamics for electrons to classical MD and from short-time electron response to more leisurely subsequent ionic dynamics, see Chapter 2 for modeling and Chapter 3 for tools to analyze the simulation results. We could happily report enormous progress at all frontiers over the past decades, see Chapter 4 for examples. The huge amount of applications and results hinders us from summarizing that here again.

In spite of great success, there remain a couple of problems to be attacked in future developments. The first difficulty may look harmless but is not: in many situations, one cannot ignore the quantum nature of electrons. Fully quantum mechanical approaches of electron dynamics are meanwhile well established and work in many cases. But they become excessively involved as soon as one considers processes at high excitation energies. These require dynamical correlations, but the accessible phase space for them grows dramatically with energy. Major approximations would be needed, but quantum mechanical benchmarks do not exist in that dynamical regime, which make such approximations hard to validate on firm theoretical grounds. On the other hand, large phase spaces allow for (semi-)classical approximations which are then naturally the method of choice in violent dynamical processes. And this raises the second difficulty: the system during the dynamical evolution crosses in energy space the border between quantum and classical regimes and in real space the border between localized excitation and large-scale energy transport. We encounter a typical multi-scale situation. This multi-scale aspect dimension implies, by nature, different relevant degrees of freedom at the various scales, hence complementing approaches relying on the basis of these relevant degrees of freedom. For each regime, we have at our disposal well-working methods. The challenge is yet to develop proper interfaces between the methods which will allow us to connect smoothly from one approach to the other in the course of the process.

We have written this book as a guidebook. This has several noticeable consequences. First we did not enter all possible details of available theories. There exist numerous books and review papers where these details can be found and are analyzed in depth. We refer the reader to these references when needed. In that respect the guidebook is thus opening doors and pointing toward relevant aspects rather than discussing them in detail. This allows us to cover a wide range of approaches at an acceptable page cost. The second aspect, also typical of a guidebook, is to offer, beyond mere opinions or tastes, a status. We have tried, as far as possible, to explicitly state what is feasible and what is not for a given theory. This has been discussed explicitly in Chapter 4 on examples and in more general terms in Chapter 5 where we also tried to summarize what standard theories can and cannot do. After all, we have tried to report the state-of-the-art in dynamical simulations of finite electron systems driving always to point out yet open ends in theoretical and computational modeling. We hope that this will motivate further investigations in this fundamental field of physics with its many practical and societal issues.

A Appendix

A.1 UNITS

We use the Gaussian system of units for electromagnetic properties. Data from [S63].

$$\text{electron mass:} \quad m_e c^2 = 510.9\,\text{keV} = 37.57 \times 10^3\,\text{Ry} = 18.78 \times 10^3\,\text{Ha}$$

$$m_e = 0.0156\,\text{eV fs}^2\text{a}_0^{-2} = 0.5\,\text{Ry}^{-1}\text{a}_0^{-2} = 1\,\text{Ha}^{-1}\text{a}_0^{-2}$$

$$\text{light velocity:} \quad c = 5670\,\text{a}_0\,\text{fs}^{-1} = 274.12\,\text{Ry a}_0 = 137.06\,\text{a}_0\,\text{Ha}^{-1}$$

$$\text{charge:} \quad e^2 = 1\,\text{Ha a}_0 = 2\,\text{Ry a}_0 = 14.40\,\text{eV Å}$$

$$\text{dielectric constant:} \quad \epsilon_0 = \frac{1}{4\pi} \equiv \text{Gaussian system of units}$$

$$\text{Bohr energy:} \quad E_B = \frac{e^4 m_e}{2\hbar^2} = \frac{\alpha^2 m_e c^2}{2} = 13.604\,\text{eV} = 1\,\text{Ry} = \frac{1}{2}\,\text{Ha}$$

$$\text{Bohr radius:} \quad \text{a}_0 = \frac{\hbar^2}{m_e c^2} = 0.5291\,\text{Å} = 0.05291\,\text{nm}$$

$$= 0.5291 \times 10^{-10}\,\text{m}$$

$$\text{Boltzmann constant:} \quad k_\text{B} = 8.6174\,10^{-5}\,\text{eV K}^{-1}$$

$$\text{energy/} \quad 1\,\text{Ha} = 2\,\text{Ry} = 27.2\,\text{eV}$$

frequency scales:

$$1\,h\,\text{GHz} = 4.136 \times 10^{-6}\,\text{eV} \; ;$$

$$hc = 0.1240 \times 10^{-3}\,\text{eVcm}$$

$$\text{time scales:} \quad 1\,\text{fs} = 10^{-15}\,\text{s} = 1.519\,\hbar\text{eV}^{-1} = 20.66\,\hbar\text{Ry}^{-1}$$

$$= 41.32\,\hbar\text{Ha}^{-1}$$

$$1\,\hbar\text{Ha}^{-1} = 0.5\,\hbar\text{Ry}^{-1} = 0.0242\,\text{fs}$$

$$\text{laser intensity:} \quad I = \frac{c}{8\pi}|E_0|^2 \; ; \; I = 27.8|E_0\;[\text{in V cm}^{-1}]|^2\text{W cm}^{-2}$$

$$\text{scale factors:} \quad \hbar c = 1.9731 \times 10^{-7}\,\text{eV m} = 1973.1\,\text{eVÅ} = 274.12\,\text{Ry a}_0$$

$$\frac{\hbar^2}{m_e} = 1\,\text{Ha a}_0^2 = 2\,\text{Ry a}_0^2 = 7.617\,\text{eV Å}^2$$

We use for energy/distance/time scales: eV, a_0 and fs; Ry, a_0, $\hbar\,\text{Ry}^{-1}$ (1 $\hbar$ $\text{Ry}^{-1} = 0.0484$ fs); and Ha, a_0, $\hbar\,\text{Ha}^{-1}$ (1 $\hbar\,\text{Ha}^{-1} = 0.0242$ fs, atomic unit).

DOI: 10.1201/9781003127949-A

A.2 NOTATIONS

- vectors with bold characters (e.g., the position $\mathbf{r}$)
- operators with a hat (e.g., the many-body Hamiltonian operator $\hat{H}$)
- Dirac notation $|\Psi>$; wavefunction $\Psi(\mathbf{r}) = <\mathbf{r}|\Psi>$
- standard atomic labeling $1s$, $2s$, $2p$ ...
- standard chemical symbol for elements: H, He, Li, ...

General quantities	
$\mathbf{r}, r$	vector position, norm thereof
x, y, z	components of $\mathbf{r}$
$d\mathbf{r}, d^3r$	volume element for 3D integration
$k, \mathbf{k}$	wave vector, wave momentum
$\mathbf{p}$	momentum conjugate to $\mathbf{r}$
∇, Δ	gradient, resp. Laplacian operator
t	time
$v, \mathbf{v}$	particle velocity
T	temperature
$F, \mathbf{F}$	force
E	energy, mostly for total systems or parts of total system
V	potential in general
System's characteristics	
$H, \hat{H}$	classical, resp. quantal Hamiltonian
$\hat{V}_{\mathrm{PsP}}$	pseudo-potential, possibly non-local
Ions	
$I, J, Q_I, M_{\mathrm{I}}, \mathbf{R}, \mathbf{P}$	labels of ions, net charge, mass, coordinates, momentum
Electrons	
$i, j, m_e, \mathbf{r}_i, \sigma$	labels of electrons, electron mass, coordinate, spin
φ, φ_i	single electron wavefunction, single electron state i
ε	energy of one-electron state
$\rho(\mathbf{r}, \mathbf{r}')$	one-electron density matrix
$\varrho(\mathbf{r})$	local one-electron density distribution
Φ	N electron wavefunction in independent particle form
Ψ	general N electron wavefunction
$\mathcal{A}$	anti-symmetrization operator
U_{back}	one-electron potential from ionic background
U_{C}	Hartree Coulomb potential
$E_{\mathrm{xc}}[\varrho], \hat{U}_{\mathrm{xc}}$	exchange-correlation energy, resp. potential (DFT)
$h, \hat{h}$	classical, resp. quantal one-electron Hamiltonian
Global characteristics	
N	particle number, atoms, ions or electrons
r_s	Wigner-Seitz radius
$\varepsilon_{\mathrm{F}}, v_{\mathrm{F}}, k_{\mathrm{F}}, p_{\mathrm{F}} = \hbar k_{\mathrm{F}}$	Fermi energy, velocity, momentum
$f(\mathbf{r}, \mathbf{p})$	electronic phase-space distribution
r_{rms}	rms radius
External fields and response	
I	laser intensity, sometimes ion label
$\mathbf{E}, \mathcal{E}$	electrical field
$f(t), F(t)$	laser profile in time, integral thereof

57bmn

ω_{las}, T_{pulse}	laser frequency, resp. pulse duration
$\mathbf{B}$, $\mathbf{A}$	magnetic field, resp. electromagnetic vector potential
ω_{Mie}	frequency of Mie surface plasmon
$\mathbf{D}(t)$, $\tilde{D}(\omega)$	dipole moment in time, resp. frequency domain
S_D, $\mathcal{P}_D$	dipole strength distribution, resp. power spectrum
ε_{kin}	kinetic energy of emitted electrons

A.3   FREQUENTLY USED ACRONYMS

Acronym	Complete name, associated section
1D, 2D, 3D	One-, two-, three-dimensional
ADSIC	Average-Density SIC, section 2.2.3.5.3
AE	All-Electron
ALDA	Adiabatic LDA, section 2.2.3.4
amu	atomic mass unit
APES	Adiabatic Potential Energy Surface, section 2.5.1
BO	Born-Oppenheimer, section 2.5.1
CI	Configuration Interaction, section 2.3.4.1
DFT	Density Functional Theory, section 2.1.5
FWHM	Full Width at Half Maximum
GGA	Generalized Gradient Approximation, section 2.2.3.2
HF	Hartree-Fock, section 2.2.2
HOMO	Highest Occupied Molecular Orbital
IP	Ionization Potential
IR	Infra-Red
KS	Kohn-Sham, section 2.2.3.3
LDA	Local Density Approximation, section 2.2.3
LUMO	Lowest Unoccupied Molecular Orbital
MC	Monte-Carlo
MCTDHF	Multi-Configurational TDHF, section 2.3.4.3
MD	Molecular Dynamics
MPI	Multi Photon Ionization
PAD	Photo Angular Distribution, section 1.2.3
PES	Photo-Electron Spectra (spectroscopy), section 1.2.3
RPA	Random-Phase Approximation, section 2.2.4.1
RTA	Relaxation Time Ansatz, section 2.3.7.2
SIC	Self-Interaction Correction, section 2.2.3.5
s.p.	single particle
STDLDA	Stochastic TDLDA, section 2.3.8
TDDFT	Time-Dependent DFT, section 2.1.5
TDHF	Time-Dependent HF, section 2.2.2
TDLDA(-MD)	Time-Dependent LDA, resp. TDLDA Molecular Dynamics, section 2.2.3.4
UV, XUV	Ultra-Violet, resp. far UV
VMI	Velocity Map Imaging
VUU(-MD)	Vlasov-Ühling-Uhlenbeck, resp. VUU Molecular Dynamics, section 2.4.1
X-ray	photons in Röntgen regime (several keV)

Bibliography

AAE⁺08.	Alonso, J. L. et al., Phys. Rev. Lett. **101**(2008), 096403
AAG14.	Abedi, A. et al., Europhys. Lett. **106**(2014), 33001
AARS96.	Abe, Y. et al., Phys. Rep. **275**(1996), 49
AB99.	Adamo, C. and Barone, V., J. Chem. Phys. **110**(1999), 6158
ABMC⁺21.	Alonso, J. L. et al., New J. Phys. **23**(2021), 063011
ACBN15.	Agapito, L. A. et al., Phys. Rev. X **5**(2015), 011006
ACER12.	Alonso, J. L. et al., in *Fundamentals of Time-Dependent Density Functional Theory* (Marques, M. A. et al., eds.), 301, Springer Berlin Heidelberg, Berlin, Heidelberg (2012)
ACG17.	Attaccalite, C. et al., Phys. Rev. B **95**(2017), 125403
ACZ⁺09.	Andrade, X. et al., J. Chem. Theo. Comput. **5**(2009), 728
ADN⁺20.	Avakyan, L. et al., Optical Materials **109**(2020), 110264
AF10.	Arbeiter, M. and Fennel, T., Phys. Rev. A **82**(2010), 013201
AGM11.	Attaccalite, C. et al., Phys. Rev. B **84**(2011), 245110
All87.	Allen, P. B., Phys. Rev. Lett. **59**(1987), 1460
AM15.	Autran, J. L. and Munteanu, D., *Soft Errors: From Particles to Circuits*, CRC Press, London (2015)
AMAG16.	Agostini, F. et al., J. Chem. Theory Comput. **12**(2016), 2127
AMG10.	Abedi, A. et al., Phys. Rev. Lett. **105**(2010), 123002
AMG12.	Abedi, A. et al., J. Chem. Phys. **137**(2012), 22A530
AP13.	Akimov, A. V. and Prezhdo, O. V., J. Chem. Theo. Comput. **9**(2013), 4959
AP14.	Akimov, A. V. and Prezhdo, O. V., J. Chem. Theo. Comput. **10**(2014), 789
APF14.	Arbeiter, M. et al., Phys. Rev. A **89**(2014), 043428
ARS02.	Andrae, K. et al., J. Phys. B **35**(2002), 1
ARS04.	Andrae, K. et al., Phys. Rev. Lett. **92**(2004), 173402
ASR⁺23.	Andersson, A. et al., Phys. Rev. A **107**(2023), 013103
AT87.	Allen, M. P. and Tildesley, D. J., *Computer Simulation of Liquids*, Oxford Univ. Press, New York (1987)
BAH⁺20.	Bonafé, F. P. et al., J. Chem. Theory Comput. **16**(2020), 4454
Bal75.	Balescu, R., *Equilibrium and Non-Equilibrium Statistical Mechanics*, Wiley, New York (1975)
Bau04.	Bauer, D., J. Phys. B **37**(2004), 3085
BB97.	Brack, M. and Bhaduri, R. K., *Semiclassical Physics*, Addision-Wesley, Reading (1997)
BBBQ20.	Baffou, G. et al., Light: Science & Applications **9**(2020), 1
BBC⁺20.	Barca, G. M. J. et al., J. Chem. Phys. **152**(2020), 154102
BBR17.	Brunger, M. J. et al., J. Phys. Chem. Ref. Data **46**(2017), 023102
BCZ⁺17.	Boriskina, S. V. et al., Adv. Opt. Photon. **9**(2017), 775
BD88.	Bertsch, G. F. and Das Gupta, S., Phys. Rep. **160**(1988), 191
BD19.	Biele, R. and DAgosta, R., Entropy **21**(2019), 752
BDJL20.	Blase, X. et al., J. Phys. Chem. Lett. **11**(2020), 7371

BDM+10.	Bär, M. et al., Phys. Stat. Sol. B **247**(2010), 1122
Bec87.	Beck, D. E., Phys. Rev. B **35**(1987), 7325
Bec88.	Becke, A. D., Phys. Rev. A **38**(1988), 3098
Bec93.	Becke, A. D., J. Chem. Phys. **98**(1993), 5648
Ben80.	Benioff, P., J. Stat. Phys. **22**(1980), 563
Ber16.	Bernardi, M., J. Phys. B **89**(2016), 1
BGG+12.	Buss, O. et al., Phys. Rep. **512**(2012), 1
BGK54.	Bhatnagar, P. L. et al., Phys. Rev. **94**(1954), 511
BH54.	Born, M. and Huang, K., *Dynamical Theory of Crystal Lattices*, Oxford Univ. Press, Oxford (1954)
BH00.	Banerjee, A. and Harbola, M. K., J. Chem. Phys. **113**(2000), 5614
BHH+09.	Bartels, C. et al., Science **323**(2009), 132
BHS82.	Bachelet, G. B. et al., Phys. Rev. B **26**(1982), 4199
BJRT06.	Bär, M. et al., Phys. Rev. A **73**(2006), 022719
BJWM00.	Beck, M. H. et al., Phys. Rep. **324**(2000), 1
BK00.	Brabec, T. and Krausz, F., Rev. Mod. Phys. **72**(2000), 545
BKP97.	Bonacic-Koutecky, V. and Pittner, J., Chem. Phys. **225**(1997), 173
BKS21.	Bedurke, F. et al., Phys. Chem. Chem. Phys. **23**(2021), 13544
BKVM01.	Bonacic-Koutecky, V. et al., J. Chem. Phys. **115**(2001), 10450
BKW+17.	Besteiro, L. V. et al., ACS Photonics **4**(2017), 2759
BLMR92.	Blum, V. et al., J. Comput. Phys **100**(1992), 364
BM92.	Bransden, B. H. and McDowell, M. R. C., *Charge Exchange and Theory of Ion-Atom Collisions*, Clarendon, Oxford (1992)
BMK+03.	Böhm, A. et al., *The Geometric phase in quantum systems: foundations, mathematical concepts, and applications in molecular and condensed matter physics*, Springer, Berlin (2003)
BMNL15.	Bernardi, M. et al., Nat. Commun. **6**(2015), 1
BMR+06.	Belkacem, M. et al., Phys. Rev. A **73**(2006), 051201
BNC+20.	Bondanza, M. et al., Phys. Chem. Chem. Phys. **22**(2020), 14433
BNM02.	Ben-Nun, M. and Martínez, T. J., *Advances in Chemical Physics*, vol. 121, 439, Wiley, New York (2002)
Bon98.	Bonitz, M., *Quantum kinetic theory*, Teubner, Leipzig (1998)
BPHH96.	Bordas, C. et al., Rev. Sci. Instrum. **67**(1996), 2257
BQ13.	Baffou, G. and Quidant, R., Laser & Photonics Reviews **7**(2013), 171
BQ14.	Baffou, G. and Quidant, R., Chem. Soc. Rev. **43**(2014), 3898
Bra93.	Brack, M., Rev. Mod. Phys. **65**(1993), 677
Bra17.	Brazil, R., Phys. World **34(7)**(2017), 30
BRvLdB06.	Berger, J. A. et al., Phys. Rev. B **74**(2006), 245117
BSN+16a.	Brown, A. M. et al., Phys. Rev. B **94**(2016), 075120
BSN+16b.	Brown, A. M. et al., ACS Nano **10**(2016), 957
BSSR07.	Botti, S. et al., Rep. Prog. Phys. **70**(2007), 357
BTA09.	Booth, G. H. et al., J. Chem. Phys. **131**(2009), 054106
BTS+20.	Bustamante, C. M. et al., J. Chem. Phys. **153**(2020), 234108
BVLN22.	Borrego-Varillas, R. et al., Rep. Prog. Phys. **85**(2022), 066401
BW39.	Bohr, N. and Wheeler, J. A., Phys. Rev. **56**(1939), 426
BYL+11.	Bondar, D. I. et al., Phys. Rev. A **83**(2011), 013420
BYW+19.	Besteiro, L. V. et al., Nano Today **27**(2019), 120

BZ83.	Bendt, P. and Zunger, A., Phys. Rev. Lett. **50**(1983), 1684
Car21.	Caruso, F., J. Phys. Chem. Lett. **12**(2021), 1734
CBA$^+$20.	Cortés, E. et al., ACS Nano **14**(2020), 16202
CBG$^+$06.	Chulkov, E. V. et al., Chem. Rev. **106**(2006), 4160
CCC$^+$12.	Chen, H. M. et al., ACS Nano **6**(2012), 7362
CDP05.	Craig, C. F. et al., Phys. Rev. Lett. **95**(2005), 163001
CDRE00.	Cazalilla, M. A. et al., Phys. Rev. B **61**(2000), 8033
Cep95.	Ceperley, D. M., Rev. Mod. Phys. **67**(1995), 279
CFB98.	Chelkowski, S. et al., Phys. Rev. A **57**(1998), 1176
CHH$^+$00.	Campbell, E. E. B. et al., Phys. Rev. Lett. **84**(2000), 2128
Chr07.	Christov, I. P., J. Chem. Phys. **127**(2007), 134110
Chu20.	Chudzinski, P., Phys. Rev. Research **2**(2020), 012048
Cir17.	Ciracì, C., Phys. Rev. B **95**(2017), 245434
CKLI96.	Chen, J. et al., Phys. Rev. A **54**(1996), 3939
CM18.	Curchod, B. F. E. and Martínez, T. J., Chem. Rev. **118**(2018), 3305
CMV$^+$09.	Castro, A. et al., C. R. Physique **10**(2009), 469
CN92.	Crank, J. and Nicolson, P., Adv. Comp. Math. **6**(1992), 207
COB18.	Crespo-Otero, R. and Barbatti, M., Chem. Rev. **118**(2018), 7026
Cor93.	Corkum, P., Phys. Rev. Lett. **71**(1993), 1994
CP85.	Car, R. and Parrinello, M., Phys. Rev. Lett. **55**(1985), 2471
CPHL$^+$17.	Ciappina, M. F. et al., Rep. Prog. Phys. **80**(2017), 054401
CR21.	Church, M. S. and Rubenstein, B. M., J. Chem. Phys. **154**(2021), 184103
CRM$^+$85.	Cusson, R. et al., Z. Phys. A **320**(1985), 475
CRS97.	Calvayrac, F. et al., Ann. Phys. (NY) **255**(1997), 125
CRSU00.	Calvayrac, F. et al., Phys. Rep. **337**(2000), 493
CTC$^+$19.	Campos, A. et al., Nature Physics **15**(2019), 275
CWH$^+$11.	Cazzaniga, M. et al., Phys. Rev. B **84**(2011), 075109
CZ68.	Cooper, J. and Zare, R. N., J. Comput. Phys. **48**(1968), 942
Dal71.	Dalal, V. L., J. Applied Phys. **42**(1971), 2274
DBG97.	Dobson, J. F. et al., Phys. Rev. Lett. **79**(1997), 1905
DCC19.	Deringer, V. L. et al., Advanced Materials **31**(2019), 1902765
DDK$^+$02.	Damrauer, N. H. et al., Euro. Phys. J. D **20**(2002), 71
DG90.	Dreizler, R. M. and Gross, E. K. U., *Density Functional Theory: An Approach to the Quantum Many-Body Problem*, Springer-Verlag, Berlin (1990)
DLM22.	Dar, D. et al., Chem. Phys. Rev. **3**(2022), 031307
DLR$^+$18.	Dinh, P. M. et al., Eur. Phys. J B **91**(2018), 246
DLRS97.	Domps, A. et al., Ann. Phys. (Leipzig) **6**(1997), 455
DMRH20.	Domaracka, A. et al., in *21st Century Nanoscience A Handbook* (Sattler, K. D., ed.), chap. 10, CRC Press, Boca Raton (2020)
DPH21.	Dienstbier, P. et al., Nanophotonics **10**(2021), 3717
DR06.	Duffy, D. M. and Rutherford, A. M., J. Phys. Condens. Matter **19**(2006), 016207
DRAJ$^+$92.	Dunlop, A. et al., *Materials under Irradiation*, Scientific Net (1992)
DRRS13.	Dinh, P. M. et al., Phys. Rev. A **87**(2013), 032514
DRS97.	Domps, A. et al., Ann. Phys. (N.Y.) **260**(1997), 171
DRS98.	Domps, A. et al., Phys. Rev. Lett. **80**(1998), 5520

DRS00.	Domps, A. et al., Ann. Phys. (N.Y.) **280**(2000), 211
DRS10.	Dinh, P. M. et al., Phys. Rep. **485**(2010), 43
DS20.	Dou, W. and Subotnik, J. E., J. Phys. chem. A **124**(2020), 757
DST00.	Durand, D. et al., *Nuclear Dynamics in the nucleonic regime*, CRC Press, London (2000)
DVC$^+$22.	Dinh, P. M. et al., Comput. Phys. Commun. **270**(2022), 108155
EB79.	Eastwood, J. W. and Brownrigg, D. R. K., J. Comput. Phys. **32**(1979), 24
EBP05.	E. B. Podgorsak, Technical Editor, *Radiation Oncology Physics: A Handbook for Teachers and Students*, IAEA (2005)
EKS00.	Eguiluz, A. et al., J. Phys. Chem. Solids **61**(2000), 383
EM16.	Elliott, P. and Maitra, N. T., Int. J. Quant. Chem. **116**(2016), 772
EVR$^+$15.	Escartn, J. M. et al., J. Chem. Phys. **142**(2015), 084118
FA34.	Fermi, E. and Amaldi, E., Accad. Ital. Rome **6**(1934), 117
Fai87.	Faisal, F. H. M., *Theory of Multiphoton Processes*, Plenum, New York (1987)
FBMB04.	Fennel, T. et al., Euro. Phys. J. D **29**(2004), 367
Fer27.	Fermi, E., Atti Accad. Naz. Lincei, Cl. Sci. Fis. Mat. Nat. Rend. **6**(1927), 602
Fer28.	Fermi, E., Z. Phys. **48**(1928), 73
Fey82.	Feynman, R. P., Int. J. Theo. Phys. **21**(1982), 1572
FFS82.	Feit, M. D. et al., J. Comput. Phys. **47**(1982), 412
FKJ13.	Fromager, E. et al., J. Chem. Phys. **138**(2013), 084101
FKT08.	Fabiano, E. et al., Chem. Phys. **349**(2008), 334
FMBT$^+$10.	Fennel, T. et al., Rev. Mod. Phys. **82**(2010), 1793
FNP$^+$08.	Feist, J. et al., Phys. Rev. A **77**(2008), 043420
FR11.	Fratalocchi, A. and Ruocco, G., Phys. Rev. Lett. **106**(2011), 105504
FRB07.	Fennel, T. et al., Phys. Rev. Lett. **99**(2007), 233401
Fre34.	Frenkel, J., *Wave Mechanics, Advanced General Theory*, Clarendon, Oxford (1934)
FSE$^+$02.	Frauenheim, T. et al., J. Phys. Condens. Matter **14**(2002), 3015
FW71.	Fetter, A. L. and Walecka, J. D., *Quantum Theory of Many-Particle Systems*, McGraw-Hill, New York (1971)
GD88.	Ghosh, S. K. and Dhara, A. K., Phys. Rev. A **38**(1988), 1149
GDGA18.	Guther, K. et al., Phys. Rev. Lett. **121**(2018), 056401
GDP96.	Gross., E. K. U. et al., in *Density Functional Theory II: Relativistic and Time Dependent Extensions* (Nalewajski, R. F., ed.), 81, Springer, Berlin (1996)
GDR$^+$17.	Gao, C.-Z. et al., Phys. Rev. A **95**(2017), 033427
GDRS15.	Gao, C.-Z. et al., Ann. Phys. **360**(2015), 98
GDRS16.	Gao, C.-Z. et al., Euro. Phys. J. D **70**(2016), 26
GHP$^+$10.	Greenman, L. et al., Phys. Rev. A **82**(2010), 023406
GKRT93.	Gherega, T. et al., Nucl. Phys. A **560**(1993), 166
GL76.	Gunnarsson, O. and Lundqvist, B. I., Phys. Rev. B **13**(1976), 4274
GL20.	Gonzalez, L. and Lindh, R., *Quantum chemistry and excited states*, Wiley, New York (2020)
GPS01.	Goldstein, H. et al., *Classical Mechanics (3rd ed)*, Addison-Wesley, Boston (2001)

GPZ10.	Granucci, G. et al., J. Chem. Phys. **133**(2010), 134111
Gre95.	Greer, J. C., J. Chem. Phys. **103**(1995), 1821
GRR86.	Goeke, K. et al., Ann. Phys. **166**(1986), 257
GRS02.	Giglio, E. et al., Ann. Phys. **4**(2002), 291
GTH96.	Goedecker, S. et al., Phys. Rev. B **54**(1996), 1703
GU97.	Goedecker, S. and Umrigar, C. J., Phys. Rev. A **55**(1997), 1765
GVM$^+$12.	Giovannini, U. D. et al., Phys. Rev. A **85**(2012), 062515
GZ15.	Govorov, A. O. and Zhang, H., J. Phys. Chem. C **119**(2015), 6181
GZS$^+$06.	Govorov, A. O. et al., Nanoscale Res. Lett. **1**(2006)
HAB$^+$20.	Hourahine, B. et al., J. Chem. Phys. **152**(2020), 124101
Hak76.	Haken, H., *Quantum field theory of solids, an introduction*, North-Holland, Amsterdam (1976)
Har11.	Hartland, G. V., Chem. Rev. **111**(2011), 3858
HBF$^+$04.	Horsfield, A. P. et al., J. Phys. Condens. Matter **16**(2004), 8251
HBF$^+$05.	Horsfield, A. P. et al., J. Phys. Condens. Matter **17**(2005), 4793
HC16.	Hui, K. and Chai, J.-D., J. Chem. Phys. **144**(2016), 044114
HCC$^+$13.	Huismans, Y. et al., Phys. Rev. A **88**(2013), 013201
Hed65.	Hedin, L., Phys. Rev. **139**(1965), A796
HF63.	Hubbard, J. and Flowers, B. H., Proceedings of the Royal Society of London. Series A. Mathematical and Physical Sciences **276**(1963), 238
HFdGC14.	Himmetoglu, B. et al., Int. J. Quantum Chem. **114**(2014), 14
HFLP20.	He, J. et al., J. Am. Chem. Soc. **142**(2020), 14664
HK64.	Hohenberg, P. and Kohn, W., Phys. Rev. **136**(1964), 864
HM76.	Hansen, J.-P. and Mac Donald, I. R., *Theory of Simple Liquids*, Academic Press, London (1976)
HOJ13.	Helgaker, T. et al., *Molecular Electronic-Structure Theory*, Wiley-Blackwell, Chichester (2013)
HRA$^+$17.	Hansen, K. et al., Phys. Rev. Lett. **118**(2017), 103001
HS99.	Haberland, H. and Schmidt, M., Euro. Phys. J. D **6**(1999), 109
HT95.	Head-Gordon, M. and Tully, J. C., J. Chem. Phys. **103**(1995), 10137
HWAL94.	Hammond, B. L. and W. A. Lester, P. J. Reynolds, *Monte Carlo Methods in Ab Initio Quantum Chemistry*, World Scientific Lecture and Course Notes in Chemistry, World Scientific, Singapore (1994)
HZR$^+$96.	Hamm, P. et al., Chem. Phys. Lett. **263**(1996), 613
IB00.	Ishikawa, K. and Blenski, T., Phys. Rev. A **62**(2000), 63204
IBL03.	Ismail-Beigi, S. and Louie, S. G., Phys. Rev. Lett. **90**(2003), 076401
IC20.	Ibele, L. M. and Curchod, B. F. E., Phys. Chem. Chem. Phys. **22**(2020), 15183
IGRS19.	Inhester, L. et al., J. Chem. Phys. **151**(2019), 054107
IM05.	Itikawa, Y. and Mason, N., J. Phys. Chem. Ref. Data **34**(2005), 1
Int.	International Irradiation Association, https://iiaglobal.com/about-us/
IZ03.	Ivanov, D. S. and Zhigilei, L. V., Phys. Rev. B **68**(2003), 064114
Jac62.	Jackson, J. D., *Classical Electrodynamics*, Wiley, New York (1962)
Jer19.	Jermyn, A. S., Phys. Rev. Materials **3**(2019), 075201
JHW$^+$20.	Jahnke, T. et al., Chem. Rev. **120**(2020), 11295
JM96.	Jäckle, A. and Meyer, H.-D., J. Chem. Phys. **104**(1996), 7974

JMTR96.	Jorgensen, W. L. et al., J. Am. Chem. Soc. **118**(1996), 11225
JRZB05.	Jungreuthmayer, C. et al., J. Phys. B. **38**(2005), 3029
JSARM+15.	Jornet-Somoza, J. et al., Phys. Chem. Chem. Phys. **17**(2015), 26599
Kap16.	Kapral, R., Chem. Phys. **481**(2016), 77
KB62.	Kadanoff, L. P. and Baym, G., *Quantum Statistical Mechanics*, Benjamin, New York (1962)
KB04.	Kurzweil, Y. and Baer, R., J. Chem. Phys. **121**(2004), 8731
KCB+19.	Krishnakanth, K. N. et al., Optical Materials **95**(2019), 109239
KCS+20.	Kotsina, N. et al., Phys. Chem. Chem. Phys. **22**(2020), 4647
KDL97.	Koonin, S. E. et al., Phys. Rep. **278**(1997), 1
KDRS13.	Klüpfel, P. et al., Phys. Rev. A **88**(2013), 052501
KE99.	Ku, W. and Eguiluz, A. G., Phys. Rev. Lett. **82**(1999), 2350
Kel65.	Keldysh, L. V., Sov. Phys. JETP **20**(1965), 1018
KGD+22.	Kasthurirangan, S. et al., Phys. Rev. A **106**(2022), 012820
KGNR84.	Kirson, Z. et al., Surface Science **137**(1984), 527
Khu19.	Khurgin, J. B., Faraday Discuss. **214**(2019), 35
KI09.	Krausz, F. and Ivanov, M., Rev. Mod. Phys. **81**(2009), 163
KJJ+10.	Kjellberg, M. et al., Phys. Rev. A **81**(2010), 023202
KK08.	Kümmel, S. and Kronik, L., Rev. Mod. Phys. **80**(2008), 3
KKK+91.	Koyama, Y. et al., Photochemistry and Photobiology **54**(1991), 433
KLI92.	Krieger, J. B. et al., Phys. Rev. A **45**(1992), 101
KLT57.	Kaganov, M. I. et al., Sov. Phys. JETP **4**(1957), 173
KLZ78.	Kmmel, H. et al., Phys. Rep. **36**(1978), 1
KMK08.	Korzdorfer, T. et al., Phys. Rev. Lett. **100**(2008), 133004
KMY+05.	Kukura, P. et al., Science **310**(2005), 1006
Koh99.	Kohn, W., Rev. Mod. Phys. **71**(1999), 1253
Koh06.	Kohanoff, J., *Electronic Structure Calculations for Solids and Molecules: Theory and Computational Methods*, Cambridge University Press (2006)
KRF12.	Köhn, J. et al., New J. Phys **14**(2012), 055011
KRMSS13.	Keller-Rudek, H. et al., Earth Syst. Sci. Data **5**(2013), 365
KS65.	Kohn, W. and Sham, L. J., Phys. Rev. **140**(1965), 1133
KS12.	Khurgin, J. B. and Sun, G., Appl. Phys. Lett. **100**(2012), 011105
KS17.	Khurgin, J. B. and Sun, G., in *Quantum Plasmonics* (Bozhevolnyi, S. I. et al., eds.), 303, Springer International Publishing, Cham (2017)
KSK92.	Krause, J. et al., Phys. Rev. Lett. **68**(1992), 3535
KSK+99.	Köller, L. et al., Phys. Rev. Lett. **82**(1999), 3783
KSKP20.	Kristiansen, H. E. et al., J. Chem. Phys. **152**(2020), 071102
Kul88.	Kulander, K. C., Phys. Rev. A **38**(1988), 778
Kul15.	Kulik, H. J., J. Chem. Phys. **142**(2015), 240901
KV95.	Kreibig, U. and Vollmer, M., *Optical Properties of Metal Clusters*, vol. 25, Springer Series in Materials Science, Berlin (1995)
Kva12.	Kvaal, S., J. Chem. Phys. **136**(2012), 194109
KvLP+21.	Karlsson, D. et al., Phys. Rev. Lett. **127**(2021), 036402
Lan50.	Lanczos, C., J. Res. Nat. Bureau of Standards **45**(1950), 255
LBB+10.	Lermé, J. et al., J. Phys. Chem. Lett. **1**(2010), 2922
LBS+15.	Lackner, F. et al., Phys. Rev. A **91**(2015), 023412

LBS$^+$17.   Lackner, F. et al., Phys. Rev. A **95**(2017), 033414

LDR$^+$15.   Lacombe, L. et al., Eur. Phys. J. D **69**(2015), 195

LDSO19.   Li, J. et al., SciPost Phys. **6**(2019), 040

Ler11.   Lermé, J., J. Phys. Chem. C **115**(2011), 14098

LGI$^+$20.   Li, X. et al., Chem. Rev. **120**(2020), 9951

LHC22.   Lassmann, Y. et al., J. Phys. Chem. Lett. **13**(2022), 12011

Lie97.   Liebsch, A., *Electronic excitations at metal surfaces*, Plenum Press, New York (1997)

Lin54.   Lindhard, J., Dan. Vid. Selsk Mat.-Fys. Medd. **28**(1954), 8

LJY$^+$19.   Li, F. et al., J. Phys. B **52**(2019), 195601

LKY17.   Loetstedt, E. et al., Prog. in Ultrafast Intense Laser Sci. VIII **13**(2017), 15

LLM$^+$20.   Lode, A. U. J. et al., Rev. Mod. Phys. **92**(2020), 011001

LMW$^+$15.   Li, H. et al., Phys. Rev. Lett. **114**(2015), 123004

LNH$^+$06.   Liu, B. et al., Phys. Rev. Lett. **97**(2006), 133401

LP88.   Lifschitz, E. M. and Pitajewski, L. P., *Physikalische Kinetik, Lehrbuch der Theoretischen Physik*, vol. X, Mir, Moscow (1988)

LR94.   Lauritsch, G. and Reinhard, P.-G., Int. J. Mod. Phys. C **5**(1994), 65

LRD16.   Lacombe, L. et al., Ann. Phys. (N.Y.) **373**(2016), 216

LS77.   Lieb, E. H. and Simon, B., Adv. Math. **23**(1977), 22

LSD$^+$20.   Loos, P.-F. et al., J. Phys. Chem. Lett. **11**(2020), 3536

LSES96.   Logunov, S. L. et al., J. Phys. Chem. **100**(1996), 18586

LSM$^+$20.   Lischka, H. et al., J. Chem. Phys. **152**(2020), 134110

LSR95.   L'Eplattenier, P. et al., Ann. Phys. (N.Y.) **244**(1995), 426

LSR06.   Legrand, C. et al., J. Phys. B **39**(2006), 2481

LSRD19.   Lacombe, L. et al., Ann. Phys. (N.Y.) **406**(2019), 233

LST94.   Lipparini, E. et al., J. Phys. Condens. Matter **6**(1994), 2025

LTdPV$^+$11.   Lopez-Tarifa, P. et al., Phys. Rev. Lett. **107**(2011), 023202

LvdHB09.   Lysaght, M. A. et al., Phys. Rev. A **79**(2009), 053411

LvL98.   Lappas, D. G. and van Leeuwen, R., J. Phys. B **31**(1998), L249

LYP88.   Lee, C. et al., Phys. Rev. B **37**(1988), 785

LZC08.   Lin, Z. et al., Phys. Rev. B **77**(2008), 075133

MAF$^+$18.   Morzan, U. N. et al, Chem. Rev. **118**(2018), 4071

Mah93.   Mahan, G. D., *Many Particle Physics*, Plenum, New York (1993)

Mai16.   Maitra, N. T., J. Chem. Phys. **144**(2016), 220901

Mak98.   Makri, N., J. Phys. Chem. A **102**(1998), 4414

Man20.   Manzano, D., AIP Advances **10**(2020), 025106

MATG17.   Min, S. K. et al., J. Phys. Chem. Lett. **8**(2017), 3048

MBC$^+$16.   Markush, P. et al., Phys. Chem. Chem. Phys. **18**(2016), 16721

MBNSGL16.   Mustafa, J. I. et al., Phys. Rev. B **94**(2016), 155105

MBW02.   Maitra, N. et al., Phys. Rev. Lett. **89**(2002), 023002

MC18.   Mignolet, B. and Curchod, B. F. E., J. Chem. Phys. **148**(2018), 134110

MCFO17.   Maldonado, P. et al., Phys. Rev. B **96**(2017), 174439

McL64.   McLachlan, A. D., Mol. Phys. **8**(1964), 39

MDL$^+$19.   Marciniak, A. et al., Nat. Commun. **10**(2019), 239

MDRS08a.   Messud, J. et al., Phys. Rev. Lett. **101**(2008), 096404

MDRS08b.   Messud, J. et al., Chem. Phys. Lett. **461**(2008), 316

MDRS11.	Messud, J. et al., Ann. Phys. (Leipzig) **523**(2011), 270
ME10.	Milonni, P. W. and Eberli, J. H., *Laser Physics*, Wiley, New York (2010)
Men12.	Mennucci, B., WIREs Computational Molecular Science **2**(2012), 386
MFH$^+$07.	Martin, F. et al., Science **315**(2007), 629
MG04.	Marques, M. A. L. and Gross, E. K. U., Ann. Rev. Phys. Chem. **55**(2004), 427
MGMS14.	Makhov, D. V. et al., J. Chem. Phys. **141**(2014), 054110
MHH$^+$03.	Moseler, M. et al., Phys. Rev. B **68**(2003), 165413
Mic83.	Micha, D. A., J. Chem. Phys. **78**(1983), 7138
Mie08.	Mie, G., Ann. Phys. (Leipzig) **25**(1908), 377
MK07.	Mundt, M. and Kümmel, S., Phys. Rev. B **76**(2007), 035413
MKH$^+$19.	Mcallister, M. et al., J. Phys. Chem. B **123**(2019), 1537
MKN$^+$12.	Marinica, D. C. et al., Nano Letters **12**(2012), 1333
MKvLR07.	Mundt, M. et al., Phys. Rev. A **75**(2007), 050501
MLKN14.	Manjavacas, A. et al., ACS Nano **8**(2014), 7630
MLL$^+$13.	Mukherjee, S. et al., Nano Letters **13**(2013), 240
MMC90.	Meyer, H.-D. et al., Chem. Phys. Lett. **165**(1990), 73
MMN$^+$12.	Marques, M. A. L. et al., *Fundamentals of Time-Dependent Density Functional Theory*, Lect. notes in Phys., vol. 837, Springer, Berlin (2012)
MN20.	Mascaretti, L. and Naldoni, A., J. Applied Phys. **128**(2020), 041101
Mor21.	Mortensen, N. A., Nanophotonics **10**(2021), 2563
MOS18.	Miao, G. et al., J. Chem. Phys. **150**(2018), 041711
MRM94.	Montag, B. et al., Z. Phys. D **32**(1994), 125
MRS10.	Maruhn, J. et al., *Simple models of many-fermions systems*, Springer, Berlin (2010)
MRSU14.	Maruhn, J. A. et al., Comput. Phys. Commun. **185**(2014), 2195
MSM21.	Meng, Q. et al., J. Chem. Theory Comput. **17**(2021), 2702
MSR$^+$21.	Medina, C. et al., New J. Phys. **23**(2021), 053011
MTOO16.	Miyamoto, Y. et al., Scientific Reports **5**(2016), 18220
MUN06.	Marques, M. A. L. et al. (eds.), *Time-dependent density functional theory*, Lect. notes in Phys., vol. 706, Springer, Berlin (2006)
MWD$^+$10.	McEniry, E. J. et al., J. Phys. B **77**(2010), 305
MZG$^+$14.	Mukherjee, S. et al., J. Am. Chem. Soc. **136**(2014), 64
NC10.	Nielsen, M. A. and Chuang, I. L., *Quantum Computation and Quantum Information*, Cambridge Univ. Press, Cambridge (2010)
NDC$^+$17.	Nisoli, M. et al., Chem. Rev. **117**(2017), 10760
New11.	New, G., *Introduction to Nonlinear Optics*, Cambridge Univ. Press, Cambridge (2011)
NHTF05.	Niehaus, T. A. et al., Eur. Phys. J. D **35**(2005), 467477
NKR$^+$18.	Naumova, M. et al., Phys. Chem. Chem. Phys. **20**(2018), 6274
NKS05.	Nest, M. et al., J. Chem. Phys. **122**(2005), 124102
NSH$^+$19.	Noda, M. et al., Comput. Phys. Commun. **235**(2019), 356
NWB$^+$20.	Nelson, T. R. et al., Chemical Reviews **120**(2020), 2215
Obo19.	Obodovskiy, I., *Radiation, fundamentals, applications, risks and safety*, Elsevier, Amsterdam (2019)

OPP⁺20.	Oliveira, M. J. T. et al., J. Chem. Phys. **153**(2020), 024117
Ott14.	Otto, F., J. Chem. Phys. **140**(2014), 014106
PA21.	Pieroni, C. and Agostini, F., J. Chem. Theory Comput. **17**(2021), 5969
Pal85.	Palik, E. D., *Handbook of optical constants of solids*, Academic Press, New York (1985)
PBE96.	Perdew, J. P. et al., Phys. Rev. Lett. **77**(1996), 3865
PDMT03.	Parker, J. S. et al., J. Phys.B **36**(2003)
PFK⁺95.	Porezag, D. et al., Phys. Rev. B **51**(1995), 12947
PFL18.	Peng, W.-T. et al., J. Chem. Theo. Comput. **14**(2018), 4129
PG96.	Pfalzner, S. and Gibbon, P., *Many-body tree methods in Physics*, Cambrigde Univ. Press, Cambridge (1996)
PG15.	Pernal, K. and Giesbertz, K. J. H., Top. Curr. Chem. **368**(2015), 125
PGK⁺21.	Pysanenko, A. et al., Phys. Chem. Chem. Phys. **23**(2021), 4317
PIT⁺12.	Passig, J. et al., New J. Phys. **14**(2012), 085020
PKZB99.	Perdew, J. P. et al., Phys. Rev. Lett. **82**(1999), 2544
PN66.	Pines, D. and Nozières, P., *The Theory of Quantum Liquids*, W A Benjamin, New York (1966)
PNDH21.	Paschen, T. et al., J. Phys. B **54**(2021), 144006
POM15.	Punwong, C. et al., J. Phys. Chem. B **119**(2015), 704
Poo83.	Poole, R. T., Physics Education **18**(1983), 280
PPS22.	Perfetto, E. et al., Phys. Rev. Lett. **128**(2022), 016801
PRM19.	Prandini, G. et al., NPJ — Comput. Mat. **5**(2019), 1
PRS00.	Pohl., A. et al., Phys. Rev. Lett. **84**(2000), 5090
PRS04.	Pohl., A. et al., Phys. Rev. A **70**(2004), 023202
Pru78.	Prugovecki, E., Ann. Phys. (N.Y.) **110**(1978), 102
PS18.	Perfetto, E. and Stefanucci, G., J. Phys. Condens. Matter **30**(2018), 465901
PSB91.	Pastore, G. et al., Phys. Rev. A **44**(1991), 6334
PT06.	Parandekar, P. V. and Tully, J. C., J. Chem. Theory Comput. **2**(2006), 229
PTVF92.	Press, W. H. et al., *Numerical Recipes*, Cambridge Univ. Press, Cambridge (1992)
PW92.	Perdew, J. P. and Wang, Y., Phys. Rev. B **45**(1992), 13244
PY89.	Parr, R. G. and Yang, W., *Density-Functional Theory of Atoms and Molecules*, Oxford Univ. Press, Oxford (1989)
PZ81.	Perdew, J. P. and Zunger, A., Phys. Rev. B **23**(1981), 5048
RC82.	Reinhard, P.-G. and Cusson, R. Y., Nucl. Phys. A **378**(1982), 418
RdB05.	Romaniello, P. and de Boeij, P. L., Phys. Rev. B **71**(2005), 155108
Rei16.	Reichl, L. E., *A Modern Course in Statistical Physics*, Wiley-VCH, Weinheim (2016)
RFS⁺13.	Rozzi, C. A. et al., Nat. Commun. **4**(2013), 1602
RG84.	Runge, E. and Gross, E. K. U., Phys. Rev. Lett. **52**(1984), 997
RG92.	Reinhard, P.-G. and Gambhir, Y. K., Ann. Phys. (Leipzig) **1**(1992), 598
RGB96.	Reinhard, P.-G. et al., Ann. Phys. (Leipzig) **5**(1996), 576
RGC83.	Reinhard, P.-G. et al., Nucl. Phys. A **398**(1983), 141

RGS06.	Rohringer, N. et al., Phys. Rev. A **74**(2006), 043420
RKT$^+$12.	Rangel, T. et al., Phys. Rev. B **86**(2012), 125125
RLJHC17.	Ronca, E. et al., J. Chem. Theory Comput. **13**(2017), 5560
Row70.	Rowe, D. J., *Nuclear Collective Motion*, Methuen, London (1970)
RPSWB97.	Rose-Petruck, C. et al., Phys. Rev. A **55**(1997), 1182
RRM10.	Rajam, A. K. et al., Phys. Rev. Lett. **105**(2010), 113002
RRMN21.	Reinhard, P.-G. et al., Phys. Rev. Lett. **127**(2021), 232501
RS80.	Ring, P. and Schuck, P., *The Nuclear Many-Body Problem*, Springer, Berlin (1980)
RS92.	Reinhard, P.-G. and Suraud, E., Ann. Phys. (N.Y.) **216**(1992), 98
RS95.	Reinhard, P.-G. and Suraud, E., Ann. Phys. (N.Y.) **239**(1995), 193
RS98.	Reinhard, P.-G. and Suraud, E., Eur. Phys. J. D **3**(1998), 175
RS03.	Reinhard, P.-G. and Suraud, E., *Introduction to Cluster Dynamics*, Wiley-VCH, Berlin (2003)
RS15.	Reinhard, P.-G. and Suraud, E., Ann. Phys. (N.Y.) **354**(2015), 183
RS21.	Reinhard, P.-G. and Suraud, E., Theo. Chem. Acc. **140**(2021), 63
RSA$^+$06.	Reinhard, P.-G. et al., Phys. Rev. E **73**(2006), 036709
RSM18.	Reinhard, P.-G. et al., J. Phys. B **51**(2018), 024007
RSU98.	Reinhard, P.-G. et al., Eur. Phys. J. D **1**(1998), 303
RT94.	Reinhard, P.-G. and Toepffer, C., Int. J. Mod. Phys. E **3**(1994), 435
RTG90.	Reynolds, P. J. et al., Comput. Phys. **4**(1990), 662
RTK17.	Rizzi, V. et al., Scientific Reports **7**(2017), 45410
RTKC16.	Rizzi, V. et al., Phys. Rev. B **93**(2016), 024306
Sö63.	Süssmann, G., *Einführung in die Quantenmechanik I*, Bibliographisches Institut, Mannheim (1963)
SAMB$^+$16.	Schütte, B. et al., Phys. Rev. Lett. **116**(2016), 033001
SBK$^+$20a.	Santiago, E. Y. et al., ACS Photonics **7**(2020), 2807
SBK20b.	Skeidsvoll, A. S. et al., Phys. Rev. A **102**(2020), 023115
SBUD19.	Sivan, Y. et al., Science **364**(2019), 9367
SCD17.	Sadasivam, S. et al., Phys. Rev. Lett. **119**(2017), 136602
Sch20.	Schröder, M., J. Chem. Phys. **152**(2020), 024108
SFM$^+$19.	Sangalli, D. et al., J. Phys. Condens. Matter **31**(2019), 325902
SFY05.	Schaadt, D. M. et al., Appl. Phys. Lett. **86**(2005), 063106
SG85.	Stoecker, H. and Greiner, W., Phys. Rep. **137**(1985), 277
SH53.	Sharp, R. T. and Horton, G. K., Phys. Rev. **90**(1953), 317
Sha09.	Shalashilin, D. V., J. Chem. Phys. **130**(2009), 244101
SI09.	Su, J. T. and W. A. Goddard III, J. Chem. Phys. **131**(2009), 244501
SI13.	Sato, T. and Ichikawa, K. L., Phys. Rev. A **88**(2013), 023402
SJL$^+$16.	Subotnik, J. E. et al., Annual Review of Physical Chemistry **67**(2016), 387
SK54.	Slater, J. C. and Koster, G. F., Phys. Rev. **94**(1954), 1498
SKvIH01.	Schlipper, R. et al., Appl. Phys. A **72**(2001), 255
Sla51.	Slater, J. C., Phys. Rev. **81**(1951), 385
SM10.	Slaviček, P. and Martnez, T. J., J. Chem. Phys. **132**(2010)
SMFH07.	Stella, L. et al., J. Chem. Phys. **127**(2007), 214104
SMHF11.	Stella, L. et al., J. Chem. Phys. **134**(2011), 194105
SNJ$^+$14.	Sundararaman, R. et al., Nat. Commun. **5**(2014), 1
SO81.	Sturm, K. and Oliveira, L. E., Phys. Rev. B **24**(1981), 3054

SOL13.	Subotnik, J. E. et al., J. Chem. Phys. **139**(2013), 214107
Sol17.	Solov'yov, A., *Nanoscale Insights into Ion-Beam Cancer Therapies*, Springer, Heidelberg (2017)
SPI10.	Solov'yov, A. V. et al., Phys. Rev. A **81**(2010), 021202(R)
SPMS91.	Schoenlein, R. W. et al., Science **254**(1991), 412
SR83.	Serene, J. W. and Rainer, D., Phys. Rep. **101**(1983), 221
SR03.	Saalmann, U. and Rost, J. M., Phys. Rev. Lett. **91**(2003), 223401
SR14.	Suraud, E. and Reinhard, P.-G., New J. Phys. **16**(2014), 063066
SR18.	Schira, R. and Rabilloud, F., J. Phys. Chem. **122**(2018), 27656
SRF$^+$11.	Schapiro, I. et al., J. Am. Chem. Soc. **133**(2011), 3354
SRL12.	Seeley, J. T. et al., J. Chem. Phys. **137**(2012), 224109
SRP15.	Sun, J. et al., Phys. Rev. Lett. **115**(2015), 036402
SRS15.	Slama, N. et al., Ann. Phys. (N.Y.) **355**(2015), 182
SRT90.	Schmitt, K.-J. et al., Z. Phys. A **336**(1990), 123
SRT09.	Shenvi, N. et al., J. Chem. Phys. **130**(2009), 174107
SS96.	Saalmann, U. and Schmidt, R., Z. f. Physik D **38**(1996), 153
SS03.	Stapelfeldt, H. and Seideman, T., Rev. Mod. Phys. **75**(2003), 543
SS04.	Stolze, J. and Suter, D., *Quantum Computing: A Short Course from Theory to Experiment*, Wiley-VCH, Weinheim (2004)
SS12.	Saita, K. and Shalashilin, D. V., J. Chem. Phys. **137**(2012), 22A506
SSH79.	Su., W. P. et al., Phys. Rev. Lett. **42**(1979), 1698
SSK$^+$14.	Schnorr, K. et al., Phys. Rev. Lett. **113**(2014), 073001
SSL07.	Schlegel, H. B. et al., J. Chem. Phys **126**(2007), 244110
SSR06.	Saalmann, U. et al., J. Phys. B **39**(2006), R39
ST12.	Shenvi, N. and Tully, J. C., Faraday Discuss. **157**(2012), 325
Ste13.	Stewart, J. J. P., J. Mol. Model. **19**(2013), 1
Ste16.	Stewart, J. J. P. MOPAC2016. Stewart Computational Chemistry, Colorado Springs (2016)
Stu82.	Sturm, K., Adv. Phys. **31**(1982), 1
Stu93.	Sturm, K., Zeit. Natur. A **48**(1993), 233
SvL13.	Stefanucci, G. and van Leeuwen, R., *Nonequilibrium many-body theory of quantum systems: a modern Introduction*, Cambridge Univ. Press, Cambridge (2013)
Sza85.	Szasz, L., *Pseudopotential Theory of Atoms and Molecules*, Wiley, New York (1985)
SZGV$^+$13.	Stella, L. et al., J. Phys. Chem. C **117**(2013), 8941
Tan07.	Tannor, D. J., *Introduction to quantum mechanics: a time-dependent perspective*, University Science Books, Melville, NY (2007)
TB21.	Tong, X. and Bernardi, M., Phys. Rev. Res. **3**(2021), 023072
The92.	Theilhaber, J., Phys. Rev. B **46**(1992), 12990
Tho27.	Thomas, L. H., Proc. Cambridge Philos. Soc. **23**(1927), 542
Tho60.	Thouless, D. J., Nucl. Phys. **21**(1960), 225
Tho61.	Thouless, D. J., *The Quantum Mechanics of Many-Body Systems*, Academic Press, New York (1961)
TJS$^+$18.	Tagliabue, G. et al., Nat. Commun. **9**(2018), 1
TKGV$^+$15.	Trabattoni, A. et al., Phys. Rev. X **5**(2015), 041053
TLA22.	Talotta, F. et al., J. Chem. Phys. **156**(2022), 184104
TNK$^+$17.	Takanashi, T. et al., Phys. Chem. Chem. Phys. **19**(2017), 19707

Tod01.	Todorov, T. N., J. Phys. Condens. Matter **13**(2001), 10125
TPSS03.	Tao, J. et al., Phys. Rev. Lett. **91**(2003), 146401
TR88.	Toepffer, C. and Reinhard, P.-G., Ann. Phys. (N.Y.) **181**(1988), 1
TS12.	Tao, L. and Scrinzi, A., New J. Phys. **14**(2012), 013021
TSAB15.	Thomas, R. E. et al., J. Chem. Theo. Comput. **11**(2015), 5316
Tse13.	Tselyaev, V. I., Phys. Rev. C **88**(2013), 054301
TSK⁺15.	Toscano, G. et al., Nat. Commun. **6**(2015), 1
TSS01.	Tatarczyk, K. et al., Phys. Rev. B **63**(2001), 235106
TTM04.	Toniolo, A. et al., Chemical Physics **304**(2004), 133
TTR07.	Tapavicza, E. et al., Phys. Rev. Lett. **98**(2007), 023001
Tul90.	Tully, J. C., J. Chem. Phys. **93**(1990), 1061
Tul98.	Tully, J. C., Faraday Discuss. **110**(1998), 407
TvLPS21.	Tuovinen, R. et al., J. Chem. Phys. **154**(2021), 094104
UG97.	Ullrich, C. A. and Gross, E. K. U., Comm. At. Mol. Phys. **33**(1997), 211
Ull00.	Ullrich, C. A., J. Mol. Struct. (THEOCHEM) **501-502**(2000), 315
Ull12.	Ullrich, C. A., *Time Dependent Density Functional Theory, Concepts and applications*, Oxford Univ. Press, Oxford (2012)
UMD⁺03.	Ullrich, J. et al., Rep. Prog. Phys. **66**(2003), 1463
UU32.	Uehling, E. A. and Uhlenbeck, G. E., Phys. Rev. **108**(1932), 1175
UY14.	Ullrich, C. A. and Yang, Z., Braz. J. Phys. **44**(2014), 154
VdMdC⁺22.	Ventura, E. et al., Phys. Chem. Chem. Phys. **24**(2022), 9403
VGGF⁺16.	Varas, A. et al., Nanophotonics **5**(2016), 409
Vig04.	Vignale, G., Phys. Rev. B **70**(2004), 201102
Vig12.	Vignale, G., in *Fundamentals of Time-Dependent Density Functional Theory* (Marques, M. A. et al., eds.), 457, Springer, Heidelberg (2012)
Vla50.	Vlasov, A. A., *Many Particle Theory and Its Applications to Plasma*, Gordon & Breach, New York (1950)
vLS13.	van Leeuwen, R. and Stefanucci, G., J. Phys. : Conf. Ser. **427**(2013), 012001
VR87.	Vignale, G. and Rasolt, M., Phys. Rev. Lett. **59**(1987), 2360
VRBL18.	Vril, M. et al., J. Chem. Theor. Comp. **14**(2018)
VSR17.	Vincendon, M. et al., Eur. Phys. J. D **71**(2017), 179
VUC97.	Vignale, G. et al., Phys. Rev. Lett. **79**(1997), 4878
VWD⁺10.	Vidal, S. et al., J. Phys. B: At. Mol. Opt. Phys. **43**(2010), 165102
VZMM22.	Vindel Zandbergen, P. et al., J. Phys. Chem. Lett. **13**(2022), 1785
WBEV16.	Waldecker, L. et al., Phys. Rev. X **6**(2016), 021003
WDRS15.	Wopperer, P. et al., Phys. Rep. **562**(2015), 1
WDSR12.	Wopperer, P. et al., Phys. Rev. A **85**(2012), 015402
WDTZ15.	Weber, W. J. et al., Curr. Opin. Solid State Mater. Sci. **19**(2015), 1
WFD⁺10a.	Wopperer, P. et al., Phys. Lett. A **375**(2010), 39
WFD⁺10b.	Wopperer, P. et al., Phys. Rev. A **82**(2010), 063416
WGB⁺15.	Wopperer, P. et al., Phys. Rev. A **91**(2015), 042514
WGR17.	Wopperer, P. et al., Eur. Phys. J. B **90**(2017), 51
WGW⁺00.	Weber, T. et al., Nature **405**(2000), 658
Woo72.	Wooten, F., *Optical Properties of Solids*, Academic Press, New York (1972)

Wor20.	Worth, G. A., Comput. Phys. Commun. **248**(2020), 107040
WW13.	Wesolowski, T. A. and Wang, Y. A., *Recent Progress in Orbital-free Density Functional Theory*, World Scientific (2013)
Yan15.	Yan, W., Phys. Rev. B **91**(2015), 115416
YH21.	Yamijala, S. S. R. K. C. and Huo, P., J. Phys. Chem. A **125**(2021), 628
YSS+12.	Yabana, K. et al., Phys. Rev. B **85**(2012), 045134
YUG+18.	Young, L. et al., J. Phys. B **51**(2018), 032003
YY19.	Yamada, A. and Yabana, K., Phys. Rev. B **99**(2019), 245103
ZBM+23.	Ziaja, B. et al., Philos. Transac. R. Soc. A 381(2023)20220216
Zew94.	Zewail, A. H., *Femtochemistry, Vol. I & II*, World Scientific, Singapore (1994)
ZFR+14.	Zhang, P. et al., Phys. Rev. B **90**(2014), 161407
ZKBS04.	Zhangellini, J. et al., J. Phys. B **37**(2004), 763
ZLJ+16.	Zhao, S.-F. et al., Phys. Rev. A **93**(2016), 023413
ZSR+19.	Zhou, L. et al., Science **364**(2019), 9545
ZSZ+18.	Zhou, L. et al., Science **362**(2018), 69

Index

For Product Safety Concerns and Information please contact our EU
representative GPSR@taylorandfrancis.com
Taylor & Francis Verlag GmbH, Kaufingerstraße 24, 80331 München, Germany

www.ingramcontent.com/pod-product-compliance
Lightning Source LLC
Chambersburg PA
CBHW060353220326
41598CB00023B/2903